D0898057

SPREAD SPECTRUM SYSTEMS
WITH COMMERCIAL APPLICATIONS

SPREAD SPECTRUM SYSTEMS
WITH COMMERCIAL APPLICATIONS

Third Edition

ROBERT C. DIXON

A Wiley-Interscience Publication

JOHN WILEY & SONS, INC.

New York · Chichester · Brisbane · Toronto · Singapore

This text is printed on acid-free paper.

Copyright © 1994 by John Wiley & Sons, Inc.

All rights reserved. Published simultaneously in Canada.

Reproduction or translation of any part of this work beyond
that permitted by Section 107 or 108 of the 1976 United
States Copyright Act without the permission of the copyright
owner is unlawful. Requests for permission or further
information should be addressed to the Permissions Department,
John Wiley & Sons, Inc., 605 Third Avenue, New York, NY
10158-0012.

Library of Congress Cataloging in Publication Data:

Dixon, Robert C. (Robert Clyde), 1932–
    Spread spectrum systems—with commercial applications / Robert C.
Dixon.—3rd ed.
      p. cm.
    "A Wiley-Interscience publication."
    Includes index.
    ISBN 0-471-59342-7
    1. Spread spectrum communications. I. Title.
    TK5102.5.D55   1994
    621.382—dc20
                                              93-8378

Printed in the United States of America

10 9 8 7 6 5 4

To
Nancy
David and Suzi
Theresa and Jeff
Rob
Matthew and Sean
Adam
Erin, Alison
and
Melissa

PREFACE

So much has changed in the spread spectrum world since the earlier editions of this book that it is hard to describe all of the changes. The primary change is that commercial applications abound, and this includes consumer-oriented products that are in development. All of this we owe to a much more far-sighted approach on the part of the Federal Communications Commission than has ever been taken before. Thank you, FCC.

Thanks also are due to all those who have been there through all the good and bad times. And thanks to Henry Moret, who wrote the program and used it to generate the table of Gold code preferred pairs in Appendix 7.

All opinions expressed here are my own. I can blame no one else. As always, my desire is to further the use of spread spectrum systems for those applications where they can work to advantage. They are not magic, and do not solve all problems. So many commercial applications are being considered today that a second part has been added to address consumer and commercial concerns, as the same rules do not necessarily apply in the commercial world as apply to military systems.

Thank you, readers, for your interest and support.

ROBERT C. DIXON

Colorado Springs, CO

CONTENTS

SPREAD SPECTRUM SYSTEMS
WITH COMMERCIAL APPLICATIONS

PART I

GENERAL SPREAD SPECTRUM SYSTEMS

1

THE WHAT AND WHYS OF SPREAD SPECTRUM SYSTEMS

Spread spectrum system techniques, applied in recent years, have produced results in communications, navigation, and test systems that are not possible with standard signal formats. In many applications the advent of high-speed transistors and/or integrated circuits was the key to practical-sized-and-powered equipment based on spread spectrum modulation. But what is a spread spectrum system? What are the advantages of spread spectrum modulation? To be sure, there are also disadvantages, but what are they?

These questions and others are posed in this chapter and in those that follow. It is hoped that the reader will find the answers he or she needs in a useful form.

1.1 WHAT IS A SPREAD SPECTRUM SYSTEM?

Before we attempt to define a spread spectrum system, let us be sure that we understand what is meant by a spectrum. Every transmitting or modulating system has a characteristic signature that includes not only the frequency at which the signal is centered, but also the bandwidth of the signal when modulated by the intended signaling waveform. A spectrum, as we speak of it here, is the frequency domain[1703] representation of the signal and, for our purposes, especially the modulated signal. We most often see signals presented in the time domain (that is, as functions of time). Any signal, however, can also be presented in the frequency domain, and transforms (mathematical operators) are available for converting frequency- or time-domain functions from one domain to the other and back again. The most basic of these operators is the Fourier transform, for which the relationship between the

3

time and frequency domains is defined by the pair of integrals

$$F(f) = \int_{-\infty}^{\infty} f(t)\exp -j\omega t \, dt,$$

which transforms a known function of time to a function of frequency, and

$$f(t) = \int_{-\infty}^{\infty} F(f)\exp j\omega t \, df,$$

which performs the inverse operation.

Fourier transforms do not exist for some functions because they require the existence of the integral

$$\int_{-\infty}^{\infty} f(t) \, dt.$$

Therefore discontinuous signals must often be transformed by use of the Laplace integral

$$L(s) = \int_{0}^{\infty} f(t)\exp -st \, dt.$$

Tables of Fourier and Laplace transform pairs for many functions have been generated and well documented (e.g., see 1714 and 1716).

Figure 1.1 illustrates some of the Fourier transforms that are most important in our considerations of spread spectrum systems. For example, we will be interested in the spectra of carriers modulated by pseudorandom binary data streams. Also of some interest will be the spectra of frequency hopped carriers—especially where those carriers are to be used in multiple-access applications, and it is necessary to restrict any interference between mutual users of the same band of frequencies. Note that the frequency spectrum produced by modulation with a square pulse is a $(\sin x)/x$ function, while modulation having a $(\sin x)/x$ envelope produces a square spectrum. Other spectra that will be of special interest to us are those that are produced by triangular or trapezoidal envelopes, and the Gaussian-shaped pulse. Still others will be introduced in the following chapters.

Even as an oscilloscope is a window in the time domain for observing signal waveforms, so is a spectrum analyzer a window in the frequency domain. The many spectrum presentations in this book are almost all spectrum analyzer representations, generated by sweeping a filter across the band of interest and detecting the power falling within the filter as it is swept. This power level is then plotted on an oscilloscope. All spectra referred to are power spectra.

	Time function		Frequency function		
sin x / x Envelope	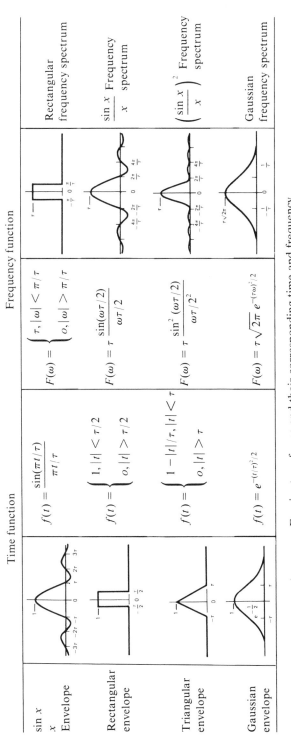	$f(t) = \dfrac{\sin(\pi t/\tau)}{\pi t/\tau}$	$F(\omega) = \begin{cases} \tau, & |\omega| < \pi/\tau \\ 0, & |\omega| > \pi/\tau \end{cases}$		Rectangular frequency spectrum
Rectangular envelope		$f(t) = \begin{cases} 1, & |t| < \tau/2 \\ 0, & |t| > \tau/2 \end{cases}$	$F(\omega) = \tau\,\dfrac{\sin(\omega\tau/2)}{\omega\tau/2}$		$\dfrac{\sin x}{x}$ Frequency spectrum
Triangular envelope		$f(t) = \begin{cases} 1 - |t|/\tau, & |t| < \tau \\ 0, & |t| > \tau \end{cases}$	$F(\omega) = \tau\,\dfrac{\sin^2(\omega\tau/2)}{\omega\tau/2^2}$		$\left(\dfrac{\sin x}{x}\right)^2$ Frequency spectrum
Gaussian envelope		$f(t) = e^{-(t/\tau)^2/2}$	$F(\omega) = \tau\sqrt{2\pi}\,e^{-(\tau\omega)^2/2}$		Gaussian frequency spectrum

Figure 1.1 Four important Fourier transforms and their corresponding time and frequency functions.

Literally, a spread spectrum system is one in which the transmitted signal is spread over a wide frequency band, much wider, in fact, than the minimum bandwidth required to transmit the information being sent. A voice signal, for example, can be sent with amplitude modulation in a bandwidth only twice that of the information itself. Other forms of modulation, such as low deviation FM or single sideband AM, also permit information to be transmitted in a bandwidth comparable to the bandwidth of the information itself. A spread spectrum system, on the other hand, often takes a baseband signal (say a voice channel) with a bandwidth of only a few kilohertz, and distributes it over a band that may be many megahertz wide. This is accomplished by modulating with the information to be sent together with a wideband encoding signal.

The most familiar example of spectrum spreading is seen in conventional frequency modulation in which deviation ratios greater than one are used. The bandwidth required by an FM signal is a function not only of the information bandwidth but of the amount of modulation. As in all other spectrum spreading systems, a signal-to-noise advantage is gained by the modulation and demodulation process. For FM signals this gain advantage (referred to as "process gain") is*

$$3\beta^2 \left(\frac{S}{N} \right)_{\text{info}},$$

where β = deviation ratio, or $\Delta f_{\text{carrier}}/f_{\text{modulation}}$,
$(S/N)_{\text{info}}$ = signal-to-noise ratio in the baseband or information bandwidth.

Wideband FM could be classified as a spread spectrum technique from the standpoint that the RF spectrum produced is much wider than the transmitted information. In the context of this book, however, only those techniques are of interest in which some signal or operation, other than the information being sent, is used for broadbanding (or spreading) the transmitted signal.

Three general types of techniques will be accepted here as examples of spread spectrum signaling methods:

1. Modulation of a carrier by a digital code sequences whose chip rate is much higher than the information signal bandwidth. Such systems are called "direct sequence" modulated systems.
2. Carrier frequency shifting in discrete increments in a pattern dictated by a code sequence. These are called "frequency hoppers." The transmitter jumps from frequency to frequency within some predetermined set; the order of frequency usage is determined by a code sequence.

---

*The factor $3\beta^2$ is the factor by which the FM signal-to-noise ratio exceeds the AM signal-to-noise ratio, given the same average signal power.

3. Pulsed-FM or "chirp" modulation in which a carrier is swept over a wide band during a given pulse interval.

Closely akin to the frequency hoppers are "time hopping" and "time-frequency hopping" systems whose direct chief distinguishing feature is that their time of transmission (usually of low duty cycle and short duration) is governed by a code sequence. In time-frequency hoppers it follows that the code sequence determines both the transmitted frequency and the time of transmission.

Figure 1.2 shows frequency spectra typical of direct sequence and frequency hopping systems. These spectra are often tens to hundreds of megahertz wide, where conventional signals are usually confined to bandwidths in the range of 10–100 kilohertz.

The basis of spread spectrum technology is expressed by C. E. Shannon in the form of channel capacity:

$$C = W \log_2\left(1 + \frac{S}{N}\right),$$

where C = capacity in bits per second,
        W = bandwidth in hertz,
        N = noise power,
        S = signal power.

This equation shows the relationship between the ability of a channel to transfer error-free information, compared with the signal-to-noise ratio existing in the channel, and the bandwidth used to transmit the information.

Letting C be the desired system information rate and changing bases, we find

$$\frac{C}{W} = 1.44 \log_e\left(1 + \frac{S}{N}\right),$$

and, for small S/N, say ≤ 0.1 (as one would wish it to be in an antijam system),

$$\frac{C}{W} = 1.44\frac{S}{N},$$

since

$$\log_e\left(1 + \frac{S}{N}\right) = \frac{S}{N} - \frac{1}{2}\left(\frac{S}{N}\right)^2 + \frac{1}{3}\left(\frac{S}{N}\right)^3 - \frac{1}{4}\left(\frac{S}{N}\right)^4 \cdots \left(-1 < \frac{S}{N} < 1\right)$$

by the logarithmic expansion.

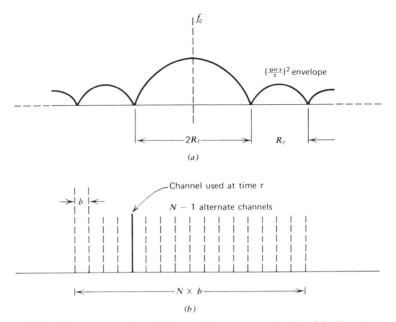

Figure 1.2 Ideal spectra of typical spread spectrum signals: (*a*) direct sequence signal spectrum; (*b*) frequency hopping signal spectrum.

From this equation we find

$$\frac{N}{S} = \frac{1.44W}{C} \approx \frac{W}{C}$$

and

$$W = \frac{NC}{S},$$

thus we see that for any given noise-to-signal ratio we can have a low information-error rate by increasing the bandwidth used to transfer the information; for example, if we want a system to operate in a link in which the interfering noise is 100 times greater than the signal and our information rate 3 kilobits per second (kbps) then our 3-kbps information must be transmitted with a bandwidth of

$$W = \frac{1 \times 3 \times 10^5}{1.44} = 2.08 \times 10^5 \text{ Hz}$$

Incidentally, the information itself may be embedded in the spread spectrum signal by several methods. The most common is that of adding the information to the spectrum-spreading code before its use for spreading modulation. This technique is applicable to any spread spectrum system that uses a code sequence to determine its RF bandwidth (either direct sequence or frequency hopping systems are good candidates). Of course, the information to be sent must be in some digital form because addition to a code sequence involves modulo-2 addition to a binary code. Alternately, information may be used to modulate a carrier before spreading it. This is usually done by some form of angle modulation, for the need in spread spectrum systems is often to output a constant-power RF envelope.

A spread spectrum system, then, must meet two criteria: (1) the transmitted bandwidth is much greater than the bandwidth or rate of the information being sent, and (2) some function other than the information being sent is employed to determine the resulting modulated RF bandwidth.

This is the essence of spread spectrum communications—the art of expanding the bandwidth of a signal, transmitting that expanded signal, and recovering the desired signal by remapping the received spread spectrum into the original information bandwidth. Furthermore, in the process of carrying out this series of bandwidth trades the purpose is to allow the system to deliver error-free information in a noisy signal environment.

It is worthy of note that another technique exists that trades bandwidth for bit-error-rate improvement. To wit: error correction coding, in which parity is added to data for the purpose of detecting or correcting errors in the composite data-plus-parity stream. The parity, however, increases the overall transmission rate, thereby increasing the bandwidth of the transmitted signal. Error correction coding could therefore be considered to be a form of spread spectrum, even though the bandwidth employed is usually much less than with other spread spectrum methods. It follows that the process gain to be expected is also much less than in other spread spectrum methods.

1.2 WHY BOTHER?

The first reaction common among those encountering spread spectrum techniques is "why bother?" The answers are varied and seldom the same. In a world beset by too little RF spectrum to satisfy the ever-growing demands of military, commercial, and private users the question "why spread spectrum" must certainly be considered valid, for spread spectrum systems have almost as many reasons for being as they have users.

Some of the properties that may be cited are the following:

1. Selective addressing capability.
2. Code division multiplexing is possible for multiple access.
3. Low-density power spectra for signal hiding.
4. Message screening from eavesdroppers.
5. High-resolution ranging.
6. Interference rejection.

These properties come about as a result of the coded signal format and the wide signal bandwidth that results. A single receiver or group of receivers may be addressed by assigning a given reference code to them, whereas others are given a different code. Selective addressing can then be as simple as transmitting the proper code sequence as modulation.

Not all of these characteristics, however, are necessarily available from the same system at the same time. It is somewhat anomalous, for instance, to expect, at the same time, a signal that is easily hidden but can also be received in the face of a large amount of interference. Signal-hiding requirements and interference are often at odds, but the same system might be used for both by using lower-power transmissions when low detectability is desired, and high-power transmissions when maximum interference rejection is needed.

When codes are properly chosen for low cross correlation, minimum interference occurs between users, and receivers set to use different codes are reached only by transmitters sending the correct code. Thus more than one signal can be unambiguously transmitted at the same frequency and at the same time; selective addressing and code-division multiplexing are implemented by the coded modulation format.

Because of the wideband signal spectra generated by code modulation, the power transmitted is low in any narrow region. At any rate, the density of a spread spectrum signal is far less than that of more conventional signals in which all the transmitted power is sent in a band of frequencies commensurate with the baseband information bandwidth.

Again, because of the coded signals employed, an eavesdropper cannot casually listen to messages being sent. Though the systems may not be "secure," some conscious effort must be made to decode the messages.

Figure 1.3 illustrates the difference in the power density of unmodulated continuous-wave and typical direct sequence spread spectrum signals.

Resolution in ranging is afforded in accordance with the code rate used, and the sequence length determines maximum unambiguous range. Ranging has been the most prominent and certainly the best known use of spread spectrum systems.

1.3   PROCESS GAIN AND JAMMING MARGIN

The most commonly used quantity in describing or specifying spread spectrum systems is that of "process gain," although it must be pointed out that what is usually intended is not process gain but "jamming margin." Process gain is a readily available quantity, if the bandwidth employed in a system is known and the information rate is available. In Section 1.1, we considered Shannon's information-rate theorem, which shows that one can send information without error, if some method can be devised that employs a wide enough bandwidth to transmit the information. The process gain that we

Figure 1.3   Illustration of power density showing equal power spread spectrum and CW signals, for spread spectrum bandwidth of (*a*) 2 MHz (0.88 MHz, 3 dB) and (*b*) 10 MHz (4.4 MHz, 3 dB).

consider here is just an embodiment of that theorem, in which the signal is spread, and the process gain produced by the spreading and despreading process is equal to the bandwidth ratio between the information and the RF bandwidth used to send it.

A spread spectrum system develops its process gain in a sequential signal bandwidth spreading and despreading operation. The *transmit* part of the process may be accomplished with any one of the band-spreading modulation methods. Despreading is accomplished by correlating the received spread spectrum signal with a similar local reference signal. When the two signals are matched, the desired signal collapses to its original bandwidth before spreading, whereas any unmatched input is spread by the local reference to the local reference bandwidth or more. A filter then rejects all but the desired, narrowband signal; that is, given a desired signal and its interference (atmospheric noise, receiver noise, or jamming), a spread spectrum receiver enhances the signal while suppressing the effects of all other inputs. This process is described in detail in Chapter 2.

The difference in output and input signal-to-noise ratios in any processor is its "process gain." A given processor, for instance, with an input signal-to-noise ratio of 10 dB and an output signal-to-noise ratio of 16 dB would have a process gain of 6 dB.

In spread spectrum processors the process gain available may be estimated by the rule of thumb:

$$\text{process gain} = G_p = \frac{\text{BW}_{\text{RF}}}{R_{\text{info}}},$$

where the RF bandwidth (BW_{RF}) is the bandwidth of the transmitted spread spectrum signal and the information rate (R_{info}) is the data rate in the information baseband channel. This does not mean, however, that a processor can perform when faced with an interfering signal having a power level larger than the desired signal by the amount of the available process gain. Another term, "jamming margin," which expresses the capability of a system to perform in such hostile environments, must be introduced.

Jamming margin is that quantity that is usually intended in the specification of spread spectrum systems, but it is less readily predicted from bandwidth and information-rate information. One can be sure, however, that jamming margin in any given system is always less than the process gain available from that system.

Jamming margin takes into account the requirement for a useful system output signal-to-noise ratio and allows for internal losses; that is (when expressed in dB),

$$\text{jamming margin} = G_p - \left[L_{\text{sys}} + \left(\frac{S}{N} \right)_{\text{out}} \right] = M_j,$$

where L_{sys} = system implementation losses,

$(S/N)_{\text{out}}$ = signal-to-noise ratio at the information output.

For example, a system with 30-dB process gain, minimum $(S/N)_{\text{out}}$ of 10 dB, and L_{sys} of 2 dB would have an 18-dB jamming margin (M_j). It could not be expected to operate with interference more than 18 dB above the desired signal.

The "jamming threshold" of a particular system is of interest in determining how well that system will operate in the presence of interference. Consider the following processor model with signal and noise inputs (the noise input may include interference):

$$\left.\begin{array}{l}\text{signal in}\\ \text{noise in}\end{array}\right\}\left(\frac{S}{N}\right)_{\text{IN}} \bullet - \boxed{G_p} \rightarrow \left(\frac{S}{N}\right)_{\text{out}} = \left(\frac{S}{N}\right)_{\text{in}} \times G_p.$$

For the region of interest, $S \ll N$. Converting to decibels, we have

$$\left(\frac{S}{N}\right)_{\text{out}}(\text{dB}) = \left(\frac{S}{N}\right)_{\text{in}}(\text{dB}) + G_p(\text{dB}),$$

but

$$\left(\frac{S}{N}\right)_{\text{in}}(\text{dB}) = -\frac{J}{S}(\text{dB}).$$

Therefore

$$\left(\frac{S}{N}\right)_{\text{out}}(\text{dB}) = G_p(\text{dB}) - \frac{J}{S}(\text{dB})$$

for the region above the jamming threshold, where the threshold for a particular system is that level for which (see Figure 1.4)

$$\left[G_p(\text{dB}) - \frac{J}{S}(\text{dB})\right]_{\text{Ideal}} - \left[G_p(\text{dB}) - \frac{J}{S}(\text{dB})\right]_{\text{meas'd}} = 1 \text{ dB}.$$

The cause of thresholding lies in such things as tracking loss, nonlinearities, and thresholding of the postcorrelation detector. No system can be designed without a jamming threshold, but with care the threshold point can be placed beyond the normal operating region. All possible thresholding sources should tend to degrade at the same point, for when one part of the system operates at interference levels beyond the others the overall system performance does not improve.

Table 1.1 compares the process gain from spread spectrum processing techniques with that provided by other methods, such as antenna pointing or electronic interference cancellation.

Antenna-cancellation techniques employ the directivity of an antenna to reject a would-be jammer. Such techniques have been used but are not always practical because of positioning restrictions. When geometry is fixed, antenna rejection of interference is limited only by the particular antenna's directivity.

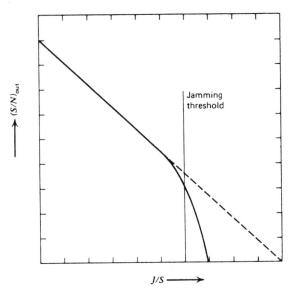

Figure 1.4   Typical process gain curve.

For effective electronic interference cancellation a duplicate of the signal to be rejected may be generated. This replica is then subtracted from the combined interfering and desired signals. The result is a residue consisting of the desired signal only. This technique has proved to be useful but is limited to those situations in which few interfering signals occur (a separate signal simulator and subtractor is required for each signal) and in which relatively noncomplex interfering signals are encountered.

TABLE 1.1   Comparison of Process Gains Available for Various Techniques, Including Signal Rejection and Cancellation

System	Process Gain
Direct sequence	$\dfrac{BW_{RF}}{R_{info}} = TW$
Frequency hopping	$\dfrac{BW_{RF}}{R_{info}} = TW =$ number of frequency choices
Time hopping	$\dfrac{1}{\text{transmit duty cycle}}$
Chirp	Compression ratio $= \tau\, dF = TW$
Antenna rejection	Antenna gain
Electronic cancellation	Depends on accuracy of replica; sometimes 40 dB
Selective rejection	High but useful only against narrowband interference.

Selective rejectors such as notch filters are useful for removing interference signals but are limited to narrowband signals such as CW. Rejection of a wideband signal with a filter could also reject much of the desired signal, which would degrade system operation and net no gain in performance, even though the interference was rejected.

In general, spread spectrum processing offers the most flexible means of providing unwanted signal rejection because it is not necessary to design for rejection of any particular kind of interference and geometric considerations are not normally important. Spread spectrum processing does not, however, offer the highest process gain (or undesired signal rejection) for every situation. The alternatives mentioned here may be combined with spread spectrum techniques to produce a compatible system with the advantages of both spread spectrum and other signal-to-noise improving methods.

Conventions in This Book

The following conventions will be employed:

Direct Sequence Bandwidth. All direct sequence signal bandwidths are assumed to be equal to the 3-dB bandwidth of the spectrum. Thus the bandwidth will vary, depending on the particular type of carrier modulation employed such as biphase, quadriphase, or minimum shift key.

To be strictly correct, we should consider the noise bandwidth, which is

$$B_N = \frac{1}{p(g)} \int_{-\infty}^{\infty} g(f)\, df.$$

For the usual waveforms, the noise bandwidth is

BPSK	R code
QPSK	R code
OQPSK	R code
MSK	0.61 R Code

Noting that for three of the modulation types, the noise bandwidth is equal to the code rate, and there is ample reason to consider process gain to be

$$10\log\frac{R\ \text{code}}{R\ \text{data}}$$

as is often done to estimate the process gain for comparison purposes. We will employ 3-dB bandwidths here unless stated otherwise.

Frequency Hopper Bandwidth. Frequency hoppers are assumed to have BW_{RF} equal to m times the channel bandwidth, where m is the number of frequency channels available.

Information Rate. The information rate is a function of the type of information transferred over the links discussed, and is the minimum bit rate necessary for satisfactory reproduction of the information at the receiver. For voice transmission 3 kHz is the analog bandwidth. Data are consistent with the minimum rate sufficient to convey analog signals* or digital data.

Chip Rate. The chip rate is equivalent to the code generator clock rate or, in frequency hopping systems, the hop rate.

Postcorrelation Bandwidth. The signal bandwidth following the receiver correlation process is the two-sided information bandwidth produced by the baseband modulating signal or by a frequency hopped carrier.

All other bandwidths are 3-dB bandwidths.

PROBLEMS

1. Given a requirement for operating in an environment in which the power of an interfering signal is 250 times the desired signal power, how much jamming margin is required?

2. If in the system described above a 10-dB output signal-to-noise ratio is required, what is the minimum process gain allowable?

3. The information rate is 6 kbps. What is the minimum RF bandwidth needed?

4. Assume that a direct sequence approach has been employed. What code rate should be used?

5. How many channels would be required by a frequency hopper for similar performance?

6. Given that chirp modulation is required and the information rate is 10 kbps, what RF sweep bandwidth is the minimum that would produce a 20-dB process gain?

7. A direct sequence system has a code rate of 1 Mcps (cps = chips per second) and an information rate of 100 kbps. Is it worth it?

---

*Data rate representing an analog signal is at least the Nyquist rate, or 2 bits per Hertz. The actual rate depends on the A/D technique used.

8. A frequency hopper has a 5-MHz RF bandwidth available and an information rate of 100 kbps. Comment.

9. How does a chirp system meet the criterion that the RF transmission bandwidth employed is not a function of the information sent?

10. How does a direct sequence system meet the same criterion?

11. Same question for the frequency hopping.

12. What is the prime limiting factor in the processing gain of a time-hopping system?

2

SPREAD SPECTRUM TECHNIQUES

Several basic spread spectrum techniques available to the communications system designer are listed in Chapter 1 and described in a general way. This chapter gives detailed descriptions of the various techniques and the signals generated. In addition to the two most important (or at least most prevalent) forms of spread spectrum modulation, other useful techniques such as chirping, time hopping, and various hybrid combinations of modulation forms are described. Each is important in that it has useful application. Curiously enough, there has been little competition between the different techniques. The historical tendency has been to confine each form to a particular field. Chirp modulation, for instance, has been used almost exclusively in radar. Here all are presented together to give the reader a reasonably complete picture of contemporary spread spectrum technology.

2.1 DIRECT SEQUENCE SYSTEMS

Direct sequence (or, to be more exact, directly carrier-modulated, code sequence modulation) systems are the best known and most widely used spread spectrum systems. This is because of their relative simplicity from the standpoint that they do not require a high-speed, fast-settling frequency synthesizer. The JPL ranging technique described by Golomb *et al.* in *Digital Communications with Space Applications*,[219] and used in various space applications, is a direct sequence approach. Today, direct sequence modulation is being used for communication systems and test systems, and even laboratory test equipment capable of producing a choice of a number of code sequences or operating modes is available. It is reasonable to expect that direct sequence modulation will become a familiar form of modulation in many areas in the foreseeable future. Even now, commercial applications of such systems are being contemplated (see part 2, Commercial Applications).

Characteristics of Direct Sequence Signals

Direct sequence modulation, as mentioned in Chapter 1, is just exactly that —modulation of a carrier by a code sequence. In the general case, the format may be AM (pulse), FM, or any other amplitude- or angle-modulation form. Very common, however, is 180° biphase phase-shift keying, though the reasons for it may not be immediately obvious. The basic form of direct sequence signal is that produced by a simple, biphase-modulated (PSK) carrier. A spectrum typical of this signal format is shown in Figure 2.1 The main lobe bandwidth (null-to-null) of the signal shown is twice the clock rate of the code sequence used as a modulating signal. Each of the sidelobes has a bandwidth from null-to-null that is equal to the clock rate; that is, if the code sequence being used as a modulating waveform has a 5-Mcps (cps = chips per second) operating rate, the main lobe null-to-null bandwidth will be 10 MHz and each sidelobe will be 5 MHz wide. This is exactly the case in Figure 2.1.

Typically, the direct sequence biphase modulator has the form shown in Figure 2.2, where a balanced mixer whose inputs are a code sequence and an RF carrier operates as the biphase modulator. In the time domain the

Vertical scale, 10 dB/cm

Horizontal scale, 5 MHz/cm

Figure 2.1 Direct sequence spread spectrum signal 180° biphase-modulated by a 5 Mcps code.

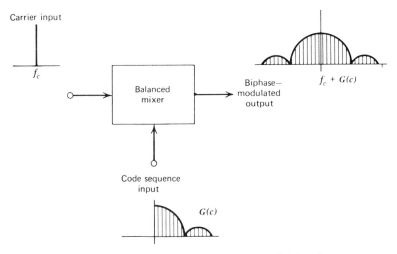

Carrier input

f_c

Balanced
mixer

Biphase—
modulated
output

$f_c + G(c)$

Code sequence
input

$G(c)$

Figure 2.2 Direct sequence modulator (biphase).

biphase-modulated carrier looks like the signal shown in Figure 2.3. There
the carrier is transmitted with one phase when the code sequence is a "one,"
and a 180° phase shift when the code sequence is a "Zero." In the usual
biphase-modulated system, which phase is which is of little consequence. It is
worthy of note that, although other modulation forms such as PAM (pulse-
amplitude modulation) could be used, balanced* modulation is the most
common in direct sequence systems. This is true for several reasons:

1. The suppressed carrier produced is difficult to detect without resorting
   to somewhat sophisticated methods. Figure 2.4 shows a typical direct
   sequence signal (as does Figure 2.1 in less detail) with carrier suppression. It is obvious that a conventional receiver would not be useful for
   detecting the carrier here because it is well below the level of the
   "noise" produced by the code modulation. For comparison, Figure 2.5
   shows the same modulator but without carrier balance; therefore, the
   carrier is no longer suppressed.

2. More power is available for sending useful information because the
   transmitter power is used to send only the code-produced signal.

3. The signal has a constant envelope level (except for incidental AM) so
   that the transmitted power efficiency is maximized for the bandwidth
   used. PAM modulation in which a carrier would be pulse modulated by
   a code as in Figure 2.6 is acceptable; it produces a similar $[\sin x)/x]^2$

---

*Biphase-balanced modulation, for example, is well adapted to code modulation in that one
carrier phase may be transmitted for a one in the code and the opposite phase for a zero. This
produces a signal $A \cos \omega_c t \cos \omega_m t = A \cos(\omega_c \pm \omega_m)t$ after filtering in which the carrier term
is absent. Balanced modulation is discussed in detail in Chapter 4.

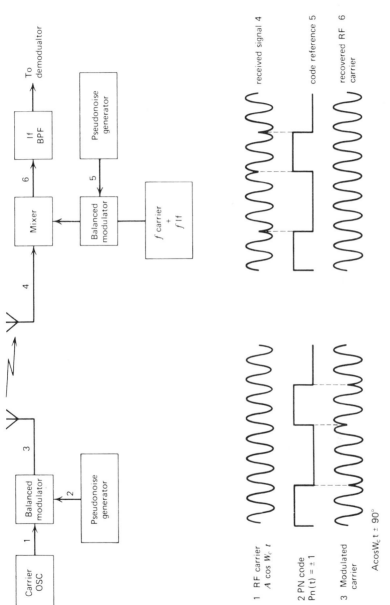

Figure 2.3 Overall direct sequence system showing waveforms.

1 RF carrier
$A \cos W_c t$

2 PN code
$Pn(t) = \pm 1$

3 Modulated carrier

$A \cos W_c t \pm 90°$

received signal 4

code reference 5

recovered RF 6 carrier

21

Figure 2.4 Direct sequence, suppressed carrier spectrum biphase, code modulated.

Figure 2.5 Direct sequence unsuppressed carrier, spectrum biphase and code modulated.

Figure 2.6   Pulse amplitude-modulated carrier, code modulated.

power spectrum but lacks the same effective power at the receiver. Thus for the same range of operation higher peak power would be necessary. PAM also does not have a suppressed carrier.

The simplified block diagram in Figure 2.3 illustrates a typical direct sequence communications link. It shows that the direct sequence system is similar to (and, in fact, can be) a conventional AM or FM communications link with code modulation overlaid on the carrier. In actual practice the carrier is not usually modulated by baseband information. The baseband information is digitized and added to the code sequence. For this discussion, however, we will assume that the RF carrier has already been info-modulated before code modulation, because this somewhat simplifies discussion of the modulation-demodulation process.

After being amplified, a received signal is multiplied by a reference with the same code and, assuming that the transmitter's code and receiver's code are synchronous, the carrier inversions transmitted are removed and the original carrier restored. This narrowband restored carrier can then flow through a bandpass filter designed to pass only the baseband-modulated carrier.

Undesired signals are also treated in the same process of multiplication by the receiver's reference, that maps the received direct sequence signal into the original carrier bandwidth. Any incoming signal not synchronous with the receiver's coded reference (a wideband signal) is spread to a bandwidth equal to its own bandwidth plus the bandwidth of the reference; that is, the receiver's multiplier output signal variance is the covariance of its two input signals.

Because an unsynchronized input signal is mapped into a bandwidth at least as wide as the receiver's reference, the bandpass filter can reject almost all the power of an undesired signal. This is the mechanism by which process gain is realized in a direct sequence system; that is, the receiver transforms synchronous input signals from the code-modulated bandwidth to the baseband-modulated bandwidth. At the same time non-synchronous input signals are spread at least over the code-modulated bandwidth.

The system information bandwidth sets the postcorrelation filter bandwidth, and this filter in turn determines the amount of power from an unsynchronized signal which reaches the information demodulator. Therefore we see that, as already stated, the multiplication-and-filtering sequence provides the desired signal with an advantage or "process gain."

Radio-Frequency Bandwidth in Direct Sequence Systems

Radio-frequency (RF) bandwidth in direct sequence systems, as discussed in the preceding sections, directly affects system capabilities. If a 20-MHz bandwidth is available, the process gain possible is limited by that 20 MHz.

Several approaches are useful in choosing the proper bandwidth in a signal-hiding application; the interest is in reducing the power transmitted per Hertz of bandwidth, and wide bandwidths are used. When a maximum process gain for interference rejection is desired, the bandwidth again should be large. When either frequency allocation or the propagation medium does not permit a wide RF bandwidth, some restraint must be applied.

A prime consideration in spread spectrum systems (and especially direct sequence systems) is the bandwidth of the system with respect to the interference furnished to other systems (that may not necessarily be spread spectrum systems) operating in the same or adjacent channels. Referring to Figures 2.4 and 2.5 will show that high sidelobe energy is present in some direct sequence formats, even though the content of the signal may not be enhanced by transmitting the sidelobes. In systems such as JTIDS (Joint Tactical Information Distribution System), for example, the band employed is shared with both IFF and TACAN systems. Therefore, a special form of direct sequence modulation called MSK (minimum shift key) is used.

There are, in fact, a number of waveforms, all closely related, that can be employed in direct sequence systems to control the level of energy contained in the unneeded sidelobes. We shall discuss these in detail in Chapter 4, but for now we simply summarize the characteristics of the available waveforms in Table 2.1. That the simple biphase or quadriphase direct sequence signal has a $[(\sin x)/x]^2$ spectrum can be simply shown: Given a rectangular pulse whose duration is T and whose amplitude is A over $T \pm T/2$ and zero

TABLE 2.1   Comparison of Direct Sequence Waveforms

Waveform	Null-to-Null Main Lobe BW	3-dB BW	First Sidelobe	Rolloff Rate
BPSK	2 × code clock	0.88 × code clock	− 13 dB	6 dB/octave
PAM	2 × code clock	0.88 × code clock[a]	− 13 dB	6 dB/octave
QPSK	2 × code clock[b]	0.88 × code clock	− 13 dB	6 dB/octave
OQPSK	2 × code clock[b]	0.88 × code clock	− 13 dB	6 dB/octave
MSK (classic)	1.5 × code clock	0.66 × code clock	− 23 dB	12 dB/octave

[a]Not including unsuppressed carrier.
[b]Requires two codes at same rate as single BPSK code.

elsewhere,

$$C_n = \int_{-T/2}^{T/2} A \exp(-j\omega t)\, dt = \frac{A}{j\omega} \exp(-j\omega t) \int_{-T/2}^{T/2}$$

$$= \frac{A \exp(j\omega\, T/2) - A \exp(-j\omega T/2)}{j\omega}$$

$$= \frac{2A}{\omega} \sin \frac{\omega T}{2} = TA \frac{\sin \omega T/2}{\omega T/2}.$$

This expression is obviously of the form $(\sin x)/x$: It is the voltage distribution for the signal whose power distribution is then $[(\sin x)/x]^2$. This shows what is readily apparent when using a spectrum analyzer—the spectrum of biphase and quadriphase signals has a $[(\sin x)/x]^2$ distribution. Pulse shaping, as we shall see, can provide significantly improved power distribution, and that is the approach used in the other forms of direct sequence modulation that we will discuss.

In any case, the direct sequence system has a main lobe bandwidth that is a function of the waveshape and the code rate used. For example, in Figure 2.1 the code rate is 5 Mcps, Figure 2.4 is generated with a code rate of 25 Mcps, and the chip rate in Figure 2.7 is a very high 100 Mcps. All of these signals are from biphase modulators, but the spectra are identical to those that would be produced by the equivalent quadriphase or offset quadriphase modulator. Comparison of these spectra will show that the spectral shape is $[(\sin x)/x]^2$ and the null-to-null bandwidth is $2R_c$.

The signal lost as a result of accepting only the main lobe is small. Only 10% of the power in a BPSK or QPSK signal is contained in the sidelobes (see Figure 2.8). Signal-power loss is not the only effect of bandwidth

Figure 2.7   High-speed code-modulated direct sequence (BPSK) spectrum.

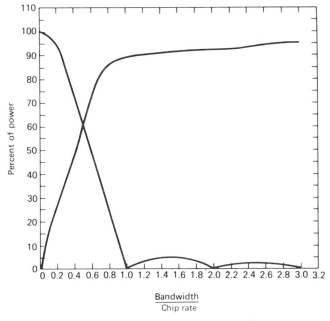

Figure 2.8   Power distribution in $[(\sin x)/x]^2$ spectrum.

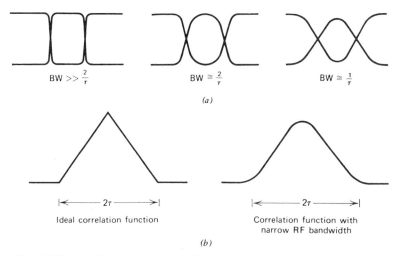

Figure 2.9 RF bandwidth restriction and its effect on typical direct sequence signals: (a) RF envelope of direct sequence signal for various RF bandwidths; (b) effect of bandwidth restriction on correlation function.

restriction, however. The sidelobes contain much of the harmonic power of the modulation. Thus restriction to a narrow RF bandwidth is the equivalent of restricting the rise and fall times of the modulating code. Also, the sharply peaked triangular correlation of a coded signal is rounded by a bandwidth restriction. Figures 2.9 and 2.10 show the effect of band restriction on the RF envelope and correlation function of a direct sequence signal.

Quadriphase (QPSK) transmission (discussed in Chapter 4) is one method of restricting RF bandwidth when a given code rate is to be used. QPSK does halve the RF bandwidth required, but processing gain is then a function of the halved bandwidth. More than four phases can also be used to restrict bandwidth further, but the same rationale holds; for example, a system with a 22.75 Mcps code rate and 10 kbps data to be sent would require a 20 MHz (3 dB) bandwidth using the standard form of BPSK modulation. Its process gain would be 20 MHz/10 kbps = 2×10^3. On the other hand, QPSK could reduce the bandwidth required to only 10 MHz but would also reduce process gain to 10 MHz ÷ 10 kbps = 1×10^3, or 3 dB less.

The choice of modulation and code rate is highly dependent on the system in which they are to be used. One must consider bandwidth availability, process gain required, and the basic data rates. RF bandwidth restriction is of more than usual importance in direct sequence ranging systems, for degradation of the correlation function, as shown in Figure 2.9, could result in a loss of ability to measure timing precisely. Bandwidth restriction, therefore, can reduce range resolution (i.e., increase the minimum measurement).

Figure 2.10 This photograph shows amplitude distortion of a bandlimited direct sequence signal.

Distribution of the power in a direct sequence signal (whether biphase or quadriphase) is plotted in Figure 2.8 along with a plot that shows the relative amplitudes of the first two sidelobes over a region equal to three times the code rate. Figure 2.8 shows that 90% of the total power is contained in a bandwidth equal to twice the chip rate, 93% of the total power in four times the chip rate, and 95% of the total power is in six times the chip rate.

Further consideration of RF bandwidth in both direct sequence and frequency hopping systems is discussed in Chapter 7 with respect to the RF transmission and reception link. Here it is sufficient to say that the direct sequence-signal bandwidth is a direct function of the code rate, modulation envelope, and number of phase positions employed.

Direct Sequence Process Gain

Process gain in a direct sequence system is a function of the RF bandwidth of the signal transmitted, compared with the bit rate of the information. The gain in question is exhibited as a signal-to-noise improvement resulting from the RF-to-information bandwidth tradeoff.

The usual assumption taken is that the RF bandwidth is that of the main lobe of the $[(\sin x)/x]^2$ direct sequence spectrum, which is 0.88 times the bandwidth-spreading code clock rate. Therefore, for a system having a 10-Mcps code clock rate and a 1-kbps information rate the process gain would be $(0.88 \times 10^7)/(1 \times 10^3) = 8.8 \times 10^3$ or 39 dB. (Such parameters are typical for practical systems.)

Let us consider the limitations that exist with respect to expanding the bandwidth ratio arbitrarily so that process gain may be increased indefinitely. (Unfortunately, physical limitations prevent this increase.) Only two parameters are available to adjust process gain. The first, the RF bandwidth, depends on the code rate used. If we wish to have an RF (null-to-null) bandwidth 100 MHz wide, the code rate would be at least 50 Mcps. On this basis, how wide is it practical to make the system RF bandwidth? At present, integrated circuits are available that allow limited code generation at rates up to two or even three Gcps. Is there a profit in going higher or, for that matter, in using these chip rates? Consider higher chip rates: doubling the present state-of-the-art code rates would increase process gain by only 3 dB, which is at best a modest gain when compared to the effort required to double the operating speed of present circuits.

Another consideration is that as code rates go higher, operating errors must go lower in inverse proportion. To be operationally useful a code generator should be able to operate for hours or even days without error. If we consider what this means in error rate, we see that a 100-Mcps code sequence generator must operate error free for 3.6×10^{11} chips, just to allow a system to operate for 1 h. On the other hand, high-speed logic circuits tend toward noise sensitivity and are more susceptible to error.

Finally, but not of least concern, is that high-speed digital circuits consume large amounts of current and power dissipation is high. High-speed code generators often consume as much as 1.0 W/stage.

The foregoing reasons, coupled with the problems of spectrum occupancy, equipment implementation (both RF and digital), and propagation constraints, tend to limit the code rates used for band spreading. For the near future, it appears that code rates of 50-100 Mcps are the highest that are practical for general use.

Reiterating an important point, we state that to improve process gain significantly beyond that obtainable with a 100-Mcps code, we would be forced to generate codes in the 1×10^9 cps region. This, at present, is not practical for the long, multifeedback sequence generators required for useful communications systems. Though codes generated at rates over 500 Mcps have been shown as laboratory curiosities, they have been extremely short and exhibited errors at rates in the 1×10^{-3} range. A 100-Mcps code generator with even a 1×10^{-8} error rate could be expected to make one mistake every second. This would prevent data transmission even if it permitted system synchronization. (It is unlikely that the system could be synchronized.)

Data rate reduction to improve process gain is limited in its extent by the willingness of a user to slow information transfer and by the overall stability of the transmission link. Once the data rate is slowed to the tens of bits-per-second area or lower, such things as local oscillator phase noise or instability in the propagation medium become significant and can cause errors. It is conceded that systems with 1 bps and lower data rates have been built. Here, however, 10 bps is recommended as a lower limit.

Now, if we consider systems with code rate limited at 100 Mcps and data rate limited at 10 bps, we see that for practical systems process gain is limited at about

$$\frac{0.88 \times 10^8}{10} = 0.88 \times 10^7, \text{ or 69 dB.}$$

It is understood that future systems will transmit data at gigabit rates and that even as this paragraph is being written techniques for building such systems are being worked out. Thus, in setting a seat-of-the-pants practical limit of 100 Mcps on spread spectrum code generators for direct sequence communications, it is understood that this limit appears paradoxical. However, there is a difference in high-speed codes used for RF spectrum spreading and in high-speed data (the high-speed data would also result in a wide RF spectrum, of course), even though both are bit streams. The difference in the two lies in the action of errors on the system; that is, an error (or, for that matter, many errors up to some correctible limit) might be tolerated in a data-modulated system, whereas an error in a code generator would completely disable the code-modulated system until resynchronization occurs.

Direct Sequence Code and Spectrum Relationships

It has been repeatedly stated here, and emphasized in illustrations, that the spectral power envelope of a direct sequence signal is of a $[(\sin x)/x]^2$ form. This is an expected result since any good set of Fourier transform pairs [1714] will show that the frequency function corresponding to a square pulse is $(\sin x)/x$ (which is a voltage distribution function), and that code modulation produces a series of pulses. Because the power envelope is a function of the voltage squared, the $[(\sin x)/x]^2$ power spectrum results. {As an aside, we point out that a spectrum-analyzer presentation of a direct sequence signal always appears to have a $[(\sin x)/x]^2$ distribution because spectrum analyzers do not display phase information, only amplitude.}

Any code sequence used to modulate a direct sequence system produces a similar power envelope, but what about the details of distribution within that envelope?

Observation of a pseudonoise code sequence will show that it is made up of a series of variable-period pulses, which could be viewed as half-period square waves. Figure 2.11 illustrates this concept. These square-wave half-periods vary in duration from one code clock chip to n chips for a linear maximal sequence $2^n - 1$ bits in length. Each of these half-periods has a $(\sin x)/x$ spectrum associated with it. As a code sequence modulates a direct sequence transmitter, the output spectrum is actually a composite made up of

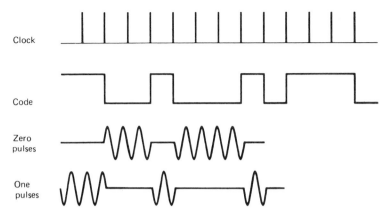

Figure 2.11   Direct sequence transmission as a pulse series.

a series or group of spectra produced by the various half-wave components of the code.

Because the pulse of shortest duration in any code sequence is equal in length to the code sequence clock period, it follows that the frequency spectrum for the composite modulation containing that code sequence must have a main-lobe bandwidth such that its first nulls fall at the code clock rate. Reinforcing this expectation is the fact that the distribution of single ones and zeros in a code sequence is such that they outnumber all other pulse lengths. Chapter 3 includes a detailed discussion of the distribution of the half-periods (or runs) in a pseudonoise code sequence. For now, it is enough that one-fourth of the runs in a maximal code are one chip long, and that the number of runs or lengths longer than one decreases by half for each chip of increase in run length.

The number of frequency sets available is a function of the length of the code (and code generator being used). For an n-chip sequence generator, there are $n + 1$ frequency sets, and the spacing of individual frequency components is as narrow as $R_c/(2^n - 1)$, where the code length is $2^n - 1$ (maximal), n is the sequence generator length, and R_c is the code chip or clock rate.

The Minimum Shift Key Technique

Minimum shift keying (MSK) is a direct sequence modulation technique that offers exceptionally good suppression of signal sidelobes and fast sidelobe rolloff. MSK has a main lobe whose power distribution is approximately the same as BPSK and QPSK, but the null-to-null bandwidth is $1.5R_c$.* Although somewhat more difficult to implement than BPSK or QPSK, MSK does offer

---

*Classic MSK.

some advantages that other forms of direct sequence modulation cannot match. We hasten to state that there are many forms of MSK modulation, each with different spectrum and characteristics. Offset QPSK is, in fact, one form of MSK modulation. We will further discuss these forms and compare their characteristics in detail in Chapter 4.

2.2   FREQUENCY HOPPING†

"Frequency hopping" modulation is more accurately termed "multiple-frequency, code-selected, frequency shift keying." It is nothing more than FSK (frequency shift keying) except that the set of frequency choices is greatly expanded. Simple FSK most often uses only two frequencies; for example f_1 is sent to signify a "mark," f_2 to signify a "space." Frequency hoppers, on the other hand, often have thousands of frequencies available. One real system[111] has 2^{20} discrete frequency choices, randomly chosen, each selected on the basis of a code in combination with the information transmitted. The number of frequency choices and the rate of hopping from frequency to frequency in any frequency hopper is governed by the requirements placed on it for a particular use.

Characteristics of Frequency Hopping Signals

A frequency hopping system or "frequency hopper" consists basically of a code generator and a frequency synthesizer capable of responding to the coded output from the code generator. A great deal of effort has been expended in developing rapid-response frequency synthesizers for spread spectrum systems.

Ideally, the instantaneous frequency hopper output is a single frequency. Practically, however, the system user must be satisfied with an output spectrum which is a composite of the desired frequencies, sidebands generated by hopping, and spurious frequencies generated as by products.

Figure 2.12 is a simplified block diagram of a frequency hopping transmission system. The frequency spectrum of this frequency hopper is shown in Figure 2.13.

Over a period of time, the ideal frequency hopping spectrum would be perfectly rectangular, with transmissions distributed evenly in every available frequency channel. The transmitter should also be designed to transmit, to a degree as close as practical, the same amount of power in every channel.

The received frequency hopping signal is mixed with a locally generated replica, which is offset a fixed amount such that $\{f_1, f_2, \ldots, f_n\} \times \{f_1 + f_{IF}, f_2 + f_{IF}, \ldots, f_n + f_{IF}\}$ produces a constant difference frequency f_{IF} when transmitter and receiver code sequences are in synchronism.

†See the Bibliography, Sections 1, 7, and 8.

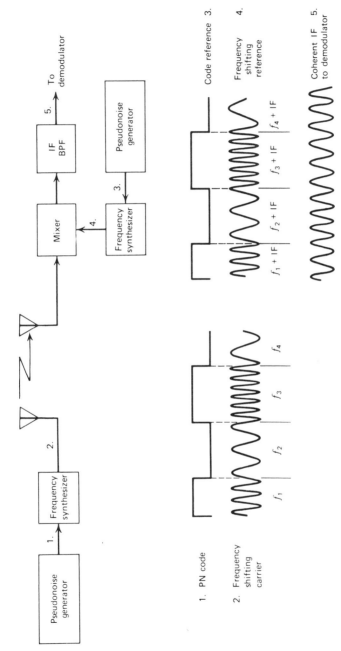

Figure 2.12 Basic frequency hopping system with waveforms.

33

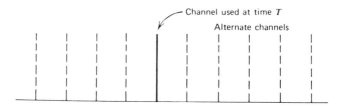

Figure 2.13   Ideal Frequency hopping signal spectrum.

As in a direct sequence system, any signal that is not a replica of the local reference is spread by multiplication with the local reference. Bandwidth of an undesired signal after multiplication with the local reference is again equal to the covariance of the two signals; for example, a CW signal appearing at the frequency hopping receiver's input would be identical to the local reference when translated to the IF frequency. A signal with the same bandwidth as the local reference (but nonsynchronous) would have twice the reference bandwidth at the IF. The IF following the correlator, then, can reject all of the undesired signal power that lies outside its bandwidth. Because this IF bandwidth is only a fraction of the bandwidth of the local reference, we can see that almost all the undesired signal's power is rejected, whereas a desired signal is enhanced by being correlated with the local reference.

In the preceding section on direct sequence systems, we saw that a direct sequence system's operation is identical from the standpoint of undesired signal rejection and remapping of the desired signal. From this general viewpoint, direct sequence and frequency hopping systems are identical; they are different, however, in the details of their operation.

Process gain for the frequency hopping system (in which channels are contiguous) is the same as that for the direct sequence system; that is,

$$G_p = \frac{\mathrm{BW}_{\mathrm{RF}}}{R_{\mathrm{info}}}.$$

If the channels are not contiguous, a better measure of process gain is

$$G_p = \text{the number of available frequency choices} = N,$$

which also holds for contiguous channels; for example, a frequency hopping system containing 1000 frequency choices could have 30-dB available process gain.

The only restriction in use of these simple calculations for process gain is that interchannel interference (crosstalk) is neglected. When IF bandwidth or channel overlap does not provide for unambiguous channel recognition, an allowance must be made for this source of error, which reduces process gain.

Figure 2.14 Ideal frequency hopper output spectrum.

Figures 2.14 through 2.17 illustrate actual frequency hopping signals typical of those that might be seen with a spectrum analyzer.

Figure 2.14 shows an ideal frequency hopping spectrum, actually far better than those normally seen from wideband, multifrequency synthesizers. This particular spectrum would be seen from a narrowband frequency hopper (only 28 frequencies are shown), so that the output level is flat over the output band. Compare this spectrum with those of Figures 2.15 and 16.

Figure 2.15 is a time exposure of a several-thousand-frequency hopper output. The bright center trace is a CW carrier equal in power to the frequency hopping signal. Note that the output spectrum of this wider-band synthesizer is not so flat in amplitude as the preceding narrowband hopper.

Figure 2.15 Multifrequency hopper spectrum with CW interference.

Figure 2.16 Multifrequency hopper spectrum with partial band multitone interference.

A wideband frequency hopper, combined with partial band interference, is shown in Figure 2.16. The interference is the larger set of signals at its center. The actual frequency hopping signal is the set of smaller, much-wider-band signals. Here again it is obvious that amplitude linearity across the band is less than perfect, as is spurious suppression.

Where spurious signal suppression is desired, it is necessary to use pulse shaping of the frequency hopped carrier. Transmission of a rectangular pulse results in (as might be expected) a $(\sin x)/x$ spectrum centered at each frequency cell. This is the origin of most of the spurious signals seen in Figure 2.16.

It is not necessarily obvious that the output spectrum for a frequency hopping system needs to be flat. If bit decisions are made on the basis of signal-power comparison in alternate channels, it becomes important that transmitted signal variation be small. When the mark/space detector has a narrow threshold band, interference must be within that band to be effective. Use of a narrow decision band depends, however, on having a transmitter that has closely controlled amplitude from channel to channel.

Determining Frequency Hopping Rate and Number of Frequencies

The minimum frequency switching rate usable in a frequency hopping system, is determined by a number of parameters:

1. The type of information being sent and its rate.
2. The amount of redundancy used, if any.
3. Distance to the nearest potential interferer.

Information in a frequency hopping system may be transmitted in any way

Figure 2.17 Frequency hopping and direct sequence ssignal comparison. (*a*) Integrated circuit synthesizer at 500 hops/sec, frequency separation 10 kHz. (*b*) Direct sequence signal compared to equivalent-power frequency hopped signal.

available to other systems. Usually, however, some form of digital signal is used, whether the information is a digitized analog signal or data. Assume for the present that some digital rate is prescribed and that frequency hopping has been chosen as the transmission medium. How, then, is the frequency hopping or "chip" rate chosen?

A frequency hopping system must have a large number of frequencies usable on demand. The number required is dependent on the system error rate; for instance, a library of 1000 frequencies could provide good operation when interference or other noise is evenly distributed at every available frequency. For equal distribution of noise in every channel the noise power required to block communications would approach 1000 times the desired signal power (in other words, the jamming margin would be 30 dB). Unless some form of redundancy that allows for bit decisions based on more than one frequency is used, however, a single narrowband interferer would cause an error rate of 1×10^{-3}, which is generally acceptable for digital data. For a simple frequency hopping system without any form of transmitted data redundancy, the expected error rate is just J/N, where J equals the number of interferers with power greater than or equal to signal power and N equals the number of frequencies available to the system. We see then that some form of data transmission other than simple binary frequency hopping is required because of the inherent high error rates caused by a small amount of interference.

Error rate for a frequency hopping system, in which we assume that binary redundant FSK is used (f_a = mark, f_b = space) may be approximated* by the expression for the cumulative binomial expansion:

$$P_e = \sum_{x=r}^{c} \binom{c}{r} p^x q^{c-x},$$

where p = error probability for a single trial = J/N,

J = the number of jammed channels,

N = the number of channels available to the frequency hopper,

q = probability of no error for a single trial = $1 - p$

c = the number of chips (frequencies sent per bit of information),

r = the number of wrong chip decisions necessary to cause a bit error.

A chip decision error is defined as occurring any time the interference power in a "space" channel exceeds the power in an intended "mark" channel (or vice versa) by some amount e that is sufficient to cause a decision in favor of what was not intended.

---

*This error-rate approximation is good to the extent that the effects of interference falling into the occupied signal channel are ignored; that is, in addition to the probability J/N of falling into the channel that signifies the opposite of what is intended, there is a probability (also J/N) that an interfering signal will fall into the intended or occupied channel. When this happens, the phase relationship of the two signals occupying the channel determines what occurs. To complicate matters further, there is the probability that an interfering signal will fall in both the intended and opposite channels, in which case the mark/space decision is based on whether the signal in the opposite channel is greater than the composite of the signals in the intended channel.

If three or more[†] frequencies (chips) are sent for each bit of information, we find that performance with interference is greatly enhanced. When the receiver's bit decision is made on the basis of two out of three being correct, a single-channel interferer would cause no more than

$$\binom{3}{2} p^2 q^{3-2} = 3p^2 q^{3-2} = 3p^2 q,$$

but $q = 1 - p$, therefore, $3p^2 q = 3(p^2 - p^3)$ errors.

Returning to our 1000-channel system, we see that p is $1/1000$ and $q = 1 - 1/1000 = 0.999$, and the error rate is now

$$3\left(\frac{1}{10^6} - \frac{1}{10^9} \right) \approx 3 \times 10^{-6},$$

which is far better than the 1×10^{-3} error rate given by the simple one-chip-per-bit system.

We have paid for the reduction in errors with a threefold increase in transmission (hop) rate; our frequency synthesizer must now hop three times as fast. Given that the synthesizer can do so, however, it is almost always advantageous to make the three-for-one trade in speed to reduce errors 3000 to 1.

The possibility of increasing redundancy (and, with it, the hop rate) to improve bit error rate depends on the system parameters. It is obvious that the more chips sent for a bit, the lower the bit error rate is. The hopping rate required and the RF bandwidth increase in direct proportion. If either the bandwidth allocated or the frequency capability of the frequency synthesizer is limited, then some tradeoff must be made between sending a larger number of chips per bit and reducing the number of frequencies available. As an example, assume that a data source at 1 kbps is to be transmitted and that a 10-MHz RF bandwidth is allowable. The 1-kb data rate suggests that frequency hopping be at least 1 khps, so that the main lobe of the dehopped $[(\sin x)/x]^2$ carrier spectrum is 2 kHz wide. Therefore, if no overlap is allowed (ignoring sidelobes), 10 MHz/2 kHz = 5000 frequency hopping channels are available. The maximum desired error rate is assumed to be 1×10^{-3}, when the interference-to-signal ratio is $100:1$.

Now if only one chip is sent per bit of data, $P_s + e$ interference power in 100 of our 5000 available channels (which would equal our desired $100:1$

[†]Three or more chips per bit of information allows a majority decision (or some other decision) to be made. Usually an odd number of bits is used. In majority decision systems (in which a decision is based on two of three, three of five, . . . , trials) an odd number of trials (r) is always better than $r + 1$ or $r - 1$ trials because another trial gives no more information but does increase the chance of error. A three-of-five decision, for instance, allows two errors in five, yet gives a correct decision. Increasing the number of trials to six requires four of the six to be correct and still allows only two errors.

interference power ratio) would have a probability of causing errors of one in 50. If three chips are sent for each bit of information, the transmission rate goes to 3000 hps and the received signal bandwidth to 6 kHz. Thus the number of nonoverlapping signal channels available is 10 MHz/6 kHz = 1666 and the 100 : 1 interference ratio gives us $J/N = 100/1666 = 0.06 = P_e$, for 100 channels containing interference among the 1666 available. The error rate could then be expected to be

$$\sum_{x=2}^{3} \binom{3}{x} 0.06^x 0.94^{3-x} = 1.2 \times 10^{-2},$$

which still does not reach our error-rate goal.

Observing Figure 2.18, we see that to reach the desired error rate (1×10^{-3}) it is necessary to decrease the ratio J/N or change the decision criteria. For a two-out-of-three (one error allowable in three) decision at a

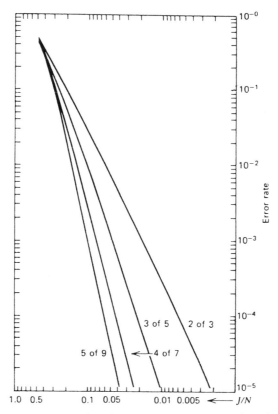

Figure 2.18 Error rate versus fraction of channels jammed (J/N) for various chip decision criteria in multichip transmissions.

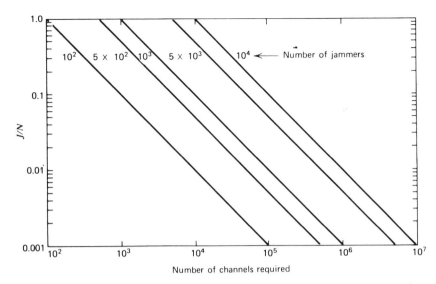

Figure 2.19   Number of channels required versus fraction of channels jammed (J/N) and number of jammers.

10^{-3} error rate we see that J/N can be no greater than 0.019 and that $100/0.019 = 5260$ transmit frequencies are the minimum allowable for $100:1$ interference rejection. At three chips per bit of information the receiver dehopped-signal bandwidth is 6 kHz. Therefore, the RF bandwidth would be $6 \times 5260 \times 10^3 = 31.56$ MHz.

Alternatively, if the number of chips is increased to five per information bit, we see that J/N is increased to 0.047 at a 10^{-3} error rate. For the same interference ratio $(100:1)$ we are required to use $100/0.047 = 2130$ channels. Because the contiguous (nonoverlapping) dehopped-signal bandwidth is now 10 kHz, the RF bandwidth required is only 21.3 MHz, or approximately one-third less than that required for decisions made on the basis of two of three chips.

Figure 2.19 plots bandwidths required for various data rates, desired jamming margin, and J/N. Note that any kind of bit decision that can be made on the basis of one error allowed in three, or two in five, etc., would give the same result. Here the use of simple majority bit decisions has been implied. Any other form of error allowance would do just as well. It is worthy of mention that there are techniques for forcing three out of three chip errors before causing a bit error. If, for example, instead of making a bit decision based on two out of three being the same, suppose that a receiver looks for the *smallest* signal out of six possibilities. (Slot 1, F_{one}/F_{zero}; slot 2, F_{one}/F_{zero}; slot 3, F_{one}/F_{zero}.) In other words, either F_{one} or F_{zero} in any of the three time slots (hop times) may have the smallest signal. Now, since transmitting a One would cause all three frequencies associated with F_{one} to

Figure 2.20 Illustration of contiguous versus overlapping channels showing gain in channels per unit bandwidth; (*a*) contiguous channel spacing; (*b*) overlapping channel spacing.

be sent, and all three frequencies associated with F_{zero} would be sent when transmitting a Zero, the smallest signal detected would be associated with the opposite from that which was intended. Therefore, the receiver's detection algorithm would choose a One to be output when the smallest signal is detected in the Zero interval, and a Zero to be output when the smallest signal is in a One interval. This means that a would be jammer must hit in three out of three intervals that are the opposite of what is intended by the transmitter.

For the frequency hopping systems discussed so far, it has been assumed that contiguous frequency spacing is used; that is, the dehopped signal appearing at the receiver is not allowed to overlap from one channel into another. This is not strictly a true picture. In many systems, depending on the receiver used, the transmit spacing can be such that significant overlap occurs. This overlap greatly reduces the RF bandwidth required for the transmitted spread spectrum signal. Figure 2.20 illustrates overlapping channels and the bandwidth savings. Between Figures 2.20*a* and 2.20*b*, there is a doubling of channels in the same bandwidth. Overlapping is such that the center of one channel falls at a null for the adjacent channels (we are assuming noncoherent carrier reception). Here then is a technique for allowing the frequency hopping system in our example to meet the RF bandwidth restrictions imposed while keeping the chip rate low. The three-chip-per-bit system of the example would have only 15.5-MHz bandwidth, whereas the five-chip-per-bit system would come close to the 10-MHz bandwidth goal with a 10.6-MHz requirement.

One other significant consideration regarding the chip rate is the effect of signals which are received at the same frequency as the desired signal but are different in phase. These signals are due to multipath or, even worse,

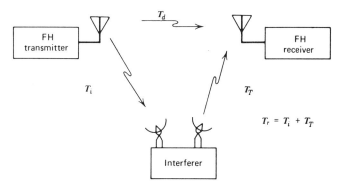

Figure 2.21   Repeater-type interference diagram.

deliberate interference. In most cases, the multipath signal arriving at a receiver is much smaller than the desired signal and is therefore not of great consequence. The deliberate interferer who receives a signal from the frequency hopping transmitter and amplifies it, modulates with noise (or, if he or she knows the key, transmits the complement), can be extremely effective with transmitter power similar to the friendly one. Against this kind of threat, the frequency hopper might have no advantage. To combat the threat the frequency hopper may hop at a rate that allows it to skip to another frequency before the interferer can respond to the last one. The desired hopping rate is then greater than $(T_r - T_d)^{-1}$, where T_r is the total propagation time from the frequency hopping transmitter to the interferer and from there to the intended receiver. T_d is direct path delay, illustrated in Figure 2.21.

As we observe Figure 2.21, we see that the minimum required hopping rate is a function of distance to the interfering station and the angle of offset from the direct path. For fixed stations, we can derive a minimum hop rate, but for mobile applications the conclusion must be to "make the system hopping rate as fast as possible."

To give an idea of the magnitude of a possible problem, assume that the transmitter and desired receiver are 10 miles apart and that the interferer is 8 miles from each of them; $T_r - T_d = 109 - 68 = 41$ μs and the hopping rate should be at least 24.2 khps. This is a readily achievable hopping rate.

Because of its importance in frequency hopping, let us recap what has been said:

Frequency hopping rate is a function of information rate.

Information error rate cannot be reduced below J/N unless some form of redundant* transmission is used.

---

*"Redundant" includes error detection and correction coding.

The number of frequency channels required is determined by the desired interference rejection capability (jamming margin) and chip error tolerance.

The correlated (dehopped) signal seen in a noncoherent frequency hop receiver is $[(\sin x)/x]^2$ in shape and has main-lobe bandwidth equal to twice the chip rate.

The chip rate is bounded on the low side by multipath and repeater interference considerations. The best rule for variable geometry situations may be to hop as fast as possible.

Avoiding the Repeat Jammer

We have mentioned that frequency hopping systems can be faced with the prospect of a jamming signal which follows the signal from frequency to frequency, and can thus have an effect on the desired signal similar to that which a single-frequency jammer would have on a nonhopping signal. It is possible to negate or at least greatly reduce the effects of the repeat jammer by relatively simple techniques, using a noncoherent detector.

One simple approach is to use frequency hopped on-off keying. That is, if a One is intended, let the transmitter send a signal selected from the library by the code. If a Zero is intended, however, the transmitter sends nothing. The repeat jammer then has no advantage since:

Transmission of a One provides a signal for the jammer to detect and repeat, but doing so will in most instances only help the receiver to detect the signal and make the correct decision that a One has been sent. The only hope that the jammer has to cause an incorrect decision is to attempt to have its signal arrive at the receiver with an amplitude that is identical to the desired signal but opposite in phase, and this is a highly improbable (if not impossible) state to maintain.

When a Zero is intended, the desired signal transmitter sends nothing, so the jammer can only make a guess at random as to the correct signal frequency if a One were being sent. In this case, the jammer's probability of causing an error is just J/N.

The repeat jammer is similarly ineffective against other frequency hopping techniques as long as the frequency hopped carrier is unmodulated, and the receiver uses noncoherent detection. Thus in many applications the repeat jammer can generally be ignored as a viable threat to the frequency hopping system. (The hopping rate, however, must be $\geq$ data rate.)

Let us turn for a moment to m-ary frequency hopping systems in which a block of $\log_2 m$-coded data bits is sent by transmitting one of a set of m frequencies. (This corresponds in an analog system to one of m levels.) When such transmissions are used, an interfering signal needs only to hit in any one

TABLE 2.2   Block Error Rate for *m*-ary Transmission of Three-, Five-, and Seven-Bit Blocks with Various Numbers of Errors Corrected

	Percent of Channels Jammed									
Corrected	5	10	15	20	25	30	35	40	45	50
	Error Rate									
1 in 7	0.04	0.15	0.28	0.42	0.56	0.67	0.76	0.84	0.89	0.93
2 in 7	0.004	0.026	0.073	0.15	0.24	0.35	0.47	0.58	0.68	0.77
3 in 7	0.0002	0.003	0.012	0.033	0.07	0.13	0.2	0.29	0.39	0.5
1 in 5	0.022	0.081	0.16	0.26	0.37	0.47	0.57	0.66	0.74	0.81
2 in 5	0.0012	0.0086	0.027	0.058	0.10	0.16	0.24	0.32	0.41	0.5
1 in 3	0.007	0.03	0.06	0.1	0.16	0.22	0.28	0.35	0.43	0.5

of the $m - 1$ channels not intended for transmission to cause a block error. In an 8-ary system this means that one-seventh the number of jamming signals is required to cause the same probability of a chip error that would be required in a simple binary system. This is a clear loss of 8.5 dB in processing gain. Of course, the reason for using an m-ary transmission format is to reduce chip rate, not to reduce errors or increase jamming margin. Chip-rate reduction, however, is not necessarily a good thing when we are faced with the "look-through" or repeating jammer.

At any rate, for m-ary frequency hopping, only N/m^* m-ary channels or sets are available, which not only reduces the processing gain available (to $10 \log N/m$ dB) but increases the probability of chip error per jammed (single) channel. Chip error rates for m-ary transmitters may then be calculated on the basis of J/N_c = number of jammed channels per number of m-ary sets. Character or block error rate for m-ary transmission would be

$$p(\epsilon) = \sum_{a=r}^{m} \binom{m}{a} p^a q^{m-a},$$

which gives the probability distribution for independent trials with probability p of a correct bit and $q = 1 - p$ incorrect bit probability for any given chip. In the case of m-ary signaling $p = J/N_c$ and $q = 1 - J/N_c = (N_c - J)/N_c$. Given a probability of error that is acceptable, we can determine the number of jamming signals that will produce this error. Table 2.2 lists the percentage of jammed channels allowable for various levels of error, correctable in blocks of three, five, and seven bits.

The prime advantage of m-ary transmissions in frequency hopping is that they are well adapted to the use of some powerful error-correction tech-

*N is the number of channels available. For contiguous channels $N = BW/R$, where BW = bandwidth available, R = chip rate.

niques without a great increase in the hopping rate being required. A good example is the TATS technique, which will be discussed in Chapter 4 in more detail. That system employs an octal Reed-Solomon error-correcting code in such a way that six-bit input data blocks are converted to seven octit output symbols. Each octit (a three-bit word) is then transmitted as one of eight contiguous frequency hopping signals. Two advantages accrue from this format: One is that up to three errors in seven symbols can be corrected, and up to four errors in seven symbols can be detected. The second advantage is that the technique (which converts from binary to octal and back again) permits the use of an extremely powerful error-correcting code while increasing the symbol and hopping rates (the symbol rate is the rate at which the system sends encoded information) by a factor of only $\frac{7}{6}$.

2.3   TIME HOPPING

Time hopping, in other words, is familiar pulse modulation; that is, the code sequence is used to key the transmitter on and off, as in Figure 2.22. Transmitter on and off times are therefore pseudorandom, like the code, which can give an average transmit duty cycle of as much as 50%. This form of signal, as spread spectrum modulation, has found its major application in combination with frequency hopping as in the RACEP systems developed by the Martin Marietta Corp. The fine point of difference separating time frequency and plain frequency hopping is that in frequency hopping systems the transmitted frequency is changed at each code chip time, whereas a

Figure 2.22   Pseudorandom code-keyed (time-hopping) signal waveform.

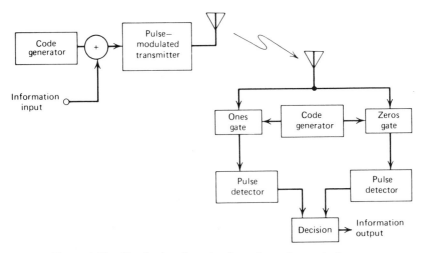

Figure 2.23   Simple time-hopping (pseudorandom pulse) system.

time-frequency hopping system might change frequency and/or amplitude only at one/zero transitions in the code sequence. Figure 2.23 shows a time-hopping system in block form. The simplicity of the modulator is obvious. Any pulse-modulatable signal source capable of following code waveforms is eligible as a time-hopping modulator.

Time hopping may be used to aid in reducing interference between systems in time-division multiplexing. However, stringent timing requirements must be placed on the overall system to ensure minimum overlap between transmitters. Also, as in any other coded communications system, the codes must be considered carefully from the standpoint of their cross-correlation properties.

Simple time-hopping modulation offers little in the way of interference rejection because a continuous carrier at the signal center frequency can block communications effectively. The primary advantage offered is in the reduced duty cycle; that is, to be really effective an interfering transmitter would be forced to transmit continuously (assuming the coding used by the time-hopper is unknown to the interferer). The power required of the reduced-duty-cycle time-hopper would be less than that of the interfering transmitter by a factor equal to the signal duty cycle.

Because of this relative vulnerability to interference, simple time-hopping transmissions should not be used for antijamming unless combined with frequency hopping to prevent single frequency interferers from causing significant losses. For ranging, multiple access, or other special uses time-hopping may be especially useful, if only because of the simplicity of generating the transmitted signal.

2.4 PULSED FM (CHIRP) SYSTEMS

Characteristics of Chirp Signals

One type of spread spectrum modulation that does not necessarily employ coding but does use a wider bandwidth than the minimum required so that it can realize processing gain is "chirp" modulation. This form has found its main application in radar, but is also applicable to communications. Chirp transmissions are characterized by pulsed RF signals whose frequency varies in some known way during each pulse period. The advantage of these transmissions for radar is that significant power reduction is possible. The receiver used for chirp signals is a matched filter, matched to the angular rate of change of the transmitted frequency-swept signal. Coding is not normally used with this type of matched filter.

The transmitted frequency-swept-signal chirp signal is just that produced by a common laboratory sweep generator, because most chirp systems use a linear sweep pattern. Any pattern, suitable to the requirement that a matching receiver filter must be built, is suitable. Here, however, we consider only linearly swept signals. Figure 2.24 illustrates typical waveforms that exist in a chirp system.

Letting $\mu = d\omega/dt$ for the frequency sweep rate, we see that the linearly swept transmitted signal (and the input to the chirp filter) is

$$F(t) = f_1 + \mu t$$

$$= A \cos(\omega_1 t + \mu t)$$

over the linear range of frequency sweep. For the larger compression ratios of interest here amplitude distribution in our linear FM pulse is quite flat. Cook[210] and Klauder et al.[223] have shown that the chirp signal is distributed according to the Fresnel integral

$$Z(u) = \int_0^u e^{i\pi} \alpha^{1/2} \, d\alpha$$

and have plotted spectral amplitude for compression ratios as high as $130:1$. Their plots confirm that ripple decreases as compression ratio increases.

The dispersive filter at the heart of a chirp receiver is a storage and summing device that accumulates the energy received over an interval, assembles it, and releases it in one coherent burst. We may visualize the filter's operation by imagining a transmitter that sends a series of signals over a time period, each of which is tagged in some way. The receiver then sorts the signals according to their tags and assembles them so that they add

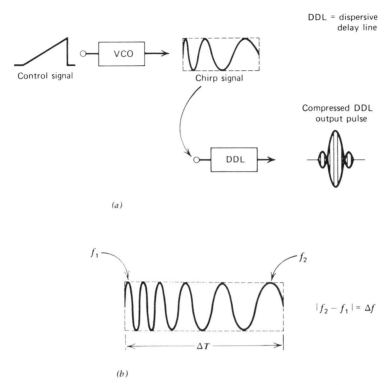

Figure 2.24 Chirp system waveform: (*a*) typical chirp waveforms; (*b*) transmitted chirp signal (one sweep).

together within a much shorter time period, reinforcing one another to produce a much stronger signal.

In actual practice, the "tagging" process used by the transmitter is its frequency sweep (which may be in either direction—low to high or high to low), with each increment in frequency denoting a new signal to be sorted and stored by the receiver. A received signal is stored until the entire ensemble of frequencies in one sweep arrives, and the summed power at all frequencies is output at one time.

The transmitted signal may be developed in several ways, the simplest of which is application of a linear voltage sweep to a voltage-controlled oscillator. An alternative to the voltage-controlled oscillator is the chirp-generating filter. This is a filter whose impulse response is such that its output is a time series of output frequency components; that is, an impulse with a "continuous" frequency spectrum causes a series of output signals whose time delay is dependent on the frequency component being delayed. If the delay slope dt/df is linear (constant over the range of interest), the output is a linear frequency sweep and the chirp-generating filter can be made to be the exact complement of the receiver's dechirping filter. In fact, if chirp filters were designed as bilateral devices, a chirp-generating filter could also be used to dechirp.

The transmitted chirp signal has two parameters of primary interest. Once these parameters are defined, system performance is also defined, and the rest of the problem is making use of the capability inherent in the chirp signal. These parameters are (1) the frequency sweep $|f_2 - f_1| = \Delta f$ and (2) the amount of time ΔT used to sweep the Δf band. We assume that any amplitude change during ΔT is negligible.

The dispersive delay line or chirp-matched filter compresses a frequency sweep, usually linear, which provides an improvement in output signal (voltage)-to-noise ratio equal to $\sqrt{D}$, where the "compression ratio" is

$$D = \Delta T \Delta f$$

Δf is defined as the frequency sweep $|f_2 - f_1|$ and ΔT is defined as the time occupied by the sweep in frequency.

All signal parameters are transformed by the factor D between the filter's input and output as given below:

	Input	Output
Frequency sweep	Δf	$\Delta f / D$
Sweep time (signal duration)	ΔT	$\dfrac{\Delta T}{D} = \tau$
Signal (peak) power	$\hat{P}_i$	$D \hat{P}_i = \hat{P}_0$
Time resolution	ΔT	$\pm \dfrac{\Delta T}{D} = \pm \dfrac{1}{\Delta f}$
Frequency resolution	Δf	$\pm \dfrac{\Delta f}{D} = \pm \dfrac{1}{\Delta T}$

Chirp systems are not the only ones that employ matched filters for reception. Matched filters made up of delay elements matched to a coded signal are also used. These are discussed in Chapter 6.

Figure 2.25 illustrates typical delay and bandwidth parameters available in contemporary delay lines of the dispersive type.

2.5 HYBRID FORMS

In addition to the more usual forms of spread spectrum modulation, there are the hybrid combinations of modulation that offer certain advantages over, or at least extend the usefulness of, the direct sequence and frequency hopping techniques. The most often used combination (or hybrid) spread

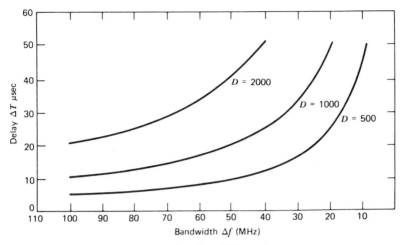

Figure 2.25   Chirp filter TW product curve.

spectrum signals are made up of (1) simultaneous frequency hopping and direct sequence modulations, (2) simultaneous time and frequency hopping, and (3) simultaneous time-hopping and direct sequence modulations.

The advantage in combining two spread spectrum modulation methods is usually that characteristics can be provided which are not available from a single modulation method. However, implementation is not necessarily increased in difficulty by the same factor. That is, a combined frequency hopping and direct sequence transmitter might be built up of readily constructed code sequence generators and frequency synthesizers, whereas a straightforward direct sequence modulator or frequency hopper could not be constructed to do the same job.

The specific construction of the various subsystems is not discussed here because it is covered elsewhere. The emphasis here is on describing the hybrid techniques in use and reasons for that use.

Frequency Hopped / Direct Sequence Modulation

As suggested by its title, frequency hopped/direct sequence (FH/DS) modulation consists of a direct sequence modulated signal whose center frequency hops periodically. Figure 2.26 illustrates the frequency spectrum from such a

Figure 2.26   Frequency spectrum of hybrid FH/DS system.

modulator. The spread spectrum signal shown is made up of a number of spread spectrum signals. A direct sequence signal covering a part of the band appears instantaneously, and the entire DS signal appears in other parts of the band as dictated by a frequency hopping pattern.

Hybrid FH/DS signals are used for various reasons:

To extend spectrum spreading capability.

For multiple access and discrete address.

For multiplexing.

To make otherwise impractical systems practical, when the maximum in reliable sequence generator clock speed has already been obtained or a limit in the number of frequency hop channels has been reached, a hybrid modulator can be especially valuable. Say, for example, that an RF spectrum 1 GHz wide is wanted. The system designer could specify a 1136-Mcps code sequence generator to support a direct sequence transmitter, a frequency synthesizer with 200,000 5-kHz-spaced outputs for frequency hopping, or a hybrid combination. By combining the direct sequence and frequency hopping approaches we could construct a signal that has a 1-GHz spread spectrum bandwidth by using a 114-Mcps code sequence generator and a frequency synthesizer with only twenty 50-MHz-spaced frequencies available. It is obvious that the difficulty of building the hybrid modulator is significantly lower than that of either of the other approaches. Both the 114-Mcps code sequence generator and the 20-frequency synthesizer could be readily constructed today, which is much more than can be said for a 1136-Mcps code sequence generator or a 200,000 frequency, 1-GHz bandwidth, high speed frequency synthesizer.

Hybrid FH/DS transmitters are straightforward superimpositions of direct sequence modulation on a frequency hopping carrier, as shown in Figure 2.27. This modulator differs from a simple direct sequence modulator mainly in that the carrier frequency is varying (hopped) rather than being at a constant frequency, as for simple DS modulation. Note also that the same code sequence generator can supply code data both to the frequency synthesizer to program its hopping pattern, and to the balanced modulator for direct modulation.

Synchronism between the FH and DS code patterns means that whenever a given sector of the DS code pattern is transmitted the frequency channel will be the same. It should be noted that the DS code rate is normally much faster than the rate of frequency hopping. Therefore, many chips of DS modulation occur in a single frequency channel. Also, the number of frequency channels available is usually much smaller than the number of code chips so that in the course of a complete code length all the frequency channels are used many times. The pattern of their use, however, is as random (or pseudorandom) as the code itself.

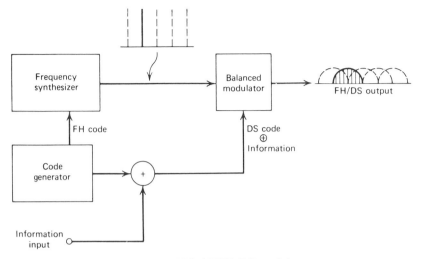

Figure 2.27 Hybrid FH/DS modulator.

The correlator used to strip off the spread spectrum modulation in a receiver, before baseband demodulation, is, for FH/DS hybrids, again a superimposition of a direct sequence correlator on a frequency hopping correlator; that is, the local reference signal becomes a hybrid FH/DS signal, which is then multiplied with all received input signals. Figure 2.28 shows a typical FH/DS receiver configuration in which the local reference generator is essentially a replica of the transmitting modulator, with two exceptions: (1) the local reference center frequency is offset an amount equal to the IF, and (2) the DS code is unmodified by baseband input.

The process gain for FH/DS hybrids is a composite of that furnished separately by the frequency hopping and direct sequence modulations, with certain restrictions; that is, from an overall view the process gain is a function of

$$\frac{BW_{RF}}{R_{info}} = FH/DS \text{ process gain.}$$

When viewed with respect to particular signals, however, we see that this does not hold true when predicting jamming margin; for a hybrid FH/DS system, the jamming margin is not necessarily related to the process gain by the relation (see Chapter 1)

$$\text{jamming margin} = G_P - \left[L_{sys} + \left(\frac{S}{N} \right)_{out} \right],$$

although from an overall standpoint the relation holds. The reason is that in any particular segment of the overall input spectrum the FH/DS receiver

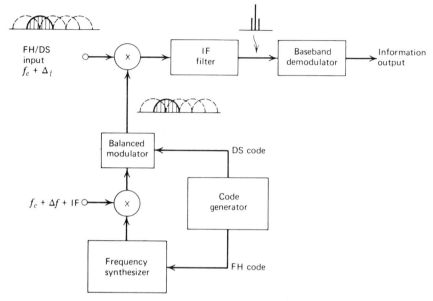

Figure 2.28   Hybrid FH/DS receiver.

can reject only that amount of interference commensurate with the spread spectrum bandwidth of the partial-band direct sequence signal; that is, if we refer to our 1-GHz bandwidth example, in which the DS signal bandwidth was 100 MHz, the process gain in that segment is limited by the 100-MHz DS bandwidth. This is true because an interfering signal (say a CW carrier) whose power falls wholly within one sector of the 1-GHz band is spread by only the bandwidth of the DS signal whenever the signal hops into that sector. This means that when the signal lies in that one sector, a narrowband interferer can cause a loss of performance, with significantly less power than is required to interfere with the entire FH/DS signal structure.

The reason for pointing out this possible loss in partial-band jamming margin is that when the separate direct sequence channels are used as access channels for multiplexing more than one DS signal on a frequency hopping carrier it is possible for a single frequency interfering signal to interfere with only one channel, with no significant effect on the others. To prevent one user (or channel) from being completely lost, while spreading the effect of interference across all channels, we need to change the order of channel use periodically so that each user employs all available frequencies equally. In this way, even though an interfering signal causes complete loss of one frequency channel, each of the multiplexed signals would be affected only a part of the time equal to

$$\frac{\text{number of jammers}}{\text{number of channels}}.$$

Thus if our 20-channel system had an interfering signal in one channel, it would be affected 1/20th of the time, and if all users shared that channel, a particular user would also be affected 1/20th of the time. It is reasonable to say, then, that process gain in decibels for a properly designed FH/DS hybrid system is the sum of the process gains produced by two overlaid spread spectrum modulations; that is,

$$G_{p(FH/DS)} = G_{p(FH)} + G_{p(DS)}$$

$$= 10\log(\text{number of channels}) + 10\log\frac{BW_{DS}}{R_{info}}.$$

Comparing the results of our 1-GHz spread spectrum example, we have process gain for a straightforward direct sequence system (assuming a 5-kbps information rate):

$$G_p = 10\log\frac{1 \times 10^9}{5 \times 10^3} = 53\text{ dB}$$

for a straightforward frequency hopper with 200,000 frequencies,

$$G_p = 53\text{ dB}$$

for the FH/DS system, and

$$G_{p(FH/DS)} = 10\log 20 + 10\log\frac{1 \times 10^8}{5 \times 10^3} = 56\text{ dB}.$$

Process gain is actually 3 dB higher for the FH/DS system because of overlap allowed in the twenty 100-MHz-wide channels. (The same degree of overlap in a straightforward frequency hopper would produce 400,000 channels for a 56-dB process gain.) We emphasize that jamming margin may be significantly less than either the simple FH or DS system, however.

Time-Frequency Hopping

Time-frequency hopping modulation has found its greatest application in those systems in which a large number of users with widely variable distances or transmitted power are to operate simultaneously in a single link. Such systems tend to employ simple coding, primarily as an addressing medium, rather than to spread the spectrum specifically. The general tendency is to design for the equivalent of a wireless telephone switching system in which random access and discrete address are the prime operational goals. For such uses time-frequency hopping is well adapted; it offers one of the few (and perhaps only) viable solutions to the near-far problem.

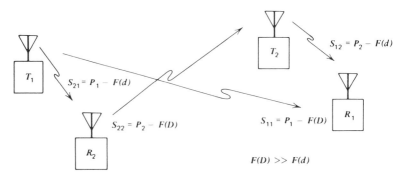

Figure 2.29 Two-link communication system illustrating the near-far problem.

Consider for a moment two receiving and transmitting links whose receivers and transmitters are spaced as in Figure 2.29 and in which the receiver for each link is positioned so that the other link transmitter causes interference. The problem in this system lies in the difference in distance from a receiver to its desired transmitter and to the nearby transmitter.

The antijamming capacity of spread spectrum systems aids in such a situation but unfortunately is not enough under most conditions to overcome the disparity seen in nearby and remote signals. A close transmitter, for instance (say within a mile), would normally produce signals attenuated by 40-plus dB (the plus is a function of transmitting frequency), whereas another transmitter 20 miles away would produce signals attenuated by 63-plus dB. To operate in such an environment a receiver would need process gain in the order of 35 dB with no allowance for fading. Processing alone is not really the answer to near-far reception difficulties. A much better solution is to time all transmissions so that the desired and undesired transmitters are never transmitting at the same time. Even better, with time-frequency hopping the two transmitters can be programmed to transmit on different frequencies as well as at different times. Many links can then be operated at once if their time slots and frequency hopping channels are properly synchronized.

Time-Hopping Direct Sequence

When direct sequence transmission is used, and code division multiplexing does not permit sufficient access to the link, time hopping has proved to be a useful way of adding time-division multiplexing (TDM) to aid in traffic control. The high degree of time synchronization between direct sequence transmit-receive terminals because of their code correlation requirements leads ideally to time-hopping; that is, because a direct sequence receiver must align its local reference code within a fraction of a *pn* code chip time, it already has timing good enough to support TDM operation.

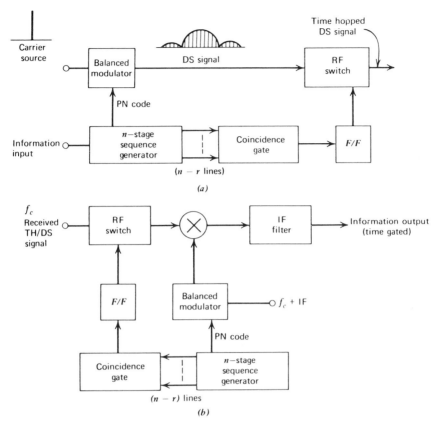

Figure 2.30 TH/DS system block diagrams: (a) time-hopping/direct sequence transmitter; (b) TH/DS receiver.

All that is required to add time-hopping TDM to a direct sequence system is on-off switching and control. For time hopping the on and off decision can be easily derived from the same code sequence generator used to derive the spectrum spreading code. An easily implemented keying control would be an n-input gate which senses some previously chosen shift register state and turns the transmitter on or off at the appearance of that state. Figure 2.30 illustrates a time-hopping direct sequence (TH/DS) transmitter. The receiver for this system would operate similarly, except that the control signal would be used to block or enable the receiver front end synchronously with the expected transmit signal.

When selecting the code state for time control, we choose a shift register state that occurs repeatedly and detect the occurrence of that state. For an n-stage maximal-connected shift register generator there are $2^n - 1$ unique states, each occurring once per repetition of the m-sequence. By choosing a number of register stages $(n - r)$ and sensing an all-ones condition in those stages, we could expect to see that condition 2^r times per code sequence and the distribution of that occurrence would be (pseudo-) random.

On the average, the state sensed would occur at a rate

$$\frac{R_c 2^r}{2^n - 1} = \text{average state occurrence for } (n - r) \text{ stage vector}$$

and the direct sequence "on" periods would occur at an average rate

$$\frac{R_c 2^r}{2^{n+1} - 2} = \text{average pulse occurrence for TH/DS transmission}$$

where R_c = code clock rate,
        n = shift register length,
        r = register stages not sensed.

We could then select the average frequency of occurrence of direct sequence output, with the actual output occurring at pseudorandom intervals for a pseudorandom duration, as dictated by the code. This pseudorandom time division can permit more channel users, improve operation over simple code division multiplex, and allow more accesses. Again, as in all other applications of codes for signal discrimination, code cross-correlation should be well defined before a TH/DS operation is attempted.

PROBLEMS

1. Why employ a balanced mixer to generate a direct sequence spread spectrum signal?

2. Remembering that a direct sequence signal has a $[(\sin x)/x]^2$ power spectrum, what is the signal's 3-dB bandwidth? What is the peak power level of the first sidelobe compared with that of the main lobe?

3. What percentage of total signal power is contained in the main lobe of a direct sequence signal?

4. Compare the power densities that would exist at a point 5 MHz removed from the center frequency of a 20 MHz wide (a) frequency hopped signal, and (b) direct sequence signal, BPSK, QPSK, classic MSK.

5. What number of channels is required by a frequency hopper whose maximum bit error rate is 0.01 and which must have a 30-dB jamming margin?

6. What RF bandwidth would this frequency hopper employ if its frequency cells were contiguous, its data rate were 1 kpbs, and three chips were transmitted per bit?

7. For the same 30-dB jamming margin, 0.01 bit error rate, and 1 kbps data, what RF bandwidth would be required by a direct sequence system?

8. The process gain expected from a hybrid FH/DS system with a 20-Mcps code rate, 100 frequency cells, and 3 kbps data rate would be how many decibels?

9. Compare the demodulator bandwidth required on coherent and noncoherent frequency hopping systems.

10. Compare demodulator bandwidth for one- and three-chip-per-bit frequency hopping systems.

11. What would be the effect of a Doppler frequency shift on an upchirp-downchirp system?

12. Compare multipath sensitivity for direct sequence and frequency hopping systems.

13. How might time/frequency hopping be employed to overcome near-far and transmitter imbalance problems?

14. Determine the minimum hop rate that should be used for a fixed receiver when an interferer is 10 miles away and the desired signal transmitter is 100 miles away at a 45° angle from the desired signal. The interferer is a repeater.

15. Compare power density for a chirp signal with that of direct sequence and frequency hopping signals.

3

CODING FOR COMMUNICATIONS AND RANGING

It is intended in this chapter to discuss the codes used in communications and ranging systems—those that act as noiselike (but deterministic) carriers for the information being transmitted. Error-correcting codes are not emphasized since they are more suited to the task of bit- or word-error correction in the information stream. The code sequences of interest are of much greater length than those considered in the usual areas of coding for information transfer, since they are intended for bandwidth spreading and not for the direct transfer of information.

This chapter does not, and indeed cannot, cover the entire field of coding for communications, even for spread spectrum communications alone. Volumes of information and speculation have been written on this subject by specialists in the coding field, and it is suggested that those who wish to explore it consult some of the works listed in the Bibliography under coding techniques.

The importance of the code sequence to a spread spectrum communications or ranging system is difficult to overemphasize, for the type of code used, its length, and its chip rate set bounds on the capability of the system that can be changed only by changing the code.

The purpose of this chapter will be well served if only enough of the rudiments of code selection for communications and ranging systems are taken into account to give a feel for the scope of the code's influence. The emphasis is placed on a description of desirable characteristics and their effects rather than on the design of codes with these characteristics.

Other code types, especially those of the nonlinear variety, are given only minimum attention here—not because of their lack of importance but because of the lack of a requirement for their use when less than true message security is desired. In some communications or ranging systems this security is not required, and in others, security is provided by processing linear codes nonlinearly. Furthermore, nonlinear encoding, in addition to not being critical to the subject at hand—that of providing codes suitable for signal bandwidth spreading in communications and ranging systems—is beyond the scope of this book. It must be stressed, however, that though the linear codes are suitable for interference rejection, ranging, and other spread spectrum applications they are not usable to secure a transmission system. The linear codes are easily decipherable once a short sequential set of chips $(2n + 1)$ from the sequence is known. This has been shown by Birdsall and Ristenbatt,[34] among others. Appendix 5 describes one simple approach to linear code breaking.

If the reader feels that the greatest emphasis is placed on linear codes, or to be more specific, on maximal linear codes, he or she is correct. The maximal linear code sequences (often called *m*-sequences or *pn* codes) are unexcelled for general use in communications and ranging. (Other codes can do no better than equal their performance.) Therefore, it is proper that they and some of their applications and variations be given adequate exposure. It is also proper that we point out here that the most advanced systems ever constructed employ linear codes internal to themselves.

Figure 3.1 illustrates linear and nonlinear code generation on a block diagram basis. To be sure that the point is understood, a spread spectrum system is not a secure system unless the codes used* are cryptographically secure, and although they are useful in spread spectrum systems, linear codes are not secure.

Having (hopefully) dispatched any misconceptions about security, let us define the concepts used in coding for nonsecure spread spectrum systems. It is in order to point out, however, that in general a nonsecure spread spectrum system may be made secure simply by replacing its nonsecure code sequence with a secure code sequence.

To summarize the properties of the codes for use in spread spectrum systems, in which we are interested:

Protection against interference. Coding enables a bandwidth trade, for processing gain against interfering signals.

Provision for privacy. Coding enables protection of signals from eavesdropping to the degree that the codes themselves are secured.

Noise-effect reduction. Error-detection-and -correction codes can reduce the effects of noise and interference.

---

*To be exact, if the band-spreading code is not a secure code, the overall system could still be secure if the information itself were encoded by a cryptographically secure technique. The point here is that encoding in some form is necessary but not necessarily sufficient for security purposes.

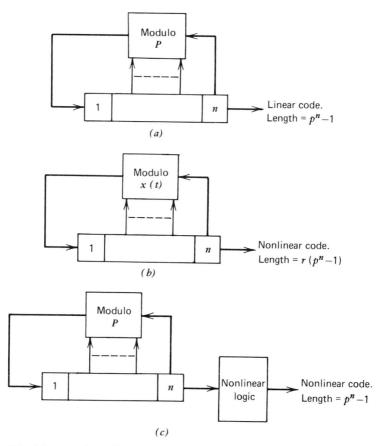

Figure 3.1 Linear and nonlinear generator configurations: (a) linear code generator for p-ary shift register (each stage has p possible states); (b) nonlinear code generator using p-ary shift register (there are $p^n - 1$ possible states for each feedback condition); (c) nonlinear code generator based on linear generator.

(Note that coding is necessary to provide all of the properties claimed for spread spectrum systems.)

3.1 MAXIMAL SEQUENCES

Maximal codes are, by definition, the longest codes that can be generated by a given shift register or a delay element of a given length. In binary shift register sequence generators, which are the only type considered here, the maximum length sequence is $2^n - 1$ chips, where n is the number of stages in the shift register. A shift register sequence generator consists of a shift register working in conjunction with appropriate logic, which feeds back a logical combination of the state of two or more of its stages to its input. The output of a sequence generator, and the contents of its n stages at any sample (clock) time, is a function of the outputs of the stages fed back at the preceding sample time. Feedback connections have been tabulated for maximal code generators from 3 to 100 stages, so that some sequences of any length from 7 through $2^{36} - 1$ chips are readily available. A number of feedback connections are listed in Table 3.6.

Properties held by all maximal code sequences are:

1. The number of ones in a sequence equals the number of zeros within one chip. For a 1023-chip code there are 512 ones and 511 zeros. Consider a code implementation in which a one is represented by a positive voltage $+V$, and a zero by a negative voltage $-V$. The amount of offset over the code length is proportional to the inverse of the code length, or $V/(2^n - 1)$. Similarly, when a code sequence biphase modulates a carrier, the residual carrier component is down by a factor $(2^n - 1)^{-1}$. Thus we see that not only is the modulator important in carrier suppression but the codes used must be capable of supporting the amount of suppression required; for example, when carrier suppression is 30 dB, the shortest code usable is 1000 chips. (For most practical purposes much longer codes are usually required.)

2. The statistical distribution of ones and zeros is well defined and always the same. Relative positions of the runs vary from code sequence to code sequence, but the number of each run length does not.

3. Autocorrelation of a maximal linear code is such that for all values of phase shift the correlation value is -1, except for the 0 ± 1 chip phase shift area, in which correlation varies linearly from the -1 value to $2^n - 1$ (the sequence length). A 1023-chip maximal code $(2^{10} - 1)$, therefore, has a peak to minimum autocorrelation value of 1024, a range of 30.1 dB. (It must be realized, however, that this value for autocorrelation is valid only for integrating over the entire sequence length.)

4. A modulo-2 addition of a maximal linear code with a phase-shifted replica of itself results in another replica with a phase shift different from either of the originals.

5. Every possible state, or n-tuple, of a given n-stage generator exists at some time during the generation of a complete code cycle. Each state exists for one and only one clock interval. The exception is that the all-zeros state does not normally occur and cannot be allowed to occur.

Each of these properties is especially useful in a communications or ranging system. Therefore, let us engage in further discussion and explanation of each one and what each means in these systems.

The fact that an almost equal number of ones and zeros exists in any maximal linear code is important in that it allows the DC component in a code or in a code-modulated signal to be neglected. [It is noted that whether a code is $2^3 - 1$ (from a three-stage generator) or $2^{99} - 1$ (from a 99-stage generator) the total number of ones always exceeds the total number of zeros by the same amount—one.] To be exact, the number of ones in any linear maximal code is

$$\frac{2^n}{2} = \text{number of ones} = 2^{(n-1)},$$

and the number of zeros is

$$\frac{2^n}{2} - 1 = \text{number of zeros} = 2^{(n-1)} - 1,$$

where n is, as usual, the number of stages in the code generator, and the code length is $2^n - 1$ chips.

When modulating a carrier with a code sequence, one-zero balance can limit the degree of carrier suppression obtainable because carrier suppression is dependent on the symmetry of the modulating signal. Therefore, the longer the code sequence, the less the effect of the code sequence on carrier balance.

Run-Length Distribution

Freymodsson[311] has shown that there are exactly $2^{n-(p+2)}$ runs of length p for both ones and zeros in every maximal code sequence (except that there is only one run containing n ones and one containing $n - 1$ zeros; there are no runs of zeros of length n or ones of length $n - 1$). A run is defined as a series of ones or zeros grouped consecutively (Figure 3.2).

An example of the distribution of runs is shown in Table 3.1 for $2^7 - 1$ chip m-sequences. The pattern exhibited here is typical of all m-sequences;

a   Run of length one one
b   Run of length two zeros
c   Run of length two ones
d   Run of length one zero
e   Run of length three ones

Figure 3.2   Illustration of runs.

TABLE 3.1   Distribution of Runs for a $2^7 - 1$-Chip m Sequence

Run Length (Chips)	Number of Runs		Number of Chips Included
	Ones	Zeros	
1	16	16	32
2	8	8	32
3	4	4	24
4	2	2	16
5	1	1	10
6	0	1	6
7	1	0	7
			$\overline{127}$ Total

that is, the number of runs of each length is a decreasing power of 2 as the run length increases. Therefore, as might be suspected, the number of runs of length p is equal to the sum of the number of runs of all lengths $p + r$, where $1 \leq r \leq n - p$. Thus, the expectation of a run of length p is twice that of a run of length $p + 1$, and this is the same as a truly random sequence.

Comparing the structure of m-sequences with truly random noise, White[335] confirms Tausworthe's theory[329] that for r-tuples such that $r \leq n$, taken as samples from the output sequence of a maximally connected generator, the distribution of chips is statistically independent. This was done by comparing the actual distribution of ones in samples of length $r \leq n$ with the binomial probability of occurrence of ones that would be evidenced by a truly random stream. Thus it was shown that for periods less than $n \times R_c$ the output of a maximal linear generator is random for all practical purposes. When randomness is a governing criteria, we should, as might be supposed even without analysis, increase the generator length (and thereby the code length.)

Even through the randomness properties can be demonstrated, the maximal linear sequences are deterministic. (Certainly, no proof is required because m-sequences repeat at intervals of $2^n - 1$ chips.) Each repetition exhibits the same one-zero distribution. As sample size is increased, the one-zero distribution becomes less and less random. Finally, when sample size equals $2^n - 1$, the number of ones per sample is $2^n/2$ and the number of zeros is $2^n/2 - 1$.

Autocorrelation

Autocorrelation in this context refers to the degree of correspondence between a code and a phase-shifted replica of itself. Autocorrelation plots show the number of agreements minus disagreements for the overall length of the two codes being compared, as the codes assume every shift in the field of interest. Such a plot can be generated easily with a pair of code generators or a computer simulation.

The plot of code correlations for the linear maximal sequences is two-valued, with a peak only at the zero shift point. This is an invaluable property because it allows the receiver to discriminate between signals on a yes-no basis. It is ideal for a communications system. To illustrate the difference between autocorrelation of a code and crosscorrelation of two different codes, consider the following example. Given a short linear maximal code 1110100 of length $2^3 - 1$, we will compare it with itself for all possible shifts, taking the number of agreements minus the number of disagreements at each shift.

Shift	Code/shifted code	A − D
0	1110100	7
	1110100	
1	1110100	−1
	0111010	
2	1110100	−1
	0011101	
3	1110100	−1
	1001110	
4	1110100	−1
	0100111	
5	1110100	−1
	1010011	
6	1110100	−1
	1101001	
7	1110100	7
	1110100	

Now let us compare this code with another code on the same basis, where the second code is 1110010.

Shift	Code 1/shifted code 2	A − D
0	1110100	3
	1110010	
1	1110100	−1
	0111001	
2	1110100	3
	1011100	
3	1110100	−1
	0101110	
4	1110100	−1
	0010111	
5	1110100	−5
	1001011	
6	1110100	3
	1100101	
7	1110100	3
	1110010	

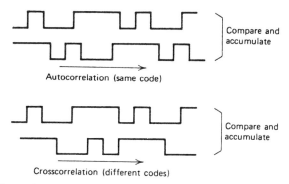

Figure 3.3 Illustration of autocorrelation and cross-correlation of binary code sequence (ψ = agreements − disagreements).

It is easy to see the difference in the autocorrelation of the first code with itself (there are only two values, and the values are $2^n - 1 = 7$, and -1), and the crosscorrelation values of two different codes. The crosscorrelation values are 3, -1, and -5, and these values may cause the receiver to falsely synchronize.

Cross-correlation of two codes is of similar importance. Cross-correlation, however, is the measure of agreement between two different codes. Again, it is plotted as a function of phase shift. Unfortunately, cross-correlation is not so well behaved as autocorrelation, and when large numbers of transmitters, using different codes, are to share a frequency band, the code sequences must be carefully chosen to avoid interference between users. The effect of a high degree of correlation between an undesired code received and a receiver reference is an increase in the receiver's false alarm rate and, under extreme circumstances, false recognition of synch by the receiver (Figure 3.3).

An even worse possible effect of poorly chosen codes with high cross-correlation is that a jammer might transmit a code from the set being used. This in turn could cause every receiver within range of the jammer to be affected by partially correlating with its reference code and thereby causing false synchronization.

Linear Addition Properties

Maximal linear code sequences possess a particularly interesting combinatorial property. This property, which allows generation of any desired code phase (e.g., 500 chips delay or any other desired delay up to $2^n - 1$ chips), is valuable any time a different code phase is required. One of its uses is in operating multiple correlators to reduce effective synchronization time. Another treats the phase-shifted sequence as a different sequence, with both the normal output and the shift-and-add generated output for separate communication links. This is possible because of the well-defined autocorrelation. An example of a typical system employing the same code for modulation, by all transmitters, is the global positioning system (GPS). In GPS, each transmitter sends the same P code (the overall code having a period of 266 days) but each transmitter's signal is offset by one week from every other transmitter, which prevents a receiver from synchronizing to more than one signal at once. The autocorrelation of the code is useful in this case to permit good effective signal orthogonality.

Another valuable property of m-sequences is the way in which two (or more) sequences add. When two m-sequences of different lengths, say $2^n - 1$ and $2^p - 1$, are linearly (modulo-2) added, the result is a composite sequence with a length $(2^n - 1)(2^p - 1)$. This composite is not maximal but may be a segment of a longer maximal sequence. The primary applications of these composite sequences have been in the JPL[219] ranging technique, and GPS systems.

Possibly the most valuable linear addition property is that the addition of two m-sequences, each of length r, produces a composite sequence also of length r but not maximal. The composite sequence itself, however, is different for each combination of delay between the two m-sequences. If we combine two sequences of the same length, modulo-2, as follows:

Shift	Code 1/Code 2	Resulting Code
0	1110100 1110010	0000110
1	1110100 0111001	1001101
2	1110100 1011100	0101000
3	1110100 0101110	1011010
4	1110100 0010111	1100011
5	1110100 1001011	0111111
6	1110100 1100101	0010001
7	1110100 1110010	0000110

note that the modulo-2 sum of the two codes is different for every shift (We have a different code for every shift), which means that this method can give us a large number of different code sequences. That is, a pair of sequence generators of length r can generate r nonmaximal linear codes, each r chips long. More important, if the component m-sequences are properly chosen, the set of r composite sequences will have low and bounded crosscorrelation; for example, a pair of ten-stage shift register generators (see Figure 3.4) would be capable of generating 1023 different 1023-chip nonmaximal linear

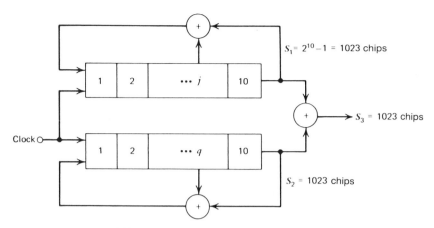

Figure 3.4 Shift register generator for 1023-chip nonmaximal codes.

codes in addition to the two basic linear maximal codes. For each change in the feedback logic to either of the two shift registers a new set of 1023 codes would be produced.

A disadvantage of the shift and add property is that the linear maximal codes are also predictable by anyone who knows the current code state, so that future operation can be anticipated.

TABLE 3.2 Binary-to-Decimal Conversion of Code n-Tuples (Four-Stage Register) Showing Pseudorandom Ordering

State	4-Tuple				Base-10 Equivalent
1	1	1	1	1	15
2	0	1	1	1	7
3	0	0	1	1	3
4	0	0	0	1	1
5	1	0	0	0	8
6	0	1	0	0	4
7	0	0	1	0	2
8	1	0	0	1	9
9	1	1	0	0	12
10	0	1	1	0	6
11	1	0	1	1	11
12	0	1	0	1	5
13	1	0	1	0	10
14	1	1	0	1	13
15	1	1	1	0	14
1	1	1	1	1	15

State Exhaustion

The number of states possible for a set of n elements, each capable of r discernible states, is r^n. A binary shift register generator with maximally connected feedback goes through $2^n - 1$ states in generating a $(2^n - 1)$-chip m-sequence. These states are n-tuples that may be employed to control a processor such as a frequency synthesizer or a Monte Carlo test generator. Table 3.2 lists a set of 15 binary 4-tuples produced by a four-stage shift register generator with its corresponding base-10 conversion. This pseudorandom ordering of n-tuples is typical of all maximal linear sequences. The particular ordering depends on the feedback used. Note that no n-tuple corresponds to zero. Selection of one particular n-tuple or code vector permits monitoring the code generation process to ensure proper operation and to use as a lower rate clock or as a code phase measuring mark.

3.2 LINEAR CODE GENERATOR CONFIGURATION

A linear code sequence generator can be made up of any set of delay elements in conjunction with linear combining elements in a feedback path such that the number of states the generator can assume is a function of the length (in time) of the delay elements and the particular combination of feedback. Typical implementations of code sequence generators have often used delay lines or even lengths of coaxial cable, but the most common contemporary technique uses digital circuits (flip-flops) in a shift register configuration.

Figure 3.5 illustrates the general form of a simple linear generator. Outputs from the last delay stage $D_n{}^*$ and an intermediate stage D_j are combined in a modulo-2 adder and fed back to the input of the first delay element. The truth table in Table 3.2 illustrates the action of feedback in generating all of the possible states of a register. In this particular case the code sequence generated, 100010011010111, is cyclic (repetitive) with a total period $2^n - 1$ times the period of a single delay element. This is the longest code sequence that can be generated by a given number of stages of delay; that is, for n delay stages the longest sequence that can be generated is $2^n - 1$, and this sequence is entitled "maximal" or, to be more specific, "linear maximal" when the generator is linear.

An equivalent and much preferable sequence generator configuration places the feedback adders between stages, as in Figure 3.6. That the same code sequence can be generated with the same number of delay stages and modulo-2 adders, as in the simple generator, is shown by Figure 3.7.

Often a large number of code sequences is needed for applications such as code-division multiplexing. Under these conditions multiple feedback points

---

*The notation D_n signifies delay element (such as flip-flop or delay line) number n; that is the nth delay element, where the convention used is to count from left to right.

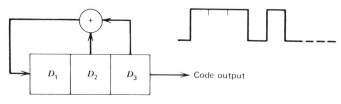

Figure 3.5 Typical code sequence generator (simple type).

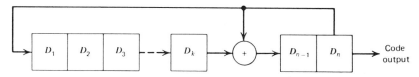

Figure 3.6 Alternate sequence generator configuration (modular type).

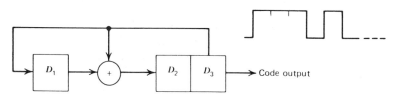

Figure 3.7 Code sequence generator equivalent to that of Figure 3.5.

are necessary, since the maximum number of codes of any length available from a set of delay elements using single-tap feedback would be only $n - 1$, or one less than the number of delay elements. Unfortunately, many of the $n - 1$ codes may be very short cyclic sequences, or the generator may halt operation altogether by entering zeros in all elements in some configurations.

Using all of the possible linear combinations of feedback taps for an n-stage register, there are $[\phi(2^n - 1)]/n$ maximal sequences* that can be generated. [Here, $\phi(2^n - 1)$ is an Euler number—the number of positive integers including 1 that are relatively prime to and less than $2^n - 1$.] As an example of the use of this formula, let us observe a five-stage register, for

---

*Here and in the remainder of the chapter, the feedback points or taps referred to do not include the last stage feedback because the last stage must always be used, otherwise the generator produces codes of a length that is a function of the last stage in the feedback network. Thus, when single-tap feedback is mentioned, two stages are actually supplying feedback—the last delay element and one intermediate point. Notation signifying feedback includes the final stage, however, and thus specifies the register. As an example (13, 9, 5, 1) specifies a 13-stage shift register with feedback (to modulo-2 adders) from stages 13, 9, 5, and 1.

which $2^n - 1$ is 31. Because 31 is itself prime, all positive integers less than 31 are relatively prime to it. The five-stage register then has maximal linear sequences available:

$$\frac{\phi(2^n - 1)}{n} = \frac{30}{5} = 6.$$

Table 3.3 lists the number of Linear Maximal codes available for many register lengths 2 through 89. Codes whose length, $2^n - 1$, is a prime number (as above) are called "Mersenne prime". If their length is not prime,

TABLE 3.3   Number of Maximal Sequences Available from Register Lengths 3 through 31, 61, and 89

n	Number of Codes	Prime Factors of $2^n - 1$
3	2	7
4	4	3;5
5	6	31
6	4	3;3;7
7	18	127
8	16	3;5;17
9	48	7;73
10	60	3;11;31
11	176	23;89
12	96	3;3;5;7;13
13	630	8,191
14	756	3;43;127
15	1,800	7;31;151
16	2,048	3;5;17;257
17	7,710	131,071
18	1,728	3;3;3;7;19;73
19	27,594	524,287
20	19,200	3;5;5;11;31;41
21	72,576	7;7;127;137
22	120,032	3;23;89;683
23	356,960	47;178,481
24	184,320	3;3;5;7;13;17;241
25	1,296,000	31;601;1,801
26	1,719,900	3;2,731;8,191
27	4,260,864	7;73;262,657
28	4,741,632	3;5;29;43;113;127
29	18,407,808	233;1,103;2,089
30	11,880,000	3;3;7;11;31;151;331
31	69,273,666	2,147,483,647
61	31,800,705,069,076,960	2,305,843,009,213,693,951
89	6,954,719,320,827,979,072,466,990	618,970,019,642,690,137,449,562,111

however, the number of codes that can be generated is

$$\frac{(2^n - 1)\left(1 - \dfrac{1}{pf_1}\right)\left(1 - \dfrac{1}{pf_2}\right)\cdots\left(1 - \dfrac{1}{pf_N}\right)}{n},$$

where $pf_1, pf_2 \ldots$ are the prime factors of $2^n - 1$, and n is the generator length. For example, let $n = 10$. Then $2^n - 1 = 1023$, and the number of linear maximal codes of length 1023 is

$$\frac{(2^{10} - 1)\left(1 - \dfrac{1}{3}\right)\left(1 - \dfrac{1}{11}\right)\cdots\left(1 - \dfrac{1}{31}\right)}{10}$$

$$= \frac{1023 \times \dfrac{2}{3} \times \dfrac{10}{11} \times \dfrac{30}{31}}{10} = \frac{600}{10} = 60.$$

We may also easily show that the number of codes is

$$\frac{(pf_1 - 1)(pf_2 - 1)(pf_3 - 1)\cdots}{n}.$$

What is not so obvious is the fact that half of the codes are the mirror image (in time) of the other half. This means that there are really only

$$\frac{(pf_1 - 1)(pf_2 - 1)(pf_3 - 1)\cdots}{2n}$$

unique codes of length $2^n - 1$.

When more than one feedback point is used, the alternate generator configuration of Figure 3.6 is of most value. Consider Figure 3.8, which shows a multitap sequence generator of the simple type. In this configuration, the delay inherent in the feedback loop is additive and consists of the sum of all the propagation delays in the path from last stage to first stage. Therefore, the limitation set by the feedback delay path sets the maximum useful speed

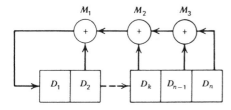

Figure 3.8 Multiple-tap simple sequence generator (SSRG).

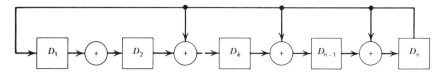

Figure 3.9 Modular multiple-tap sequence generator (MSRG).

of the sequence generator. Figure 3.9 extends the idea of Figure 3.6 to illustrate a sequence generator with a modulo-2 adder between each delay stage and its neighbor. This configuration reduces the delays inherent in the feedback path so that the total delay is never greater than that of a single-tap sequence generator. Another important, though less evident, property of the configuration in Figure 3.9 is that it lends itself more easily to implementation, being readily divisible into units consisting of a delay element (flip-flop) and a modulo-2 adder. For this reason sequence generators of this type have been dubbed "modular," and shift register sequence generators made up in this way are called "modular shift register generators" or MSRG's.

In recent years, first with the advent of high-speed switching transistors and then integrated circuits (IC's), it has become progressively simpler to design high-chip-rate code generators. One particular TTL integrated circuit, whose block diagram is shown in Figure 3.10, has been specifically designed to act as a flip-flop and gated modulo-2 adder module in MSRG's with code chip rates up to or higher than 20 Mcps. It is expected that entire code generators, with all stages selectable for feedback, will be available as

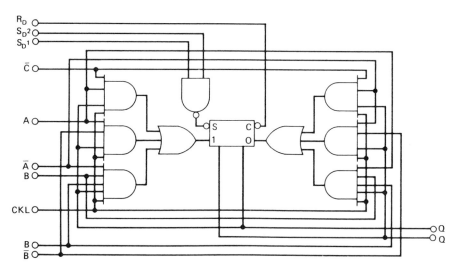

Figure 3.10 Single package flip-flop and gated modulo-2 adder for code sequence generation.

Figure 3.11 MC8504 integrated circuit.

single-chip monolithic integrated circuits. Already, shift registers are available with up to 100 chips storage and need only have feedback added to perform as sequence generators.

One particularly applicable integrated circuit capable of acting as an MSRG is the MC8504 developed by Motorola. This IC includes four D flip-flops and four modulo-2 adders, plus a gate that can be used to sense the nonallowable all zeros condition. For longer generators, the output of one

Figure 3.12 Thirteen-stage MSRG design for MC8504. ($*\frac{1}{2}$ 74 74) Power and grounds not shown.

MC8504 can be cascaded as an input to the next, and code generators that operate to 17 Mcps can be readily constructed. Figure 3.11 is a block diagram of the MC8504, and Figure 3.12 shows a typical 13-stage MSRG constructed with these devices. Note that the internal all-zeros gate is used in this case to sense an all-zeros hangup condition (and to correct) as well as to generate an $n - 1$ zeros output. (The $n - 1$ zeros state is equivalent to the all-ones state for test or other purposes, since it also occurs once and only once in each complete cycle of a linear maximal code.)

A shift register configuration worthy of mention here is the "toggle" shift register generator. The toggle SRG is a special type of code sequence generator that takes advantage of the logic included in the generally available integrated circuit J-K flip-flops; that is, toggle-type code sequence generators do not employ modulo-2 adders that are separate from the shift register stages themselves. This configuration is advantageous in that it provides for

TABLE 3.4   Toggle Registers which are Maximal[a]

Stages	Number of Toggles	Associated Polynomial
3	1	$1 + x + x^3$
3	2	$1 + x^2 + x^3$
4	1	$1 + x + x^4$
5	2	$1 + x^2 + x^5$
5	3	$1 + x + x^2 + x^3 + x^5$
6	1	$1 + x + x^6$
6	5	$1 + x + x^4 + x^5 + x^6$
7	1	$1 + x + x^7$
7	3	$1 + x + x^2 + x^3 + x^7$
7	4	$1 + x^4 + x^7$
7	6	$1 + x^2 + x^4 + x^6 + x^7$
9	4	$1 + x^4 + x^9$
9	5	$1 + x + x^4 + x^5 + x^9$
10	7	$1 + x^2 + x^3 + x^4 + x^5 + x^6 + x^7 + x^{10}$
11	2	$1 + x^2 + x^{11}$
15	1	$1 + x + x^{15}$
15	4	$1 + x^4 + x^{15}$
15	8	$1 + x^8 + x^{15}$
15	11	$1 + x + x^2 + x^3 + x^8 + x^9 + x^{10} + x^{11} + x^{15}$
17	3	$1 + x + x^2 + x^3 + x^{17}$
17	5	$1 + x + x^4 + x^5 + x^{17}$
17	6	$1 + x^2 + x^4 + x^6 + x^{17}$
17	11	$1 + x + x^2 + x^3 + x^8 + x^9 + x^{10} + x^{11} + x^{17}$
17	12	$1 + x^4 + x^8 + x^{12} + x^{17}$
17	14	$1 + x^2 + x^4 + x^6 + x^8 + x^{10} + x^{12} + x^{14} + x^{17}$
18	11	$1 + x + x^2 + x^3 + x^8 + x^9 + x^{10} + x^{11} + x^{18}$

[a]We note that there are no maximal toggle registers of 8, 12, 13, 14, 16, or 19 stages.

code generation at the highest rates that the flip-flops are capable of operating without their being slowed down by external logic. On the minus side, however, not all codes normally available for a given shift register length are possible with the toggle configuration.

Table 3.4 lists polynomials for linear maximal codes from toggle registers up through a length of 18 stages (degree 18). There are no available linear maximal toggle codes of degree 8, 12, 13, 14, 16, or 19.

Any binary code may be represented by a polynomial, where the degree of the polynomial is equal to the number of stages in the generating register. The terms in the polynomial (i.e., $x, x^2, x^4, \ldots, x^n$) represent the stages in the register, and the coefficients that are 0 or 1 determine which stages are included in the feedback network. (A 1 is always included as the x^0 stage, with coefficient 1.) As an example, $1 + x + x^2 + x^4 + x^7$ is a seven-stage code generator with feedback from stages 1, 2, and 4, with of course, 7.

3.3 AUTOCORRELATION AND CROSS-CORRELATION OF CODES

The correlation properties of the code sequences used in spread spectrum communications depend on code type, length, chip rate, and even the chip-by-chip structure of the particular code being used. Both autocorrelation and cross-correlation are of interest in communication system design.

Autocorrelation, in general, is defined as the integral

$$\psi(\tau) = \int_{-\infty}^{\infty} f(t) f(t - r) \, dt,$$

which is a measure of the similarity between a signal and a phase-shifted replica of itself. An autocorrelation function is a plot of autocorrelation over all phase shifts $(t - r)$ of the signal, where Δt is in one-chip intervals.

Autocorrelation is of most interest in choosing code sequences that give the least probability of a false synchronization. In a communications system designed for maximum sensitivity it is no mean task to discriminate between correlation peaks in a poorly chosen code. Therefore, the designer should investigate the code he or she uses carefully, even if that code is one of the relatively safe m-sequences. Statements such as "our extremely long 127-bit chip code sequence assures noiselike properties," which have been observed in the literature on spread spectrum systems, exhibit a lack of investigation.

Cross-correlation is of interest in several areas such as (a) code division multiple access systems (or any code addressed system) in which receiver response to any signal other than the proper addressing sequence is not allowable, and (b) antijamming systems that must employ codes with extremely low cross-correlation as well as unambiguous autocorrelation.

Cross-correlation is the measure of similarity between two different code sequences. The only difference between autocorrelation and cross-correlation is that in the general convolution integral for autocorrelation a different term is substituted:

$$\psi_{(cross)} = \int_{-\infty}^{\infty} f(t)g(t-r)\, dt.$$

Cross-correlation for different code sequences can be tabulated by generating a comparison table and curve of agreements minus disagreements, just as in autocorrelation (Figure 3.16).

It is well, at this point, to introduce a term for the property of a code sequence, pair of sequences, or a sequence and other signal that determines a receiver's ability to recognize the proper point of code synchronization. This property is called the index of discrimination (ID) and denotes the difference in correlation between the fully correlated (perfectly synchronized) code and the peak of minor autocorrelation or of cross-correlations. A particular code will then have separate ID values for autocorrelation and cross-correlation with noncoded signals. The higher the ID value, the better the code.

Code sequence autocorrelation and cross-correlation are expressed as the number of agreements minus the number of disagreements when the code or codes are compared chip by chip. The following example shows autocorrelation for all shifts of a three-stage shift register generator, generating a seven-chip maximal linear code:

$$S = S_1 S_2 S_3, \; S_3 \oplus S_2 \to S_1, \; L(S) \to R(S).$$

Reference sequence: 1110010

Shift	Sequence	Agreements(A)	Disagreements(D)	$A - D$
1	0111001	3	4	-1
2	1011100	3	4	-1
3	0101110	3	4	-1
4	0010111	3	4	-1
5	1001011	3	4	-1
6	1100101	3	4	-1
0	1110010	7	0	7

Note that the net correlation $A - D$ is -1 for all except zero-shift or synchronous conditions, and $2^n - 1 = 7$ for the zero-shift condition. This is typical of all m-sequences.

In the region between the zero and plus or minus one chip shifts, correlation increases linearly so that the autocorrelation function for an m-sequence is triangular as shown in Figure 3.13. This characteristic autocorrelation is used to great advantage in communication and ranging systems. Two communicators may operate simultaneously, for instance, if their codes

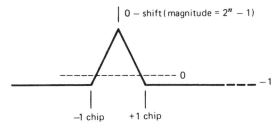

Figure 3.13 *m*-sequence autocorrelation function.

are phase shifted more than one chip. In a ranging system a range measurement is ensured of being accurate within one chip by using the correlation peak as the marker for measurement. This may be accomplished by setting the correlation detector in such a way that it recognizes the level associated with ± 1-chip synchronization and does not recognize a lower level.

When codes other than *m*-sequences are used, their autocorrelation properties may be markedly different from those of the *m* sequences. Figure 3.14 illustrates a typical autocorrelation function for a nonmaximal code. The minor correlation peaks are dependent on the actual code used and are caused by partial correlations of the code with a phase-shifted replica of itself. When such minor correlations occur, a receiving system's ability to synchronize may be impaired because it must discriminate between the major (± 1 chip) and minor correlation peaks, and the margin of discrimination is reduced.

For purposes of illustration, let us consider the five-stage shift register generator shown in Figure 3.15. If feedback is taken from stages five and three, the code sequence output is

$$\cdots 1111100011011101010000100101100 \cdots \text{ (31 chips)}.$$

The autocorrelation of this sequence is shown in Figure 3.13; its maximum value is $2^n - 1 = 31$ and its $ID_{auto} = 32$. This ID value is, as expected, typical of all linear maximal sequences (of which this is an example) for which ID_{auto} is always equal to 2^n.

Figure 3.14 Typical nonmaximal code autocorrelation function.

Now if we modify the feedback to come from stages five and four, one possible output sequence is only 21 chips long:

$$\cdots 111110000100011001010 \cdots .$$

This is an example of a nonmaximal linear sequence that is less than $2^n - 1$ chips long. There are two other nonmaximal linear sequences available from this same feedback configuration whose lengths are seven and three chips:

$$\cdots 1001110 \cdots \quad \text{and} \quad \cdots 101 \cdots .$$

The initial start vector contained in the register determines which of the sequences is generated. For this reason greater care is necessary when nonmaximal sequences are used, both to ensure that the initial start vector is correct (or at least is one of the allowable states) and that noise does not cause the register to go to a state outside the desired set. (In such a case the output code could suddenly change from one sequence to another.)

The three sets (not counting the images) of sequence generator states for the available nonmaximal sequences are

	Set 1 $Q_1Q_2Q_3Q_4Q_5$	Set 2 $Q_1Q_2Q_3Q_4Q_5$	Set 3 $Q_1Q_2Q_3Q_4Q_5$
Initial condition	1 1 1 1 1	1 1 0 0 1	0 1 1 0 1
Next states	0 1 1 1 1	1 1 1 0 0	1 0 1 1 0
	0 0 1 1 1	0 1 1 1 0	1 1 0 1 1
	0 0 0 1 1	1 0 1 1 1	0 1 1 0 1
	0 0 0 0 1	0 1 0 1 1	(3-chip cycle)
	1 0 0 0 0	0 0 1 0 1	
	0 1 0 0 0	1 0 0 1 0	
	0 0 1 0 0	1 1 0 0 1	
	0 0 0 1 0	(7-chip cycle)	
	1 0 0 0 1		
	1 1 0 0 0		
	0 1 1 0 0		
	0 0 1 1 0		
	1 0 0 1 1		
	0 1 0 0 1		
	1 0 1 0 0		
	0 1 0 1 0		
	1 0 1 0 1		
	1 1 0 1 0		
	1 1 1 0 1		
	1 1 1 1 0		
Starting point (21-chip cycle)	1 1 1 1 1		

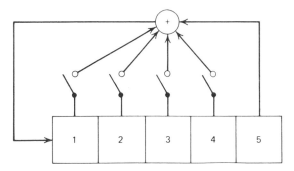

Figure 3.15 Variable tap five-stage SRG.

Note that a total set of only 31 $(2^5 - 1)$ states exists in all of these nonmaximal sequences—the same number that exists in a single maximal sequence. It is typical of linear sequence generators that for every feedback point that produces a subset of length $(2^n - 1) - k$ there are one or more other nonmaximal feedback connections whose subsets (in combination with the original set) have a total length $(2^n - 1)$. Nonmaximal sequences often have high minor autocorrelation peaks. For this reason, the use of nonmaximal codes or even sectors of maximal codes for communications should be approached with caution.

Code sequences available from the five-stage generator of Figure 3.15 are the following:

Feedback	Sequence	Length
[5, 3]	$\cdots$ 1111000110111010100001001011100 $\cdots$	31
[5, 2]	$\cdots$ 1111001101001000010101011011000 $\cdots$	31
[5, 4, 3, 2]	$\cdots$ 1111001001100001011010100011110 $\cdots$	31
[5, 3, 2, 1]	$\cdots$ 1111101110001010110100001100100 $\cdots$	31
[5, 4, 3, 1]	$\cdots$ 1111101000100101011000011100110 $\cdots$	31
[5, 4, 2, 1]	$\cdots$ 1111101100111000011010100100010 $\cdots$	31
[5, 4]	$\cdots$ 111110000100011001010 $\cdots$	21
[5, 1]	$\cdots$ 111110101001100010000 $\cdots$	21
[5, 4]	$\cdots$ 1101001 $\cdots$	7
[5, 1]	$\cdots$ 1001011 $\cdots$	7
[5, 4]	$\cdots$ 011 $\cdots$	3
[5, 1]	$\cdots$ 011 $\cdots$	3

Six of these longer sequences are maximal $(2^5 - 1)$ in length, whereas six are nonmaximal. Observation of pairs ([5, 4, 3, 2], [5, 3, 2, 1]), ([5, 4, 3, 1], [5, 4, 2, 1]), and ([5, 4], [5, 1]) will show that they are paired inverses. None of these codes is useful as the spectrum spreading element for a practical system because of their short length, but they are listed here to provide models. Let us examine these sequences for their autocorrelation and cross-correlation properties;

Figure 3.16   Autocorrelation for 21-chip maximal code [5, 4].

autocorrelation of all six maximal codes is the same (i.e., equal to -1 for all except the 0 ± 1 chip shift). The zero shift produces an autocorrelation value of 31 for all of them.

Autocorrelation of one of the 21-chip nonmaximals is shown in Figure 3.16 (for sequence [5, 4]), which is also typical for the other 21-chip sequence (although backward). The ID_{auto} value for the 21-chip sequences is 19, which could cause a reduction of more than 40% in a receiver's synchronization capability below that for the 31-chip maximal code.

Even the linear maximal sequences are not immune to cross-correlation problems, though they are, in general, the best available. It is also of some interest to note, even when the codes used exhibit excellent cross-correlation properties when averaged over their entire length, that short-term cross-correlations, which are quite effective in disrupting communications, can (and do) occur.

Note that we have restricted our considerations in this chapter to integration over a long period ($-\infty$ to ∞). This is essentially (for our simple case) the same as integrating over the code length, for the codes spoken of here repeat at intervals of $2^n - 1$ chips. We hasten to repeat that integration (as in a synchronization detector) over a period less than that of the code used allows short-term correlations; that is, a short pattern occurring in two different codes or twice in the same code could appear as a legitimate code synchronization when the integration period does not significantly exceed the pattern period.

Figures 3.17 and 3.18 are given as illustrations of cross-correlation and autocorrelation for maximal sequences. The autocorrelation curve of the [5, 3] code shows a zero-shift correlation value of 31. For the [5, 3] and [5, 2] codes cross-correlated, however, the peak value is 11, which gives an index of discrimination of 20, or 35% less than the autocorrelation value. The [5, 3] and [5, 2] codes are images; that is, one is the same as the other, but generated in reverse order. Cross-correlation of the [5, 3] and [5, 4, 3, 2] codes is lower than that for the image codes, but is still such that the peak cross-correlation value is seven, a value that occurs at 10 different shift positions.

The significant point is that these particular pairs of code sequences are not capable of operating in the same link if the transmitted power from

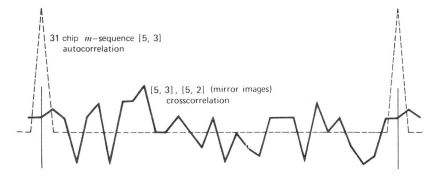

31 chip m-sequence [5, 3] autocorrelation

[5, 3], [5, 2] (mirror images) crosscorrelation

Shift		Agreements (A)		Disagreements (D)		$A - D$	
0	17	17	15	14	16	3	−1
1	18	18	13	13	18	5	−5
2	19	17	12	14	19	3	−7
3	20	11	17	20	14	−9	3
4	21	17	17	14	14	3	3
5	22	19	17	12	14	7	3
6	23	11	11	20	20	−9	−9
7	24	19	19	12	12	7	7
8	25	19	15	12	16	7	−1
9	26	21	17	10	14	11	3
10	27	15	13	16	18	−1	−5
11	28	15	11	16	20	−1	−9
12	29	17	12	14	19	3	−7
13	30	15	17	16	14	−1	3
14	31	13	17	18	14	−5	3
15		17		14		3	
16		11		20		−9	

Figure 3.17 Comparative autocorrelation and cross-correlation for 31-chip mirror image m-sequences.

either transmitter exceeds the other enough to raise the peak cross-correlation to a value near peak autocorrelation. Of course, such short codes should not be used, but the comparison is reasonably representative of operation even with much longer sequences used in code division multiplexing or other multiple-access applications.

Judge[222] has considered code division multiplexing by using "quasiorthogonal binary functions" (linear maximal sequences) and states that for two equal-power signals multiplexed together signal-to-noise is

$$\frac{S}{N_{(2)}} = \frac{S}{\left(K_1^2 + \dfrac{T_0}{T} \right)^{1/2}}$$

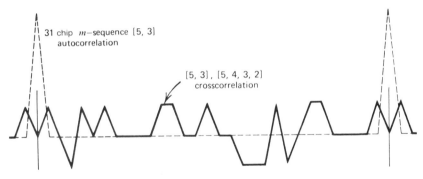

Shift		Agreements (A)		Disagreements (D)		A − D	
0	17	15	15	16	16	−1	−1
1	18	19	11	12	20	7	−9
2	19	15	11	16	20	−1	−9
3	20	11	11	20	20	−9	−9
4	21	19	19	12	12	7	7
5	22	15	11	16	20	−1	−9
6	23	19	15	12	16	7	−1
7	24	15	19	16	12	−1	7
8	25	15	19	16	12	−1	7
9	26	15	15	16	16	−1	−1
10	27	15	15	16	16	−1	−1
11	28	19	15	12	16	7	−1
12	29	19	15	12	16	7	−1
13	30	15	19	16	12	−1	7
14	31	15	15	16	16	−1	−1
15		19		12		7	
16		15		26		−1	

Figure 3.18 Comparative autocorrelation and cross-correlation for 31 chip *m*-sequences (not images).

in each receiver. For *b* signals,

$$\frac{S}{N_{(b)}} = \frac{S}{\left[b\left(K_1^2 + \dfrac{T_0}{T} \right) \right]^{1/2}},$$

where T = the crosscorrelation integration period,
  T_0 = the code bit period,
  K_1 = value of DC correlation.

Judge's result shows that some Mersenne prime sequences exhibit cross-correlation values superior to others, sometimes even for nonprime sequences longer than prime sequences.

The composite code sequences already mentioned are of great utility when cross-correlation is a prime consideration. Their real advantage lies in that for every code in a set of $2^n - 1$ codes, each of length $2^n - 1$, cross-correlation values are well defined, and a system can be designed to operate within this definition. Unless some similar coding scheme is used, each code employed must be separately analyzed with respect to every other code to ensure noninterfering operation.

3.4 COMPOSITE CODES

Though unwavering in the assertion that for communications and ranging the maximal linear sequences cannot be bettered, we must admit that composite code sequences generated by a combination of linear maximal sequences should be considered. Composite codes constructed in this way have special properties that are most advantageous under proper circumstances; for instance, the JPL ranging codes[219] and the Gold[312-314] codes, though constructed from maximal sequences, are not maximal. The JPL ranging codes have special correlation properties that permit rapid synchronization, whereas the Gold codes allow construction of families of $2^n - 1$ codes from pairs of n-stage shift registers in which all codes have well-defined correlation characteristics.

Gold Code Sequence Generators (See Appendix 7)

Gold code sequence generators are useful because of the large number of codes they supply, although they require only one pair of feedback tap sets. A bonus awarded on the basis of use of these codes is that few sets of feedback taps are needed. Thus the possibility of using a pair of single-tap feedback, simple SRG's while retaining the capacity to generate a large number of codes is present. The single-tap simple SRG is the fastest configuration possible. Thus the Gold code sequences are potentially available at rates equal to the capability of the fastest simple SRG.

The Gold codes are generated by modulo-2 addition of a pair of maximal linear sequences as shown in Figure 3.19. The code sequences are added chip by chip by synchronous clocking. The codes themselves are the same length. Thus the two code generators maintain the same phase relationship, and the codes generated are the same length as the two base codes which are added together, but are nonmaximal. Figure 3.20 is a specific example.

The shift-and-add property of maximal sequences tells us that any maximal sequence added to a phase-shifted replica of itself (any integral number of bits) produces a different phase shift as an output. Here the same

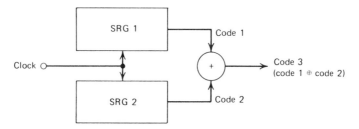

Figure 3.19   Gold code sequence generator configuration.

operation is performed, with the new sequence having the same length as those being added, and nonmaximal. Furthermore, every change in phase position between the two generators causes a new sequence to be generated. To show this advantage, consider the following example.

Given a five-stage sequence generator, we choose a set of feedback taps from Table 3.7. Note that there are only six feedback sets available for the five-stage register, and half are images of the other half. If more than six 31-chip codes are needed, we cannot get them from our five-stage register.

The solution, as stated, is to use two five-stage sequence generators connected in the Gold configuration, as shown in Figure 3.20. This figure also shows the modulo-2-combined Gold codes produced by combining the two output maximal codes with different initial offsets; that is, the two code generators are started with initial conditions offset by various amounts to give different output codes. The all-ones vector is set into both registers as an initial condition. In addition, one- and five-chip shifts (from the all-ones vector) are also shown in initial conditions:

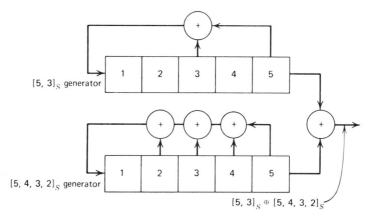

Figure 3.20   Illustration of Gold code generation with $[5,3]_s$ and $[5,4,3,2]_s$ generators.

Zero-shift combination:

1 1 1 1 1 0 0 0 1 1 0 1 1 1 0 1 0 1 0 0 0 0 1 0 0 1 0 1 1 0 0
1 1 1 1 1 0 0 1 0 0 1 1 0 0 0 0 1 0 1 1 0 1 0 1 0 0 0 1 1 1 0
───
0 0 0 0 0 0 0 1 1 1 1 0 1 1 0 1 1 1 1 1 0 1 1 1 0 1 0 0 0 1 0

One-chip-shift combination:

1 1 1 1 1 0 0 0 1 1 0 1 1 1 0 1 0 1 0 0 0 0 1 0 0 1 0 1 1 0 0
1 1 1 1 0 0 1 0 0 1 1 0 0 0 0 1 0 1 1 0 1 0 1 0 0 0 1 1 1 0 1
───
0 0 0 0 1 0 1 0 1 0 1 0 1 1 1 1 0 0 0 0 1 0 1 0 0 0 0 1 1 0 0 0 1

Five-chip-shift combination:

1 1 1 1 1 0 0 0 1 1 0 1 1 1 0 1 0 1 0 0 0 0 1 0 0 1 0 1 1 0 0
0 0 1 0 0 1 1 0 0 0 0 1 0 1 1 0 1 0 1 0 1 0 0 0 1 1 1 0 1 1 1 1 1
───
1 1 0 1 1 1 1 0 1 1 0 0 1 0 1 1 1 1 1 0 0 0 0 1 1 1 1 0 0 1 1

Any shift in initial conditions from zero to 30 chips can be used. (A 31-chip shift is the same as the zero shift.) Thus, from this Gold sequence generator, 33 maximal-length codes are available. Extending this demonstration, we can show that any two-register Gold code generator of length n can generate $2^n - 1$ maximal-length sequences (length $2^n - 1$) plus the two maximal base sequences. A multiple-register Gold code generator can generate $(2^n - 1)^r$ nonmaximal sequences of length $2^n - 1$ plus r maximal sequences of the same length, where r is the number of registers and n is register length.

In addition to their advantage in generating large numbers of codes, the Gold codes may be chosen so that over a set of codes available from a given generator the cross-correlation between the codes is uniform and bounded. Thus the Gold codes are attractive for applications in which a number of code-division-multiplexed signals are to be used. The same guarantee of bounded cross-correlation is impossible for maximal sequences of the same length.

Gold has presented a method[312] for choosing the linear maximal codes used as components in Gold sequences to give a set of sequences, each of whose members has cross-correlation and autocorrelation side lobes bounded by $|\theta(r)| \le 2^{(n+1)/2} + 1$ for n odd, and by $|\theta(r)| \le 2^{(n+2)/2} - 1$ for n even.

An equivalent result is given by Anderson[32] for the Gold codes; that is, Anderson's expression for the cross-correlation bound is

$$|\theta(\tau)|_G \le \left(\frac{\sqrt{2}\sqrt{1 + 1/L} + 1/\sqrt{2}}{\sqrt{L}} \right)^{1/2}.$$

It is apparent from this expression that as $L \rightarrow \infty$, $|\theta(\tau)| \rightarrow \sqrt{2}/\sqrt{L}$. Convergence is sufficiently rapid that for any code sequence length of interest $|\theta(\tau)| = \sqrt{2}/\sqrt{L}$ percent.

Notice that one expression gives cross-correlation in chips, whereas the other gives a percentage of maximum correlation. By normalizing maximum correlation to 1, $2^{(n+1)/2} + 1 \approx \sqrt{2L}/\sqrt{L}$ for large L.

Anderson also states that the cross-correlation function for maximal sequences is bounded by

$$|\theta(\tau)| \leq \left(\frac{1 + 1/L - 1/L^2}{L} \right)^{1/2}.$$

Now, as $L \rightarrow \infty$, $|\theta(\tau)| \rightarrow 1/\sqrt{L}$. For a given value of L the Gold codes exhibit cross-correlation that is $(\sqrt{2}/\sqrt{L})/(1/\sqrt{L}) = \sqrt{2}$ greater than maximal length sequences of the same length.

Appendix 7 describes an algorithm for Gold code sequence design. A convenient method for selection of Gold codes in hardware is that employed in the Global Positioning System (GPS) for switching from one C/A code to another. This method makes use of the shift-and-add property of linear maximal codes in a different way, and is illustrated in Figure 3.21. (The Gold code generator from Figure 3.20 is used as an illustration.) Both generators shown could be started by setting them to the all-ones vector. The lower generator's output, however, is shifted by modulo-2 adding selected outputs together. Here the number of codes selectable by the two switches shown is

$$5 + \overset{2}{\underset{5}{C}} = 15.$$

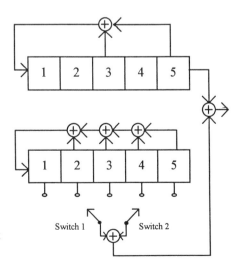

Figure 3.21 Gold code generator using shift-and-add for switching codes.

Syncopated-Register Generators

A technique that multiplexes two or more slower generators to generate high-rate sequences was used in complex form as early as 1958 by H. G. Posthumus. The method is similar to that used in Gold code sequence generators in that two separate sequences are modulo-2 added to produce a composite output. The difference lies in that separate clocks, phase shifted by $360/P$ degrees, each at a rate R/P, are used. (Here R is the desired output code chip rate and P is the number of registers used or the rate reduction desired.) Figure 3.22 shows a two-register generator of the syncopated type, which would allow clock rate reduction by a factor of 2. Timing for this code generator is also shown. Note that the chip rate of code 1 $\oplus$ code 3 is twice that of either code 1 or code 2 before their combination.

Codes generated by syncopation are subject to the same correlation requirements as others used for communications. Unfortunately, work in this specific area is unknown, if it exists, so that the designer who must resort to syncopation techniques must also spend some time in code analysis of correlation properties.

There is no difficulty in extending the syncopation technique. Five registers, for instance, each operating at 50 Mcps, could be used to generate a

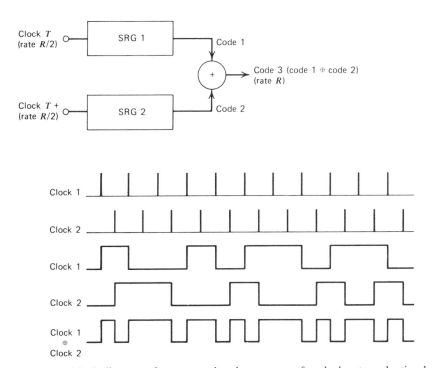

Figure 3.22 Block diagram of syncopated code generator for clock rate reduction by a factor of 2.

250-Mcps code sequence with a five-phase clock. Theoretically, the syncopation technique could achieve a rate multiplication as great as desired. The rapid increase in hardware needed places a limit on what is practical, however. The overall length of any number of syncopated codes is the product of the lengths of all the composite codes; for example, a pair of codes $(2^r - 1)$ and $(2^s - 1)$ chips long produces a composite code $(2^r - 1)(2^s - 1)$ chips in overall length. We assume here that $r \neq s$. When $r = s$, the composite code is $2^r - 1$ chips long.

Syncopation techniques, though useful for increasing the speed of marginal designs (e.g., increasing low-speed MOS large-scale integrated circuits to give effective 10-Mcps or higher output speeds), will probably find less use in the near future because of the high-speed capability of new integrated circuits, which can operate directly at logic rates that in the past only techniques like syncopation could reach.

JPL Ranging Codes

The JPL (Jet Propulsion Laboratory) ranging codes are constructed by modulo-2 addition of two or more maximal linear sequences whose lengths are relatively prime to one another. Often a clock term is also modulo-2 added to the other sequences (the clock is a short maximal sequence in itself). There are several advantages to such a technique: (a) very long codes useful for unambiguous ranging over long ranges are available; (b) these long codes are generated by a relatively small number of shift register stages; and (c) synchronization of a receiver can be accomplished by separate operations on the component codes. This can greatly reduce the time required for synchronization.

Figure 3.23 shows a typical JPL code generator configuration, which has three basic maximal-length shift register generators, each with a different

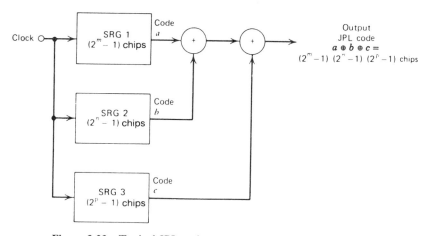

Figure 3.23   Typical JPL code sequence generator configuration.

number of stages. This configuration is identical to the Gold and syncopated code generators except for the difference in the individual code lengths. The JPL codes and their application are discussed further in Chapter 6, and details of their selection and construction may be found in Reference 219.

3.5 CHIP RATE AND CODE LENGTH

Code chip rates in spread spectrum systems affect their systems in many ways. The most obvious is in a BPSK direct sequence system, in which the transmission bandwidth is a direct function of the code chip rate (i.e., main lope null-null RF bandwidth is twice the code chip rate). Code repetition rate is also a function of the chip rate; that is, code repetition rate is simply

$$R_{rep} = \frac{\text{clock rate in chips per second}}{\text{code length in chips}}.$$

This repetition rate determines the line spacing in the RF output spectrum and is an important consideration in a system design.

One criterion for selecting code repetition rate is that the period of the code must exceed the length of any mission in which it is to be used. In most aircraft, for instance, an eight-hour code period would exceed the flight capability. Table 3.5 lists the various code lengths for a 1-Mcps chip rate to acquaint the reader with the range of code lengths. Other considerations that

TABLE 3.5 Code Sequence Periods for Various m-Sequence Lengths, 1-Mcps Rate

Register Length n	Sequence Length	Sequence Period
7^a	127	1.27×10^{-4} sec
8	255	2.55×10^{-4} sec
9	511	5.11×10^{-4} sec
10	1023	1.023×10^{-3} sec
11	2047	2.047×10^{-3} sec
12	4095	4.095×10^{-3} sec
13^a	8191	$9,191 \times 10^{-3}$ sec
17^a	131071	1.31×10^{-1} sec
19^a	524287	5.24×10^{-1} sec
23	8388607	8.388 sec
27	134217727	13.421 sec
31^a	2147483647	35.8 min
43	879609302207	101.7 days
61^a	2305843009213693951	7.3×10^4 yr
89^a	618970019642690137449562111	1.95×10^9 yr

[a]Mersenne prime generators. All codes of length $2^n - 1$.

TABLE 3.6 Chip Rates for Various Basic Range Resolutions

Basic Accuracy	Nautical Miles	Statute Miles
0.1 mile	1.61875 Mcps (608 ft/chip)	1.86333 Mcps (528 ft/chip)
0.05 mile	3.23750 Mcps (304 ft/chip)	3.72666 Mcps (264 ft/chip)
0.02 mile	8.09375 Mcps (121.6 ft/chip)	9.19665 Mcps (105.6 ft/chip)

bear on the choice of code rate and length are the relationship of the repetition rate to the information baseband and use of the system for ranging.

It is advisable that a direct sequence system's code repetition rate* be adjusted, by choosing a satisfactory code length, so that it will not lie in the information band. Otherwise, unnecessary noise will be passed into the information demodulators, especially under jammed conditions. In frequency hopping systems the equivalent of having a code repetition rate that falls within the baseband is to have a chip rate that falls there. Both conditions should be avoided whenever possible.

Code chip rates for frequency hoppers usually bear an integral relationship to the information sent; that is, 1 to p frequency chips are sent for each bit of enclosed information. Sending multiple information bits at a single frequency is more conductive to errors than multiple chip/bit transmission, although a lower frequency hopping rate is possible and the synthesizer may be simpler to build.

When ranging is important, a properly chosen code rate can ease the measurement problem and sometimes even improve resolution. If the chip rate is chosen in such a way that an integral number of code chips is accumulated for each mile of delay (propagation time), a simple count of code offset can be used to measure range without elaborate correction. Clock rates that are integral multiples of 161,875 Hz produce code at rates integrally related to the speed of propagation of the RF signal. This relationship may be derived as follows:

Speed of propagation of light (in a vacuum) is

$$c = 2.997926 \times 10^8 \text{ m/sec.}$$

There are 1.852×10^3 m/nautical mile; therefore, for one-chip per mile resolution, the clock rate should be (setting clock wavelength = 1 n mi):

$$\frac{c}{R} = \frac{2.997925 \times 10^8}{1.852 \times 10^3} = 161,875 \text{ Hz.}$$

The clock rate for one-statute mile resolution would be 186,333 Hz. Table 3.6

---

*Repetition rate is usually adjusted by choosing code length, for chip rate must be chosen to give a particular RF bandwidth in a direct-sequence system.

lists the rates that should be used for various range resolutions (numbers of chips per mile) for nautical and statute mile measurements.

3.6 CHOOSING A LINEAR CODE

Among the biggest problems in code generation are (1) finding feedback logic that gives the desired code length, and (2) checking the code once a sequence generator has been constructed to ensure that it is operating properly. It is of little use to know how to construct a shift register generator unless one also knows the feedback connections necessary to generate a useful code sequence. Of course, it is possible to find a set of feedback connections experimentally, but this requires not only constructing a shift register generator but also taking the time to check the length of codes generated until one chances on a maximal connection.

Happily, tables of feedback connections and tables of irreducible polynomials have been generated,[326, 320, 335] which make the job much easier. Table 3.7 was compiled from information given in many sources, and cross checked to ensure accuracy as far as was practical. An example of the use of the table follows.

TABLE 3.7   Feedback Connections for Linear m-Sequences

Number of Stages	Code Length	Maximal Taps
2^a	3	[2, 1]
3^a	7	[3, 1]
4	15	[4, 1]
5^a	31	[5, 2] [5, 4, 3, 2] [5, 4, 2, 1]
6	63	[6, 1] [6, 5, 2, 1] [6, 5, 3, 2]
7^a	127	[7, 1] [7, 3] [7, 3, 2, 1] [7, 4, 3, 2] [7, 6, 4, 2] [7, 6, 3, 1] [7, 6, 5, 2] [7, 6, 5, 4, 2, 1] [7, 5, 4, 3, 2, 1]
8	255	[8, 4, 3, 2] [8, 6, 5, 3] [8, 6, 5, 2] [8, 5, 3, 1] [8, 6, 5, 1] [8, 7, 6, 1] [8, 7, 6, 5, 2, 1] [8, 6, 4, 3, 2, 1]
9	511	[9, 4] [9, 6, 4, 3] [9, 8, 5, 4] [9, 8, 4, 1] [9, 5, 3, 2] [9, 8, 6, 5] [9, 8, 7, 2] [9, 6, 5, 4, 2, 1] [9, 7, 6, 4, 3, 1] [9, 8, 7, 6, 5, 3]
10	1023	[10, 3] [10, 8, 3, 2] [10, 4, 3, 1] [10, 8, 5, 1] [10, 8, 5, 4] [10, 9, 4, 1] [10, 8, 4, 3] [10, 5, 3, 2] [10, 5, 2, 1] [10, 9, 4, 2]
11	2047	[11, 1] [11, 8, 5, 2] [11, 7, 3, 2] [11, 5, 3, 5] [11, 10, 3, 2] [11, 6, 5, 1] [11, 5, 3, 1] [11, 9, 4, 1] [11, 8, 6, 2] [11, 9, 8, 3]

TABLE 3.7   *(Continued)*

Number of Stages	Code Length	Maximal Taps
12	4095	$[12, 6, 4, 1] [12, 9, 3, 2] [12, 11, 10, 5, 2, 1]$ $[12, 11, 6, 4, 2, 1] [12, 11, 9, 7, 6, 5]$ $[12, 11, 9, 5, 3, 1] [12, 11, 9, 8, 7, 4]$ $[12, 11, 9, 7, 6, 5] [12, 9, 8, 3, 2, 1]$ $[12, 10, 9, 8, 6, 2]$
13[a]	8191	$[13, 4, 3, 1] [13, 10, 9, 7, 5, 4]$ $[13, 11, 8, 7, 4, 1] [13, 12, 8, 7, 6, 5]$ $[13, 9, 8, 7, 5, 1] [13, 12, 6, 5, 4, 3]$ $[13, 12, 11, 9, 5, 3] [13, 12, 11, 5, 2, 1]$ $[13, 12, 9, 8, 4, 2] [13, 8, 7, 4, 3, 2]$
14	16, 383	$[14, 12, 2, 1] [14, 13, 4, 2] [14, 13, 11, 9]$ $[14, 10, 6, 1] [14, 11, 6, 1] [14, 12, 11, 1]$ $[14, 6, 4, 2] [14, 11, 9, 6, 5, 2]$ $[14, 13, 6, 5, 3, 1] [14, 13, 12, 8, 4, 1]$ $[14, 8, 7, 6, 4, 2] [14, 10, 6, 5, 4, 1]$ $[14, 13, 12, 7, 6, 3] [14, 13, 11, 10, 8, 3]$
15	32, 767	$[15, 13, 10, 9] [15, 13, 10, 1] [15, 14, 9, 2]$ $[15, 1] [15, 9, 4, 1] [15, 12, 3, 1] [15, 10, 5, 4]$ $[15, 10, 5, 4, 3, 2] [15, 11, 7, 6, 2, 1]$ $[15, 7, 6, 3, 2, 1] [15, 10, 9, 8, 5, 3]$ $[15, 12, 5, 4, 3, 2] [15, 10, 9, 7, 5, 3]$ $[15, 13, 12, 10] [15, 13, 10, 2] [15, 12, 9, 1]$ $[15, 14, 12, 2] [15, 13, 9, 6] [15, 7, 4, 1]$ $[15, 4] [15, 13, 7, 4]$
16	65, 535	$[16, 12, 3, 1] [16, 12, 9, 6] [16, 9, 4, 3]$ $[16, 12, 7, 2] [16, 10, 7, 6] [16, 15, 7, 2]$ $[16, 9, 5, 2] [16, 13, 9, 6] [16, 15, 4, 2]$ $[16, 15, 9, 4]$
17[a]	131, 071	$[17, 3] [17, 3, 2, 1] [17, 7, 4, 3]$ $[17, 16, 3, 1] [17, 12, 6, 3, 2, 1]$ $[17, 8, 7, 6, 4, 3] [17, 11, 8, 6, 4, 2]$ $[17, 9, 8, 6, 4, 1] [17, 16, 14, 10, 3, 2]$ $[17, 12, 11, 8, 5, 2]$
18	262, 143	$[18, 17] [18, 10, 7, 5] [18, 13, 11, 9, 8, 7, 6, 3]$ $[18, 17, 16, 15, 10, 9, 8, 7]$ $[18, 15, 12, 11, 9, 8, 7, 6]$
19[a]	524, 287	$[19, 5, 2, 1] [19, 13, 8, 5, 4, 3]$ $[19, 12, 10, 9, 7, 3] [19, 17, 15, 14, 13, 12, 6, 1]$ $[19, 17, 15, 14, 13, 9, 8, 4, 2, 1]$ $[19, 16, 13, 11, 19, 9, 4, 1] [19, 9, 8, 7, 6, 3]$ $[19, 16, 15, 13, 12, 9, 5, 4, 2, 1]$ $[19, 18, 15, 14, 11, 10, 8, 5, 3, 2]$ $[19, 18, 17, 16, 12, 7, 6, 5, 3, 1]$

TABLE 3.7 *(Continued)*

Number of Stages	Code Length	Maximal Taps
20	1, 048, 575	[20, 3] [20, 9, 5, 3] [20, 19, 4, 3]
		[20, 11, 8, 6, 3, 2] [20, 17, 14, 10, 7, 4, 3, 2]
21	2, 097, 151	[21, 2] [21, 14, 7, 2] [21, 13, 5, 2]
		[21, 14, 7, 6, 3, 2] [21, 8, 7, 4, 3, 2]
		[21, 10, 6, 4, 3, 2] [21, 15, 10, 9, 5, 4, 3, 2]
		[21, 14, 12, 7, 6, 4, 3, 2] [21, 20, 19, 18, 5, 4, 3, 2]
22	4, 194, 303	[22, 1] [22, 9, 5, 1] [22, 20, 18, 16, 6, 4, 2, 1]
		[22, 19, 16, 13, 10, 7, 4, 1] [22, 17, 9, 7, 2, 1]
		[22, 17, 13, 12, 8, 7, 2, 1] [22, 14, 13, 12, 7, 3, 2, 1]
23	8, 388, 607	[23, 5] [23, 17, 11, 5] [23, 5, 4, 1]
		[23, 12, 5, 4] [23, 21, 7, 5] [23, 16, 13, 6, 5, 3]
		[23, 11, 10, 7, 6, 5] [23, 15, 10, 9, 7, 5, 4, 3]
		[23, 17, 11, 9, 8, 5, 4, 1] [23, 18, 16, 13, 11, 8, 5, 2]
24	16, 777, 215	[24, 7, 2, 1] [24, 4, 3, 1]
		[24, 22, 20, 18, 16, 14, 11, 9, 8, 7, 5, 4]
		[24, 21, 19, 18, 17, 16, 15, 14, 13, 10, 9, 5, 4, 1]
25	33, 554, 431	[25, 3] [25, 3, 2, 1] [25, 20, 5, 3] [25, 12, 5, 4]
		[25, 17, 10, 3, 2, 1] [25, 23, 21, 19, 9, 7, 5, 3]
		[25, 18, 12, 11, 6, 5, 4] [25, 20, 16, 11, 5, 3, 2, 1]
		[25, 12, 11, 8, 7, 6, 4, 3]
26	67, 108, 863	[26, 6, 2, 1] [26, 22, 21, 16, 12, 11, 10, 8, 5, 4, 3, 1]
27	134, 217, 727	[27, 5, 2, 1] [27, 18, 11, 10, 9, 5, 4, 3]
28	268, 435, 455	[28, 3] [28, 13, 11, 9, 5, 3] [28, 22, 11, 10, 4, 3]
		[28, 24, 20, 16, 12, 8, 4, 3, 2, 1]
29	536, 870, 911	[29, 2] [29, 20, 11, 2] [29, 13, 7, 2]
		[29, 21, 5, 2] [29, 26, 5, 2] [29, 19, 16, 6, 3, 2]
		[29, 18, 14, 6, 3, 2]
30	1, 073, 74, 1, 823	[30, 23, 2, 1] [30, 6, 4, 1]
		[30, 24, 20, 16, 14, 13, 11, 7, 2, 1]
31[a]	2, 147, 483, 647	[31, 29, 21, 17] [31, 28, 19, 15] [31, 3]
		[31, 3, 2, 1] [31, 13, 8, 3] [31, 21, 12, 3, 2, 1]
		[31, 20, 18, 7, 5, 3] [31, 30, 29, 25] [31, 28, 24, 10]
		[31, 20, 15, 5, 4, 3] [31, 16, 8, 4, 3, 2]
32	4, 294, 967, 295	[32, 22, 2, 1] [32, 7, 5, 3, 2, 1]
		[32, 28, 19, 18, 16, 14, 11, 10, 9, 6, 5, 1]
33	8, 589, 934, 591	[33, 13] [33, 22, 13, 11] [33, 26, 14, 10]
		[33, 6, 4, 1] [33, 22, 16, 13, 11, 8]
61[a]	2, 305, 843, 009, 213, 693, 951	[61, 5, 2, 1]
89[a]	618, 970, 019, 642, 690, 137, 449, 562, 112	[89, 6, 5, 3]

[a]Mersenne prime length generator.

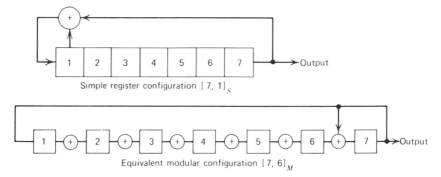

Figure 3.24 Feedback configurations for 127-chip linear m sequences.

After selecting a given code length, say 127 chips, we select a seven-stage shift register generator from the table, together with a set of maximal taps for a seven-stage register. For a minimal number of feedback taps we choose [7, 1]. Generators implemented with this feedback appear in Figure 3.24, which shows both SSRG and MSRG configurations. SSRG configurations are denoted $[n, r, q, p]_s$.

For modular shift register generators, the $[n, r, q, p]_m$ notation would be used, and this signifies an n-stage register with feedback from the nth-stage modulo-2 added with the output of the rth, qth, and pth stages and fed to the inputs of the $r + 1$st, $q + 1$st stages, respectively.

The two configurations in Figure 3.24 illustrate a point—that in order to generate an identical code sequence, simple shift register and modular shift register generators must be implemented differently. A second example will perhaps reinforce the point and better explain the use of Table 3.7.

Suppose now that another 127-chip code sequence is needed. This time we choose feedback set [7, 3, 2, 1] and construct either $[7, 3, 2, 1]_s$ or $[7, 6, 5, 4]_m$. Again the two configurations shown are equivalent.

As an example of the equivalence of a $[n, p]_s$ and a $[n, n - p]_m$ generator, we generate the pair of subsequences shown in Table 3.8. Observing the first 15 states of stage one in the $[7, 1]_s$ register and of stage seven in the $[7, 6]_m$ register, we see that both generate an identical subsequence of 15 chips:

$$101010100110011 \cdots$$

For the seven-stage register this is $2n - 1$ chips. It can be readily shown that a sequence of $2n + 1$ chips from a linear sequence generator can be used to derive the feedback in the sequence generator (see Appendix 6). Therefore, because both configurations produce an identical subsequence for $2n + 1$ chips, and $2n + 1$ chips is sufficient to establish feedback for a linear sequence generator, the two configurations must be equivalent.

Finally, we must point out that for every set of $[n, \cdots, p]_s$ feedback taps listed in Table 3.6 there is an image set; that is, there is a set of feedback taps

TABLE 3.8   Comparison of States in $[7,1]_s$ and $[7,6]_m$ Sequence Generators

		$[7,1]_s$	$[7,6]_m$
State	1	1 1 1 1 1 1 1	1 1 1 1 1 1 1
	2	0 1 1 1 1 1 1	1 1 1 1 1 1 0
	3	1 0 1 1 1 1 1	0 1 1 1 1 1 1
	4	0 1 0 1 1 1 1	1 0 1 1 1 1 0
	5	1 0 1 0 1 1 1	0 1 0 1 1 1 1
	6	0 1 0 1 0 1 1	1 0 1 0 1 1 0
	7	1 0 1 0 1 0 1	0 1 0 1 0 1 1
	8	0 1 0 1 0 1 0	1 0 1 0 1 0 0
	9	0 0 1 0 1 0 1	0 1 0 1 0 1 0
	10	1 0 0 1 0 1 0	0 0 1 0 1 0 1
	11	1 1 0 0 1 0 1	1 0 0 1 0 1 1
	12	0 1 1 0 0 1 0	0 1 0 0 1 0 0
	13	0 0 1 1 0 0 1	0 0 1 0 0 1 0
	14	1 0 0 1 1 0 0	0 0 0 1 0 0 1
	15	1 1 0 0 1 1 0	1 0 0 0 1 0 1
Stage		1 2 3 4 5 6 7	1 2 3 4 5 6 7

that generates an identical code reversed in time sequence. Furthermore, this mirror image of a feedback set $[n, r, \cdots, p]_s$ is $[n, n - r, \cdots, n - p]_s$.

Recalling that $[n, p]_s$ is equivalent to $[n, n - p]_m$, we see that $[n, p]_m$ is the mirror image of $[n, p]_s$, $[n, p]_m$ is equivalent to $[n, n - p]_s$, and $[n, p]_s$ is the mirror image of $[n, p]_m$.

To give a numerical example we select a nine-stage, 511-chip code feedback set $[9, 8, 5, 4]_s$ from our table. If we want the mirror image, we convert to

$$[9, 9\text{-}8, 9\text{-}5, 9\text{-}4]_s = [9, 1, 4, 5]_s$$

$$= [9, 5, 4, 1]_s.$$

The same conversion holds for a modular shift register; the modular feedback set would be

$$[9, 5, 4, 1]_m.$$

Here the code sequence would not be a mirror image but would be generated by the modular generator as an identical sequence to $[9, 8, 5, 4]_s$.

Caution. No repeating maximal code sequence is produced when two feedback taps (or any even number) besides the nth stage are used. This is obvious if we consider that any three ones (or other odd number) of ones modulo-2 added together produces a one. Thus the all-ones vector could not be permitted and a maximal sequence cannot be generated. If nonmaximal sequences are desired, however, even numbers of feedback taps may be acceptable.

3.7 GENERATING HIGH RATE CODES

High chip rate code sequences are desirable for many applications. Notable for the purposes of this book are those applications in which high code rates are advantageous for spreading a signal over wide frequency bands. Such is the case when a high data rate signal (wide baseband bandwidth) is to be expanded, or when more interference rejection margin is required. Data to be transmitted are often in the megabits-per-second range, and it is obvious that to gain any significant advantage from band spreading the code sequences used must be in the hundreds of megachips range.

Shift register sequence generators have been constructed from most of the known delay elements by those striving for faster code rates, smaller size, or some other parameter of interest. Tunnel diodes, delay lines (coaxial, acoustic, stripline, etc.), silicon-controlled rectifiers, transistors, vacuum tubes, and even relays have been used. Recent developments in microelectronics have produced logic elements with such flexibility, speed, and general utility that it is now difficult to justify generating code sequences by any other method. Three-hundred-megachip operation has been exceeded by shift registers made up of microelectronic elements. There is little reason to doubt that future circuits and subsystems will include complete sequence generators capable of 100-Mcps clock rates in the same size presently occupied by a single flip-flop.

For the purposes of this book the import of microelectronics is such that we have not (for digital circuits) even considered the use of discrete elements (i.e., transistors). Instead, flip-flops, gates, and other elements are treated as blocks and are assembled to make up the desired code sequence generators.

In a preceding section, we discussed the advantages of modular shift register connections for high-speed code generation. Here we consider modular, simple, and partially modular construction (i.e., with modulo-2 adders inserted between some stages rather than at every stage). Toggle registers are not specifically considered because they generate codes at the operating rate of the flip-flops themselves without special provision for high-speed capability; they are special cases of the modular SRG's and they lack the repertoire of codes available that either the simple or modular generators are able to produce.

Figure 3.25 is a diagram of a seven-stage simple shift register generator made up of ECL elements capable of generating 127-chip $(2^7 - 1)$ codes well in excess of 200 Mcps, depending on the layout of the circuit. Figure 3.26 shows a sector of the 127-chip code at a 150-Mcps rate. In high-speed sequence generators layout is extremely critical, because the propagation time in wiring approaches that in the circuit elements. Extending the length

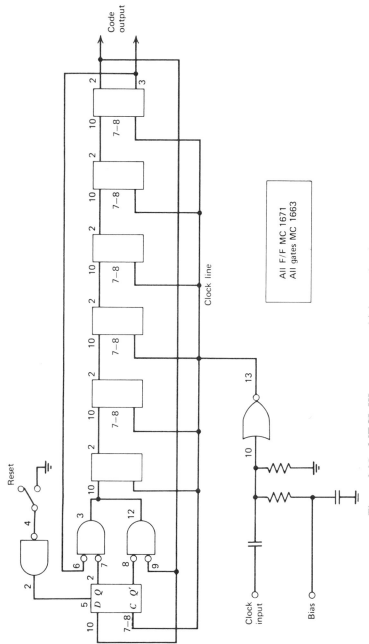

Figure 3.25 MECL III, seven-stage high-speed code sequence generator.

99

Figure 3.26   127-chip $(2^7 - 1)$ code sequence at 150 Mcps.

of the generator in Figure 3-25 to a useful length and redesigning to provide a larger number of code sequences could slow down the code rate significantly.

Emitter-coupled-logic elements that have been used to construct code sequence generators capable of operating to well in excess of 200 Mcps are available. Some instances are known in which simple-type ECL generators have operated beyond 300 Mcps. Figure 3.27 shows a 200-Mcps 15-chip code sequence generated by a four-stage MECL III generator.

The maximum chip rate of a code generator is determined not only by the rate of its shift register stages but by any delay in its feedback network. Because the feedback network output is the next-state information for some of the register stages, both the flip-flops used as feedback points and all modulo-2 adders must complete their operation before the next shift time; that is, the maximum code rate for a shift register generator is the rate at which

$$R_{\max} \approx \frac{1}{\tau_R + \tau_M} = \frac{1}{\tau_K},$$

and τ_R = time for a shift register stage to reach its next state,

   τ_M = propagation time through the feedback network,

   τ_K = period of the clock pulse.

The simple SRG with one modulo-2 adder in the feedback network is the fastest possible configuration.* The modular SRG in its simplest form is

---

*This includes the toggle registers. Toggle SRG's are no faster than simple registers with single-layer feedback logic. After all, a toggle register is a single-layer feedback configuration in which the logic is a part of the delay elements, because of their construction.

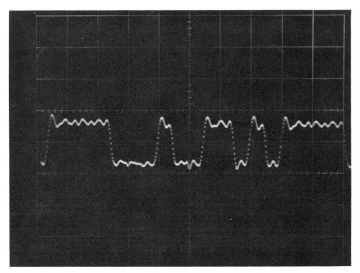

Figure 3.27   15-chip $(2^4 - 1)$ code sequence at 200 Mcps.

equivalent to it. When it is necessary to use three, five, or more (always odd) feedback taps, however, the modular register quickly becomes the fastest. (Observation of the table of codes given in this chapter will show that single-tap SRG's are the exception rather than the rule.) The multitap simple SRG again has

$$R_{\max} \approx \frac{1}{\tau_R + \tau_M},$$

but $\tau_M = \tau_{M_1} + \tau_{M_2} + \cdots + \tau_{MK}$. Proper arrangement of the feedback network can speed up the operation of the simple SRG by allowing parallel-series addition, as in Figure 3.28, in which the feedback network of configuration (b) is the equivalent of (a), but only two levels of logic are in series in the network. Writing the feedback equations

$$\lambda_a^0 = (\lambda^5 \oplus \lambda^4) \oplus \lambda^3 \oplus \lambda^2 = \lambda^5 \oplus \lambda^4 \oplus \lambda^3 \oplus \lambda^2,$$

$$\lambda_b^0 = (\lambda^5 \oplus \lambda^4) \oplus (\lambda^3 \oplus \lambda^2) = \lambda^5 \oplus \lambda^4 \oplus \lambda^3 \oplus \lambda^2,$$

$$\lambda_a^0 = \lambda_b^0$$

shows their equivalence.

A modular SRG has a maximum operating speed that is a function of the switching time of one flip-flop stage plus one modulo-2 adder, no matter how many feedback points are used. The equivalence of modular and simple

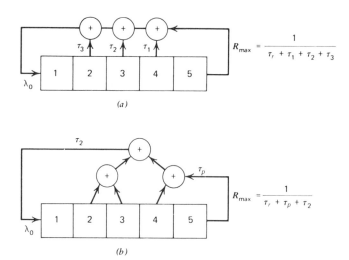

$$R_{max} = \frac{1}{\tau_r + \tau_1 + \tau_2 + \tau_3}$$

$$R_{max} = \frac{1}{\tau_r + \tau_p + \tau_2}$$

Figure 3.28   Comparative shift rates for two simple SRG configurations: (a) series connection; (b) series-parallel construction.

SRG's has been shown (also see Appendix 3). A parallel-series feedback configuration is not available, so that one modulo-2 adder is required for each feedback tap; however, the modular SRG is inherently capable of higher code rates in multitap configurations.

Because of the criticality of direct high-speed code generators, several techniques for multiplexing codes have been worked out with less complex

Figure 3.29   Biphase-modulated 300-MHz carrier modulated by a 200-Mcps code. (Code length 127 chips $= 2^7 - 1$.)

SRG configurations. Two of these composite code generators, which offer advantages for high-speed operation, are the Gold and syncopated coders.

A BPSK-modulated signal using the seven-stage high-speed MECL III code generator in Figure 3.25 is shown in Figure 3.29. Here the code rate is 100 cps and the main lobe null-to-null bandwidth is 200 MHz. A system employing this RF bandwidth and having a 10-kpbs information rate would have an inherent process gain of 17,600 or 42 dB.

We must reemphasize here that the 127-chip code shown is not capable of supporting a high process gain. A 42-dB process gain system would require a much longer code. To show the inadequacy of a short code such as this it is necessary only to compare Figures 2.7 and 3.28.[9] These spectra were photographed on two different days with different code rates, yet they are identical. This illustrates the lack of randomness in a short sequence and reinforces the conclusion that just any code will not do. The code sequence and the code rate must be fitted to the task at hand.

3.8 SOFTWARE-IMPLEMENTED CODES

It is, of course, quite practical to use a more general purpose machine such as a calculator or a computer to generate codes. In some cases, it may not be

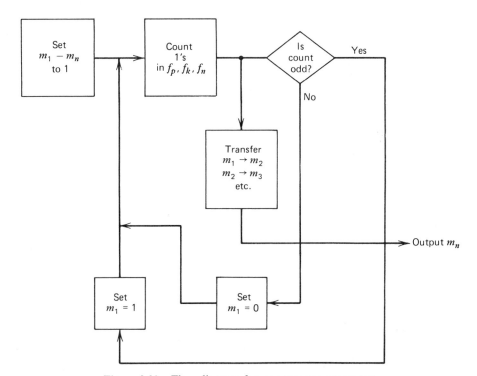

Figure 3.30 Flow diagram for m-sequence generator.

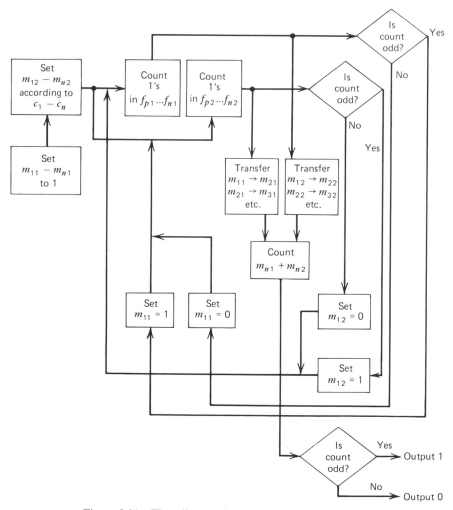

Figure 3.31   Flow diagram for Gold sequence generator.

practical to do so because of other loading, but where machine time is available the flow diagrams shown in Figures 3.30 and 3.31 illustrate techniques for generating maximal and Gold codes, respectively.

The simpler of the two diagrams, for maximal codes, requires only that a number of memory locations equal to the degree of the code be set aside (one location per register stage.) Then, with all locations set to one, count the number of ones in the stages from which feedback is taken. If there is an odd number of ones, set the first stage to one and shift all to the right. If the number of ones is even, set the first stage to zero and again shift to the right. Repetition of this procedure will produce a code at a rate equal to the algorithm repetition rate.

Figure 3.32   BPSK-modulated carrier, 8191-chip linear maximal code. (Note symmetry around center frequency.)

Figure 3.31 shows a dual maximal sequence generator configuration capable of generating Gold codes.

3.9   CODE SELECTION AND SIGNAL SPECTRA

The following series of photographs illustrates some of the effects of code selection on direct sequence (BPSK and QPSk) spectra. Both "good" and "bad" spectra are shown, so that the reader can recognize the effects of generator failure and other phenomena. The first photograph, Figure 3.32, is of the spectrum of a BPSK-modulated carrier, where the code length is $2^{13} - 1$ (8191 chips). The code used is $[13, 4, 3, 1[_m$. A point of interest in Figure 3.32 is the precise symmetry of the spectrum around the center frequency, which shows that BPSK direct sequence signals are no different from other signals. That is, they generate symmetrical upper and lower sidebands.

An illustration of the possible effects of failure in a feedback element is given in Figures 3.33 and 3.34. If a feedback element fails, or a connection is wrong, a shift register generator may output a much shorter code than the linear maximal code desired, and this can cause the modulated spectrum to be quite different from that which is expected.

Figure 3.33 BPSK modulation by 14-chip nonmaximal code from 13-stage generator (period 14 μsec).

Figure 3.34 BPSK modulation by 105-chip nonmaximal code from 13-stage generator (period 105 μsec).

Figure 3.35   QPSK modulation by same code as in Figure 3.32. Codes taken from stages 13 and 4. Marker added at center frequency.

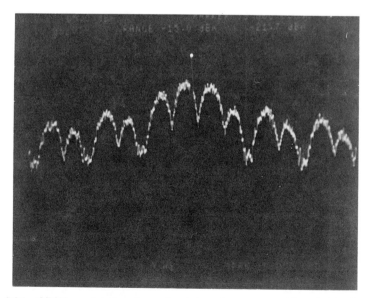

Figure 3.36   QPSK modulation by codes taken from stages 13 and 11 of 13-stage generator.

Figure 3.37 QPSK modulation by codes taken from stages 13 and 6 of 13-stage generator.

For quadriphase modulation, where two code sequences are required, it is possible to take both sequences from the same generator. Some care must be exercised when doing so, however, as is shown in Figures 3.35 through 3.38. The first of the QPSK-modulated carrier spectrum photographs is shown in Figure 3.35. Here the codes driving the QPSK modulator are taken from stages 13 and 4. The spectrum produced has the same shape as the previous BPSK spectrum (the same code generator connections are also used). The QPSK spectrum no longer displays the precise symmetry of the BPSK spectrum however.

The asymmetry of the QPSK spectrum is dramatically illustrated in Figures 3.36 through 3.38. In these photographs, the codes used have been taken from stage 13 together with various other taps on the generator. In none of these is the modulation symmetrical around the center frequency. Note, however, that the number of minor lobes in the one-sided main-lobe bandwidth is equal to the number of stages separating the code output points on the shift register. This illustrates the effects of the strong correlation between the codes, which are delayed replicas of one another. In Figure 3.38, the two individually unsymmetrical but mirror-image replica spectra are those produced by QPSK modulating the carrier with codes separated in time by ± 1 chip time (i.e., in Figure 3.38a, stage 12 is one chip time ahead of stage 13, and in Figure 3.38b, stage 13 is one chip time ahead of stage 1).

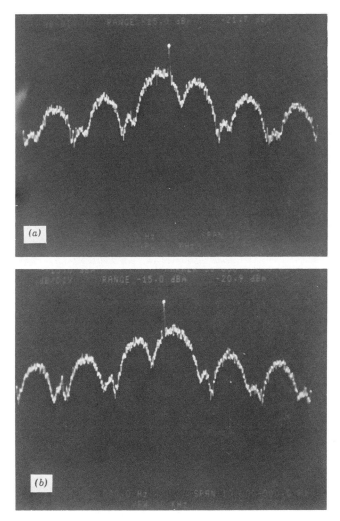

Figure 3.38   QPSK modulation showing complementary asymmetry when codes are taken from $[n, n-1]$ and $[n, 1]$ stages: (a) codes taken from 13 and 12; (b) codes taken from 13 and 1.

3.10   CODE BASEBAND SPECTRA

The code baseband spectrum can be observed, and is seen to have exactly the same one-sided distribution of frequency components as the two-sided, modulated carrier. A spectrum analyzer having sufficient capability can show not only the overall envelope of the frequency components, but also the fine structure of the code spectrum. Figure 3.39 shows the baseband spectrum of

Figure 3.39   Baseband code spectrum of 8191-chip code at 1.5 Mcps.

an 8191-chip code $[13, 4, 3, 1]_m$ at approximately a 1.5-Mcps rate. This spectrum has lobes from 0 to 20 MHz (and beyond), and code imbalance which is shown by the even harmonic content exhibited by the stationary signals at the nulls.

Fine structure of this same code's spectrum is shown in Figure 3.40. Figure 3.40*a* shows the main lobe between 0 and 10,000 Hz, with a 100-Hz resolution bandwidth. This structure is expanded further in Figure 3.40*b*, with resolution bandwidth of 30 Hz and frequency span of 0–1 kHz. The lines shown are separated by the repetition rate of the code itself. In this case $(1.5 \times 10^6)/8191 = 183$ Hz. This demonstrates very aptly that the codes used in spread spectrum systems are noiselike to a casual observer but are deterministic and decidedly not Gaussian.

3.11   ERROR DETECTION AND CORRECTION CODES

Error detection and correction (EDAC) codes are mandatory for use in frequency-hopping systems in order to overcome the high error rates induced by partial band jamming and the efficiency (to the jammer) of such a jamming strategy. The code's use, however, must be carefully defined, as error detection and correction codes do not necessarily work well under all conditions. That is, many EDAC codes have thresholds that must be exceeded before satisfactory performance is achieved.

In direct sequence systems, error detection and correction coding may not be advisable in some applications. This is because of the effect of the coding

Figure 3.40 Expanded baseband spectra of 8191 chip code at 1.5-Mcps rate: (*a*) span, 0–10 kHz; resolution BW = 100 Hz. (*b*) Span, 0–1 kHz; resolution BW = 30 Hz.

overhead, which increases the signal's apparent data transmission rate and may increase the jamming threshold level. Some demodulators (CVSD, for example) can operate to bit error rates of approximately 1×10^{-2} without EDAC, and since the threshold of EDAC decoders is often near 1×10^{-2}, it may not be worthwhile to include a complex coding and decoding scheme in a system.

Table 3.9 gives results for a rate $\frac{1}{2}$ convolutional code with Viterbi decoding at various bit error rates between 1×10^{-2} and 1×10^{-6}. At these

TABLE 3.9   Measured Performance of Rate $\frac{1}{2}$, Constraint Length 7, Convolutional Code / Viterbi Decoder

Bit Error Rate	Ed/NO (dB)		Coding Gain (dB)	Loss in Jamming Margin	Net
	Without Coding	With Coding			
10^{-2}	4.2	2.0	2.2	3.0	−0.8
10^{-3}	6.8	3.0	3.8	3.0	0.8
10^{-4}	8.4	3.8	4.6	3.0	1.6
10^{-5}	9.6	4.5	5.1	3.0	2.1
10^{-6}	10.5	5.2	5.3	3.0	2.3

bit error rates, the improvement (or coding gain) seen varies between 2.2 and 5.3 dB. The problem for spread spectrum systems is that unless RF bandwidth is increased when an error-correcting code is added, the doubled transmission rate halves the available processing gain. Since the processing gain (and jamming margin) is reduced by 3 dB by doubling the information rate while holding the RF bandwidth constant, any EDAC technique employed must have coding gain of 3 dB or more, or it is worse than useless. Table 3.8 shows that the net coding gain, for a spread spectrum system using rate $\frac{1}{2}$ convolutional codes and Viterbi decoding, is a loss of 0.8 dB when the bit error rate is 1×10^{-2} and only 0.8 dB when the bit error rate is 1×10^{-3}. Owing to the complexity of adding such coders and decoders, the modest net of 2.3-dB gain when the bit error rate is 10^{-6} may not be enough to justify their use in spread spectrum applications. Nevertheless, coding for error detection and correction has found application in some spread spectrum systems. Coding that has been employed is basically:

Frequency Hopping (or Frequency Hopping/ Direct Sequence)	Direct Sequence
BCH	Golay
Reed-Solomon	BCH
Convolutional	Convolutional
Redundant Transmission	Hamming
Hamming	

PROBLEMS

1. Given a shift register of 23 stages, what is the longest code sequence that can be generated with a single linear feedback network?

2. How many maximal linear code sequences are available from a 17-stage sequence generator?

3. What code sequence would be produced by a code sequence generator having the equation $D_0 = D_4 \oplus D_2$ and initial conditions $D_1 = 1$, $D_2 = 1$, $D_3 = 0$, $D_4 = 1$?

4. For a Gold code sequence generator, how many nonmaximal codes are available if the basic registers employed are each 11 stages long?

5. What are the maximum and minimum autocorrelation values for a $[5, 2]_s$ code?

6. What code sequences could be produced by a $[7, 2]_s$ generator?

7. Draw block diagrams for $[9, 6, 4, 3]_s$ and $[9, 8, 4, 1]_m$ sequence generators.

8. Construct a cross-correlation plot of a $[5, 3]_s$ and a $[5, 4, 3, 2]_s$ code.

9. If a $[17, 3]_s$ code sequence generator is operated at a 5-Mcps clock rate, its repetition period would be how many seconds? Line spacing in a direct sequence spectrum produced by modulation with this code would be how many hertz?

10. How many runs of three ones would be expected in a $2^{13}-1$ chip m sequence?

11. How many runs of three zeros would be expected in the same sequence?

12. What alternate Gold configurations might be employed if at least 1000 4095-chip code sequences are needed?

13. What is the shortest generator that could be used to provide the needed sequences if all must be maximal?

14. Show that the shift-and-add operation is closed for any 2^5-1 chip m sequence.

15. Given a maximal shift register sequence generator, show that the all-zero stage cannot be allowed.

16. Show the effect of an even number of feedback taps (in addition to the last stage) on a shift register sequence generator.

17. How may one synchronize to a pseudorandom code sequence to check its correctness?

4

MODULATION AND MODULATORS: GENERATING THE WIDEBAND SIGNAL

The wideband signals characteristic of spread spectrum systems are generated in different ways in the basic techniques of interest here. In a direct sequence system the encoding signal is used to modulate a carrier, usually by phase-shift keying at the code rate; that is, for a "one" in the code pattern, one carrier phase is transmitted, and for a "zero" a different carrier phase is sent. More complex phase-shift mappings are also used (such as quadriphase), but in each there is a one-to-one correspondence between the carrier phase transmitted and the code reference or references. It is of some interest to note that balanced modulation* is often used, because this suppresses carrier components and helps to hide the signal. In addition, no power is wasted in transmitting a carrier that would contribute nothing to interference rejection or information transfer.

Frequency hopping systems generate their wideband signal spectra in a different manner from direct sequence systems, although the modulation is still dictated by a code sequence. The relation of the code sequence to the RF output is such that it commands the frequency of a carrier signal rather than directly modulating the carrier; that is, for a given code state the system transmits at one frequency, and for any other code state transmits at a different frequency. In each state the code commands the system output frequency. Therefore, the carriers "hops" from one frequency to another in synchronism with code state changes.

---

*Balanced modulation multiplies a carrier $A \cos w_c t$ with a modulation term $B \cos w_m t$. The resulting signal is $AB \cos w_c \cos w_m t = AB \cos w_r \pm w_m t$, for which there is theoretically no carrier term. In practice carrier suppression for such modulators is between 20 and 60 dB.

Typically, a frequency hopping system may have a few thousand available frequencies to choose from, and a code sequence generator with at least that number of states; for example, a system with a frequency synthesizer capable of outputing any one of 10,000 frequencies separated by 1 kHz would have an RF bandwidth of 10 MHz. Its code generator should then be able to generate a code sequence at least 10,000 chips long, so that each of the frequencies available can be used. A 14-stage shift register sequence generator could be used for this purpose because it can generate sequences up to $2^{14} - 1 =$ 16,383 chips. The wideband signal spectrum from a frequency hopper is not generated by carrier modulation like the direct sequence signal. Instantaneously, the system has only one output rather than a set of symmetrically distributed outputs.

Information is normally embedded in the code sequence that determines frequency for a frequency hopper, or spectrum spreading for a direct sequence system. Therefore, it does not matter which type of spread spectrum system one is dealing with, from the standpoint of information transmission, if the information to be sent is a part of the code sequence. The same code might be used for a direct sequence biphase or quadriphase modulator or at a lower rate to determine the frequencies over which a frequency hopper spreads its output.

The point is that both direct sequence and frequency hopping systems generate wideband signals under the control of a code sequence generator. In one case, the code is used as direct carrier modulation (direct sequence) and in the other (frequency hopping) the code commands the carrier frequency. This chapter shows spectrum spreading subsystems and circuits typical of those used in practical systems and discusses the problems most often encountered.

4.1 BALANCED MODULATION

An important tool in any suppressed carrier system is the balanced modulator used to generate the transmitted signal. Balanced modulation is employed in many applications in which either one or both of the signals applied to a modulator must be suppressed at its output. Balanced modulation for carrier suppression is at least 60 years old; a patent for this purpose was filed by J. R. Carson in 1915.

Here we intend to show the specific application of balanced modulation in direct sequence spread spectrum systems. We also illustrate the effects of various circuit problems on the output signal.

Almost as many kinds of balanced modulator (or balanced mixer, depending mainly on who is using it) circuits exists as there are applications. Recent advances in semiconductor diodes (in particular, hot carrier diodes and wideband transformers) and their application to ring-configured diode mod-

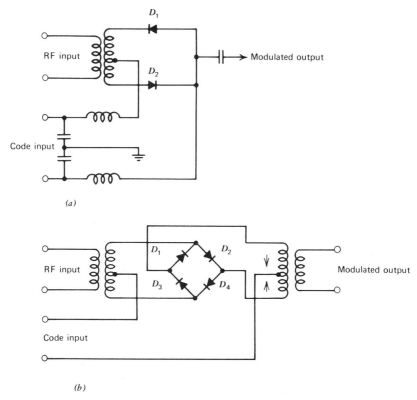

Figure 4.1 Diode-balanced-modulator configurations: (*a*) typical single-balanced mixer/modulator; (*b*) typical double-balanced mixer/modulator schematic diagram.

ulators[1019, 1022, 1023] (see Figure 4.1) have generally made obsolete all others for most applications. Therefore, our emphasis is on the diode modulator. Commercially available single- and double-balanced diode modulators cover most frequency ranges and offer excellent performance from the standpoint of input signal suppression, low loss, and stable operation.

A simple form of balanced mixer/modulator is a double throw switch, connected so that it switches between two 180°-separated input signals at a modulating signal rate. This configuration would work quite well for digital data modulation and in fact is the same in function as a modulo-2 adder. Figure 4.2 shows a balanced mixer output signal as modulated by a binary switching signal. (The binary switching signal is for our purposes a code sequence, with embedded information. This is true for biphase or quadriphase signals. Note that since the code includes the transmitted information, the single spectrum spreading, balanced-modulation operation satisfies both information transmission and RF signal spreading requirements.)

180° carrier
phase shift

Code transition

Figure 4.2   An actual balanced mixer output signal.

For code-modulated RF signals the balanced-modulator output is a square-wave-modulated carrier rather than the familiar sine-wave-modulated balanced signal. For sine wave modulation, the signal may be expressed as

$$A \cos \omega_c t \, B \cos \omega_m t.$$

Similarly, for data modulation the signal is

$$A \cos \omega_c t \, B \cos \omega_s t,$$

where ω_s is a time-varying set of frequencies harmonically related to the data bit rate and run lengths.

Output waveforms from sine-wave- and data-modulated balanced modulators are shown in Figure 4.3.

Spectral outputs of the two modulating signals are $AB \cos(\omega_c \pm \omega_m)t$ for sine wave modulation in which only two sidebands are represented, and $AB \cos(\omega_c \pm \omega_s)t$ for data. Those familiar with single-sideband systems will recognize the signal in Figure 4.3a.

The code-modulated spectrum is not a single pair of sidebands but a set of symmetrical $[(\sin x)/x]^2$-distributed sidebands due to the many code frequency components. In any case, the carrier signal is suppressed an amount that depends on the design of the balanced modulator and on the carrier waveform.

When a code is used to generate a spread spectrum output from a balanced mixer/modulator, a great deal of care must be exercised to provide both carrier and code clock suppression. Figures 4.4 through 4.10 are actual

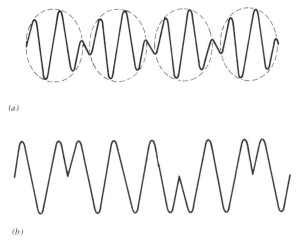

(a)

(b)

Figure 4.3 Balanced modulator output waveforms: (a) sine-wave-modulated balanced mixer output signal; (b) data (code)-modulated balanced mixer output signal.

photographs of spread spectrum output signals. The single-balanced mixer of Figure 4.1a suppresses only the RF input by inverting the phase of the RF carrier at a code-determined rate. A code input turns on either diode one (D_1) or diode two (D_2) and turns off the other diode, depending on the relative polarity of the input. Because the code input is complementary, one input is always at a higher DC level than the other, and if the difference

Figure 4.4 Satisfactory balanced mixer/modulator output.

between the two exceeds a diode's forward voltage drop, one of the two will conduct. (In practice, the code input voltage should be higher than the carrier input level.) Diode conduction then allows one phase of the RF carrier to pass to the output.

For improved overall input suppression, the double-balanced configuration of Figure 4.1b is recommended. Here a diode bridge is used and an output transformer is added. Operation is similar to the single-balanced mixer/modulator in that the polarity of the code input lines determines whether D_1 and D_4 or D_2 and D_3 are conducting, and thereby determines which RF carrier phase appears at the output. In addition, the balanced windings on the transformers allow suppression of the code input signal at the output. This suppression occurs because code-related current is in opposition in both input and output transformers and thus tends to cancel.

The spectrum of a properly operating biphase-modulated signal is shown in Figure 4.4. The power distribution of this signal is $[(\sin x)/x]^2$, with line separation within this distribution equal to the repetition rate of the code modulating signal; for example, if a 1023-chip code with a 1-Mcps rate were used, the code would repeat $(1 \times 10^6)/1023 = 977$ times per second and the line separation in the modulated signal would be 977 Hz. Because of the pseudorandum character of the code, however, the appearance of any particular spectral component in time is also pseudorandom. Therefore, the direct sequence modulated signal appears to be noise with a $[(\sin x)/x]^2$ distribution.

Figure 4.5 illustrates the effect of poor carrier balance, which could be caused by imbalance in the transformers of Figure 4.1, an open diode, or any other cause that might give the effect of unequal 180° carrier inputs. The two

Figure 4.5 Balanced mixer/modulator output showing lack of carrier balance.

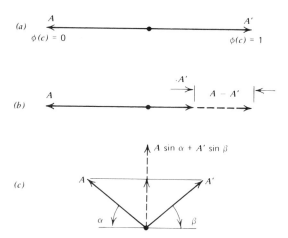

Figure 4.6   Vector diagram of biphase-balanced modulation.

phases of the carrier must also be well matched to provide satisfactory balance. (In the diode mixer/modulator of Figure 4.1 the two carrier phases are provided by the input transformer.) Either amplitude inequality or less than 180° phase shift can cause the carrier to be present in the modulated output.

Figure 4.6 is a phasor diagram of the modulated output, which is recognized as antipodal. The carrier output that results is zero when the $\phi(c)$ corresponding to the one and zero code inputs are balanced and in phase opposition.

A difference in amplitude of the two output phases results in imbalance $A - A'$ as illustrated in Figure 4.6b, and a less than 180° phase relationship is shown by Figure 4.6c.

Carrier suppression may be expressed in decibels below the output desired by the expression

$$V = 10 \log \frac{B}{A \sin \alpha + A' \sin \beta} \text{ dB,}$$

where B = amplitude of the correct output signal (i.e., $A = A'$ and $\alpha = \beta = 0$).

A = zero degree signal amplitude,

A' = 180° signal amplitude,

$\alpha = A$ phase offset,

$\beta = A'$ phase offset.

In some cases a controlled amount of carrier signal is deliberately sent for use in carrier tracking without code demodulation as in the Air Force's SGLS space ground link subsystem system. For such an application the system is usually designed so that $\alpha = \beta$, $A = A'$, and $V = (10 \log B)/(2 \sin \alpha)$.

Phase modulation by less than 180° is also used when the modulated signal is multiplied by an even number. Unfortunately, this is not practical for 180° biphase modulation because even-order multiplication of any signal that is symmetrical in the frequency domain removes the symmetrical sidebands and produces an unmodulated output at the frequency multiple desired. This problem is overcome by phase modulating by $\pm 90/k$ degrees before multiplication by the factor k. When carrier suppression is critical, however, the technique is not recommended, as modulation at odd phase angles can be quite difficult to maintain and any departure from the desired $90/k$ modulation angle would significantly reduce carrier suppression after multiplication.

Alternatives to less than $\pm 90°$ modulation in which multiplication is involved are the following:

1. Separately multiply the carrier, then modulate. In general, this is a practical solution because balanced mixers and modulators are commercially available to several gigahertz.
2. Multiply by an odd order, because odd-order multiplication preserves symmetrical modulation.

A typical direct sequence output signal in which the carrier signal is poorly suppressed is shown in Figure 4.5. In most instances poor balance might not be so clearly evidenced. Often, however, a constant signal component is detectable in the output spectrum when carrier balance is not as it should be. The best method for checking carrier suppression is to remove the code modulation and substitute a square wave. This permits observation of the carrier component in comparison to the square-wave-modulation sidebands without noiselike code components (see Figure 4.7). Varying the square wave frequency over the range of code frequency components (code rep rate^{-1} to clock rate/2) also allows a check of modulator balance with code modulation.

Figures 4.8 and 4.9 show the effects of (1) nonsymmetry in the code-modulating sequence and (2) simultaneous code nonsymmetry and lack of carrier

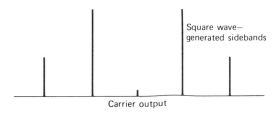

Figure 4.7 Typical square-wave-balanced modulation test spectrum for checking carrier suppression.

Figure 4.8 Balanced mixer/modulator output showing poor code balance.

balance. Code symmetry is more difficult to provide in a practical system than carrier balance because of the characteristics of the circuits used to drive the balanced modulator with a code sequence. Many integrated-circuit flip-flops are capable of driving a diode-balanced mixer/modulator directly and offer the best general symmetry; that is, both discrete circuits and separate drivers tend to exhibit more differential change in output signal than complementary outputs from a single integrated-circuit flip-flop.

Figure 4.9 Balanced mixer/modulator output showing both carrier and code imbalance.

The problem is this: to exhibit good code-rate suppression a balance-modulated output signal must weight the signal corresponding to a One at the same value as a signal corresponding to a Zero; that is,

$$A \int_0^{\tau_1} G_1(c) \, dt$$

must equal

$$A' \int_{\tau_1}^{\tau_2} G_0(c) \, dt,$$

where $G_1(c)$ is the signal envelope corresponding to a One and $G_0(c)$ is the signal envelope corresponding to a Zero. Also, the periods $\{\tau_1 - 0, \tau_2 - \tau_1\}$ and amplitudes $\{A, A'\}$ must equal one another.

Now, in flip-flop output stages the rise and fall times of the output signals are not the same. Furthermore, temperature change causes characteristic changes and the two outputs may not maintain the same relationship. Therefore, we caution the designer that no matter how good the balanced modulator may be some attention must be paid to code symmetry, or his or her work will be of little value for code-rate suppression.

Effects of Carrier and Code-Rate Spurious (Lack of Suppression)

What is the effect of having a modulated output like that in Figure 4.9? In the transmitter the most significant effect is that after having gone to the trouble of code modulation our output spread spectrum signal has stationary, easily detected signals in its midst, which defeat one purpose of spread spectrum modulation by causing narrowband interference, transmission of wasted power, and relatively easy detection.

In spread spectrum receivers a narrowband signal due to carrier or code imbalance is even less desirable. These signals can render the receiver unusable because undesired narrowband input signals may be translated directly into the IF (see Section 5.2). Figure 4.10 shows the effect of excessive code clock leakage through or around the balanced mixer/modulator so that the output is not only balanced-modulated by the code but amplitude-modulated by the code clock. This is usually simple to prevent, but the undesired output signals can cause the same kinds of spurious signals already discussed.

Quadriphase Modulation

Quadriphase modulation is an extension of biphase-balanced modulation and is often used in direct sequence systems. The main reasons for quadriphase, despite its extra complexity, are that (1) a quadriphase signal is not so seriously degraded as a biphase signal when passed through a nonlinearity

Figure 4.10   Balanced mixer/modulator output showing code clock leakage.

simultaneously with interference, and (2) the RF bandwidth for a quadriphase signal is half that required by a biphase signal, given the same data rate. Alternatively, twice as much data might be transmitted in the same bandwidth with a quadriphase signal as with a biphase signal. A quadriphase modulator shifts between one of four carrier phases spaced at 90° intervals. Thus two bits of binary data are required to select the phase to be transmitted. A typical scheme for such selection might be

Data Bits		Transmitted Phase
A	B	
0	0	0°
0	1	90°
1	0	270°
1	1	180°

Of course, any other one-to-one correspondence between the data and the transmitted phase is just as good. The only requirement is that a transmitter and receiver both employ the same data-to-carrier phase relationship.

The block diagram in Figure 4.11 shows a common quadriphase modulator configuration that consists of a combined pair of biphase modulators. An input RF carrier is phase shifted by 90° and applied in the original and quadrature form to a pair of biphase modulators. Separate binary inputs drive the two modulators and their outputs are summed (Figure 4.12). Any combination of carrier phases from the two biphase modulators (a, a' and

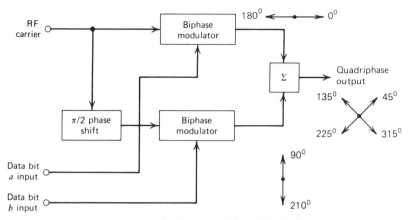

Figure 4.11 Quadriphase modulator block diagram.

b, b' are not allowed) results in a summed output that is one of the four desired carrier phases.

The capability of quadriphase signals to pass through limiting with minimal processing loss has a great deal of importance but is left for discussion in Chapter 7.

The power envelope of a quadriphase signal has the same shape as a biphase-modulated signal; that is, $(\sin^2 x)/x^2$. The main-lobe bandwidth, however, is only half that of the biphase signal for the same code rate; that is, the first null of the $(\sin^2 x)/x^2$ spectrum occurs at chip rate$/2$, and successive nulls occur at the same interval. We recall that the biphase-modulated signal has a main lobe whose first null occurs at the chip rate, with succeed-

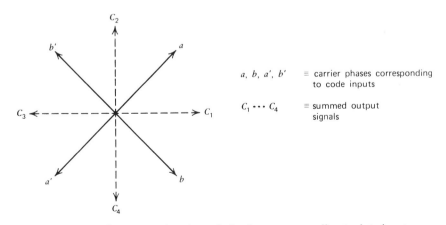

Figure 4.12 Output carrier phase derivation corresponding to data input.

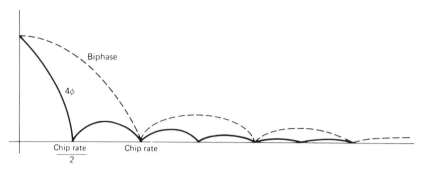

Figure 4.13 Comparison of biphase and quadriphase signal spectra for the same code rate (one code only).

ing nulls also separated by the chip rate. Figure 4.13 illustrates this relationship.

In typical QPSK-modulated spread spectrum systems the spectrum employed has the same bandwidth as a BPSK-modulated system using the same code rate. This is accomplished by using two codes in the QPSK systems, that have the same rate as the single code used in a BPSK system. This is done to avoid the loss in processing gain that would occur if the RF bandwidth were reduced.

Carrier Modulation Trade Offs

The major concerns in choosing a modulation waveform that must be taken into account in spread spectrum are:

Effect on synch acquisition technology.
Sidelobe energy output (splatter) and bandwidth.
Complexity of modulators and demodulators.
Effects of jamming on the system.
Impact of the technique on size, weight, power requirements, and reliability.

Effect on Synch Acquisition

The waveform employed is a primary force in determining the synch acquisition capability of a spread spectrum system. Direct sequence modulation, for example, employs high-chip-rate code sequences that force a receiver to synchronize to within a very narrow time window (often as narrow as a few nanoseconds), while frequency hopping with its lower code rates permits timing uncertainties as great as a few milliseconds. This means that the techniques employed for synchronizing in the various systems are quite varied, depending on the waveform employed. Representative acquisition and tracking techniques are (see also Chapter 6):

1. Direct Sequence:
   a. Initial acquisition:
      Serial and/or sequential search—slowest
      Sequential estimation—fast but poor AJ performance
      Universal timing—depends on small time uncertainty
      Matched filters/convolvers—fastest, but still has problems
   b. Tracking:
      Delay lock loops—classic or dithered
2. Frequency Hopping:
   a. Initial acquisition:
      Serial and/or sequential search—works well because of reduced uncertainty
      Universal timing—works well due to low drift rates
   b. Tracking:
      Split-bit tracking—similar to delay lock

(Some frequency hopping systems operate without tracking loops, due to their relative stability and long chip time.)

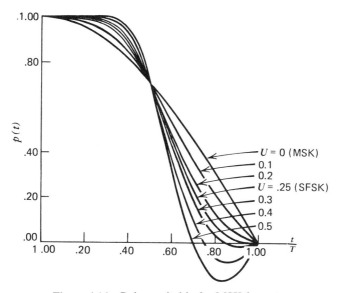

Figure 4.14   Pulses suitable for MSK format.

Sidelobe Energy Output

Production of unwanted sidelobe energy is much higher in some common direct sequence modulating approaches than in others, and is typically high in frequency hopping, although frequency hopping sidelobe energy is well contained inside the signal band. (With frequency hopped signals, one must remember that the signal is a pulsed carrier, and that the spectrum of the signal produced by pulsing the carrier is strictly dependent on the pulse envelope.) Typical rectangular-pulse frequency hopped carriers have $[(\sin x)/x]^2$ power spectra, with sidelobes at -13 dB, and that roll off at a 6-dB/octave rate. Techniques for pulse shaping exist, however, that can reduce the sidelobe energy, and these may be employed in a system design, where they are applicable.

Also of interest are the techniques needed for use in reducing sidelobe energy output in direct sequence modulated systems. Of the modulation techniques available—biphase phase shift keyed (BPSK), quadriphase shift keyed (QPSK), offset quadriphase shift keyed (OQPSK), and minimum shift keyed (MSK)—the MSK-like signals appear to be most promising.

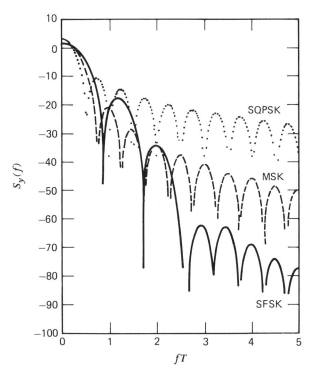

Figure 4.15 Comparison of the equivalent low-pass power spectral densities for OQPSK, MSK, and SFSK.

MSK, or minimum shift keying, is actually a family of signal structures that must satisfy two conditions:

1. The modulating pulse $g(t)$ is symmetrical about $t + T/2$ and 0 otherwise.

2. The modulated carrier has a constant envelope. That is,

$$g^2(t) + g^2(t + .T/2) = 1$$

Some examples of such waveforms are:

a.
$$g(t) = \cos\left(\frac{\pi t}{2T} - U \sin \frac{2\pi t}{T}\right) \quad - T \le t \le T$$

$$= 0 \text{ otherwise}$$

(For classical MSK, $U = 0$ and for SFSK, $U = 0.25$.)

b.
$$g^{(n)}(t) = \left(\frac{\sin(\pi t/2T)}{\pi t/2T}\right)^n \quad - 2T \le t \le 2T.$$

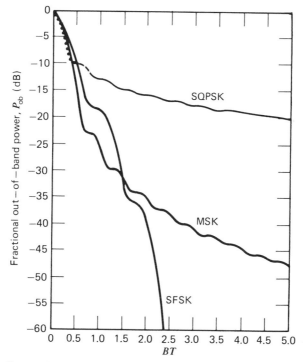

Figure 4.16 Comparison of fractional out-of-band powers for OQPSK, MSK, and SFSK.

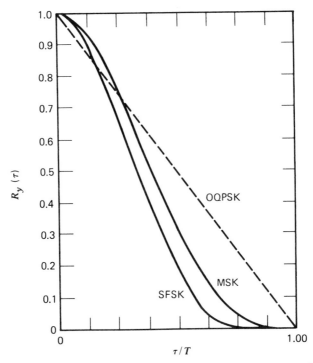

Figure 4.17 Comparison of the equivalent low-pass autocorrelation functions for OQPSK, MSK, and SFSK.

Figure 4.14 shows the (half-pulse) envelope of some signals satisfying equation (a), and Figures 4.15 and 4.16 show the spectral density and out-of-band power, respectively, for three forms of MSK. (OQPSK, though often listed separately, is actually a form of MSK, where the pulse used is square, instead of shaped as in the more usual forms.) Figure 4.17 compares the autocorrelation functions of OQPSK, and two other forms of MSK modulation (classic MSK with $U = 0$, and SFSK with $U = 0.25$).

The OQPSK form does have a more sharply peaked autocorrelation function (this is true only for unfiltered OQPSK), and rolls off linearly. This makes it optimum for ranging systems. MSK and SFSK, on the other hand, have rounded correlation functions, which make them less synchronization-critical.

Complexity of the basic modulators is compared in Figure 4.18. The BPSK and MSK modulators are least complex, and appear identical on a block-diagram basis. The filters employed are very different, however, although the balanced modulators are the same.

BPSK modulation, though simple, is too readily vulnerable, and should not be processed in a limiting channel. Figure 4.19 illustrates the problem seen when passing a BPSK-modulated signal $B \cos \omega_0 t \pm 90°$ through a limiter,

Figure 4.18 Modulator comparison.

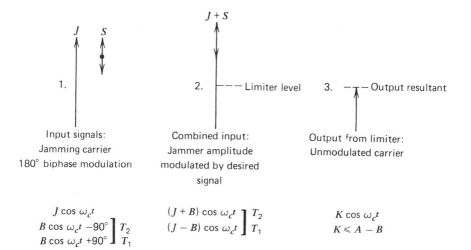

Figure 4.19 CW jammer coherent with BPSK-modulated carrier signal.

TABLE 4.1 Waveform Trade Off Summary

	Synch Technology	Sidelobe Energy	Nonlinear Tolerance	Moderate Complexity	Autocorrelation
BPSK	M.F. or Convolver	High	Poor	Best	Triangular
QPSK	M.F. or Convolver	High	Poor	2 × BPSK plus two summers	Triangular
OQPSK	Convolver	High, but filterable	Good	Same as QPSK, plus filter	Triangular
MSK	Convolver	Lowest	Best	Same as BPSK	Rounded

together with a strong CW jamming signal $J \cos \omega_0 t$. The result of the process, when the frequency of the jammer is exactly at center frequency, is that the small desired signal amplitude modulates the jamming signal. This results in amplitude modulation that is completely removed by the limiter, so that nothing remains but an unmodulated carrier.

Even in the best case, desired signals are suppressed by larger jamming signals by up to 6 dB, and all this must be considered when designing a spread spectrum system. Table 4.1 summarizes the significant characteristics of the various carrier modulation techniques useful for direct sequence modulation.

4.2 FREQUENCY SYNTHESIS FOR SPREAD SPECTRUM MODULATION

The frequency synthesis section of a frequency hopping system is at once both the heart and the most critical part of the system. The number of frequencies generated and the hopping rate are the quantities that determine system capabilities. They, in turn, depend on the type and configuration of the synthesizer itself. There is no difference, in principle, between the synthesizers used in standard communication transceivers and those in frequency hopping. The frequency hopping synthesizer, however, is designed to switch quickly from one frequency to another, whereas a conventional synthesizer may be required to change frequencies only at long irregular intervals. The rapid-switching requirement does force some difference in the details of synthesizer design. Therefore, some discussion of frequency synthesis for frequency hopping is in order.

Frequency synthesizers are divided into two general types. The first includes all those that generate the desired output frequencies from a source (or set of sources) by mixing, multiplying, and dividing, or any other means not including phase-locked loops. This class of synthesis is called "direct." Any synthesizer employing a phase-locked loop in its frequency-generation elements is called "indirect." Both are of interest here.

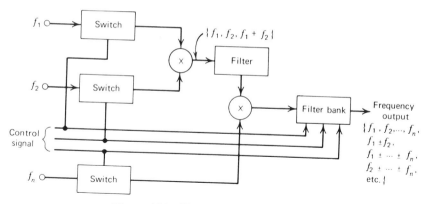

Figure 4.20   Simple direct synthesizer.

The simplest type of direct synthesizer is shown in Figure 4.20. Here a number of frequencies are mixed together in various combinations to give all the sum and difference frequencies. The utility of this simple synthesis approach quickly becomes limited when a large number of frequency choices is desired. The size and weight of the filters required are major factors in themselves.

A great variety of frequency synthesizers has been developed. In general, every new piece of communications equipment has a new synthesizer configuration, and although many commercial frequency synthesizers are available, no two have a great deal of similarity beyond being in the same class (direct or indirect). This is pointed out only to show that there are as many ways to implement a synthesizer as there are synthesizers. One direct synthesis approach has been successfully used in the TATS (tactical transmission system).[111] It is well adapted to generating large numbers of frequencies and can switch from frequency to frequency rapidly. This is the add-and-divide (or mix-and-divide) approach suggested by Hastings and Stone.[814]

Add-and-divide synthesizers are made up of a series of add-and-divide modules, which are most often made to be identical to facilitate construction. Figure 4.21 is a diagram of an add-and-divide module. The number of frequencies that a direct synthesizer of the add-and-divide type can generate is a function of K, the number of mixing frequency choices, and A, the number of times the mixing operation is performed. Frequency separation is determined by the base frequency separation, divided by the number of output frequencies, times the number of base frequencies; that is, the number of frequency outputs is K^A and the output frequency separation is (base separation$/K^A$) × number of bases. As an example, let us postulate a synthesizer with at least 4000 output frequencies and 5-kHz frequency separation. Our output center frequency f_0 is arbitrarily chosen as 50 MHz. Because we need at least 4000 frequencies, we can use $4096 = 2^{12}$ as a goal. Conveniently enough, $2^{12} = (2^2)^6$, so that if we let $K = 2^2 = 4$, then A, the

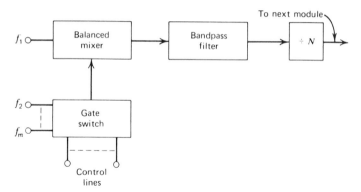

Figure 4.21   Basic add-and-divide frequency synthesizer module.

number of mixer stages, is 6; that is $K^A = 4^6 = 4096$. We need $K = 4$ base frequencies separated by K^{A-1} output separation) $= 4^6 \times 5000/4 = 5.12$ MHz. (If 4.88-kHZ spacing is acceptable, then exact 5.0-MHz base separation is satisfactory.) For simplicity let us assume 5.0-MHz base separation.

Now that we know that our base frequencies must be separated by 5.0 MHz, we can define the remaining signals. Because we have decided to use a divider (M) of 4 and an output center frequency of 50 MHz, we see that the first divider output (referring to Figure 4.22) is 12.5 MHz. Preserving the 50-MHz center frequency, we choose the base frequencies f_1 through f_K such that

$$\frac{1}{K} \sum_{n=1}^{K} f_n + 12.5 \text{ MHz} = 50 \text{ MHz},$$

and the f_K are symmetrically distributed about the mean

$$\frac{1}{K} \sum_{n=1}^{K} f_n.$$

Therefore, the mean frequency is $50 - 12.5 = 37.5$ MHz and the base frequencies are

$$f_1 = 30 \text{ MHz},$$

$$f_2 = 35 \text{ MHz},$$

$$f_3 = 40 \text{ MHz},$$

$$f_4 = 45 \text{ MHz},$$

Figure 4.22 4096 frequency direct add-and-divide synthesizer.

135

TABLE 4.2   Frequency Table for the Direct Add-and-Divide Synthesizer in Figure 4.22

Point	Band (MHz)	Δf (MHz)	Frequencies
1	42.5–57.5	5	4
2	10.625–19.375	1.25	4
3	40.625–59.375	1.171875	16
4	10.15625–14.59375	0.277343	16
5	40.15625–59.59375	0.303710	64
6	10.034062–14.898437	0.076005	64
7	40.034062–59.898437	0.077595	256
8	10.008515–14.974609	0.019399	256
9	40.008515–59.974609	0.019498	1024
10	10.002128–14.993652	0.004874	1024
11	40.002128–59.993652	0.004880	4096

which is the set of four frequencies symmetrically distributed about 37.5 MHz at 5.0-MHz spacing.

Our basic design is now complete. Figure 4.22 shows the block diagram of the synthesizer we have postulated. It has 4096 output frequencies spaced at 4.88-kHz intervals across the 40.0–59.9-MHz band. Table 4.2 shows the development of these frequencies.

Summarizing the steps we have the following:

1. Determine the number of output frequencies needed and their spacing.
2. Select the center frequency (f_0) around which they are to be supplied. (This frequency is often bounded by the maximum frequency capability of the dividers available.)
3. As a trade off between the number of add-and-divide operations (A) and the number of frequencies (K) in the base set ($f_1 \cdots f_K$), choose A and $K : K^A \geq$ the number of output frequencies required.
4. Select a divisor (M) that gives enough separation between f_0/M and the nearest of the ($f_1 - f_K$) frequencies to allow a reasonable filter design. [Because f_0 is known and the extent of the set ($f_1 \cdots f_K$) is also known, the separation between f_0/M and $f_0 - (f_1 \cdots f_K)$ is available.]

$$f_m = (f_1 - f_K) = f_0 - \frac{f_0}{M} = \frac{1}{K} \sum_{n=1}^{K} f_n$$

5. Base set ($f_1 \cdots f_K$) may now be determined from f_m and the base frequency spacing (see Figure 4.23).

Figure 4.23   Base frequency relationships.

So far we have not mentioned spurious output signals or switching speed. These must be considered and to some extent traded off against one another because the bandpass filter following the mixer in each section of this synthesizer configuration affects both. Spurious signal output is a function of the level of the unwanted sidebands produced by mixing at the output of an interstage filter. Each unwanted signal is mixed in the succeeding section and produces new unwanted frequencies. Thus the bandwidth and skirt response of the interstage filters is important in a determination of the output spurious signal levels. The filter is also important in determining frequency switching speed for the synthesizer because its response limits the rate of change of phase of any signal passed through it. Noting that several filters are cascaded in the add-and-divide synthesizer, we find it obvious that these filters ultimately limit switching speed. Also, if not well matched between synthesizers, these cascaded filters can cause a loss of frequency-to-frequency phase coherence from synthesizer to synthesizer.

One not-so-obvious source of switching-speed improvement is in the way that a new code command is applied to the frequency gating switches. We have assumed in the discussion so far that all code inputs are applied at the same instant to these selectors. For high-speed frequency hopping, however, it is advantageous to apply new code commands to the synthesizer sequentially. Referring once again to Figure 4.22, we note that the desired output signal is generated in a series of add-and-divide steps and that the propagation through these steps is additive. Thus, when a frequency change is desired, the new signal must travel from the first stage through the last before appearing at the output. Therefore, by applying the new command sequentially, so that each stage in the chain receives its command with a time delay equal to the signal propagation, all stages switch simultaneously with respect to the output signal. The switch from one frequency to another is faster; that is, less "settling time" is involved and the duration of a signal at one particular frequency may be reduced in proportion to the reduction in settling time.

The usual rule of thumb for determining the maximum frequency hopping rate for a frequency synthesizer is that the period of transmission for one frequency should be ten times the time required for setting to that new frequency. Thus a reduction in settling time is effective in increasing hopping rate, because a setting-time reduction of 1 μsec would reduce a frequency period by 9 μsec.

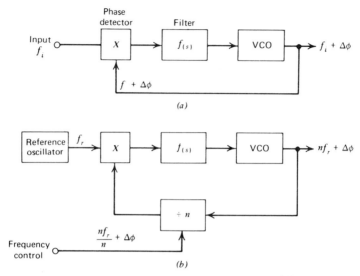

Figure 4.24   Basic indirect frequency synthesizer components: (a) simple phase-locked loop; (b) simple phase-locked loop modified for multiple frequency output capability.

Indirect Synthesis

The second type of frequency synthesizer is the indirect (any synthesizer employing phase- or frequency-locked loops is so termed). The simplest type of indirect frequency synthesizer is a phase-locked oscillator whose feedback path is modified to include a variable counter as in Figure 4.24b. Addition of this counter enables the phase-locked loop to generate any frequency multiple of its reference signal; that is, given a 1-MHz reference oscillator, a 1-to-1000 counter, and a phase-locked oscillator, theoretically we could generate 1000 frequencies spaced at 1-MHz intervals.

How does it operate? The reference input signal is compared in a phase detector that generates an output signal proportional to the phase difference between the reference and the counted-down VCO. The signal has a DC component that is applied, after filtering, to the VCO, closing the loop and causing the VCO to be locked to the input signal (or a multiple thereof) in frequency. Thus, inserting the variable counter between the VCO and the phase detector causes the VCO to operate at a frequency that is n times the reference frequency (where n is also the division ratio of the counter).

From this point, once the phase-locked loop is built and the variable counter inserted, all that remains is to add a coded input to the variable counter (to command the count) and we have a frequency hopper. In general, however, even the simplest of frequency synthesizers is not quite so easily come by, for a particular design must be tailored specifically to meet the hopping rate and stability requirements of a given communications system. Commonly, short-term stability of the output signal must be balanced against the hopping rate, because the loop filter has a direct bearing on both; that is, a narrow loop filter reduces VCO jitter due to noise pick up, reference noise, and so on. That same narrow filter can also significantly increase the time required to settle at a new frequency.

Common techniques for reducing frequency switching time while preserving the low-noise characteristics in indirect frequency synthesizers include the following:

1. Use of a sampling loop filter.
2. Digital-to-analog (D-A) conversion of the frequency command word, the analog signal is used to coarse-set the controlled VCO to a frequency near the desired output.
3. Use of more than one loop with alternate outputs taken in sequence.

Sampling Loop Filter. Although offering little to speed up loop frequency acquisition (in fact, the loop acquisition rate is limited by the sampling rate), a sample-and-hold type of loop filter helps to reduce jitter induced by the phase detector. This, in turn, permits greater freedom in selecting the loop reference frequency and filter parameters. Typical loop filters of the sampling type use field-effect transistors as switches and often employ tandem sampling filter sections to provide for rapid sampling while retaining high effective integrating time constants. Because of its sampling action, the sampling filter may also act as a phase detector.

D-A Frequency Command Conversion. A phase-lock frequency synthesizer can be locked to a new frequency by using the frequency command word (after conversion to a DC control signal) to set the loop VCO close to the desired frequency. Because frequency uncertainty is reduced, lock time is also reduced. D-A command word conversion can also reduce output frequency jitter because it allows open loop gain to be reduced.

When the output frequency range of a phase-lock loop is $|f_n - f_1| = f_0$, open-loop gain is required to be such that AoL $\geq f_0$ (Hz). Open-loop gain can be derived from only two sources: (1) phase detector gain (in volts per radian offset), or (2) VCO gain (in radians per volt). Amplification within the loop is equivalent to increasing one of these parameters. Dividing the lock range of the loop by setting the loop VCO in several increments across the desired frequency band therefore allows loop gain to be reduced to cover only the range of a sector (i.e., setting the VCO to ten intermediate frequencies within a frequency band could allow loop gain to be one-fifth of that otherwise required). The advantage of reduced loop gain is, then, that the VCO can be made less sensitive to noise occurring at its inputs; it appears as a direct reduction in output frequency jitter and noise.

Multiple Loop Techniques. Two frequency synthesizers, alternating their outputs, can increase the composite hopping rate by as much as ten times when settling time for a single synthesizer is held to 10% of a chip. Figure 4.25 is a

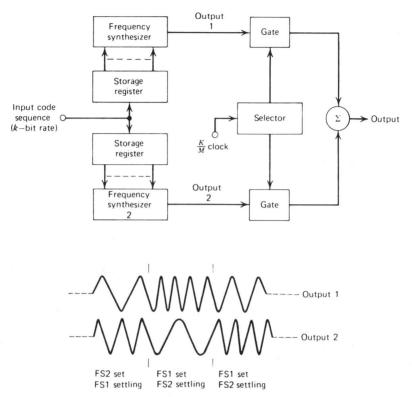

Figure 4.25 Multiple-synthesizer frequency hopper.

block diagram of a multiple loop synthesizer; waveforms show the possibility of speedup.

Especially well adapted to use in multiple-loop synthesis are the integrated-circuit frequency synthesizers described in Appendix 9. These devices (depending on the particular unit in question) are capable of generating large numbers of frequencies at output frequencies up to 2.5 GHz. Their small size permits more than one synthesizer to be included in systems that could employ only one with discrete-circuit techniques. By including two synthesizers in a system, and switching between them to allow for one to output a signal while the other settles to new frequencies, the system's hopping rate can be increased by a factor of 10 over a single synthesizer that allows for 10% settling time out of a hopping period. That is, when going from one to two synthesizers where the fraction of a hop time allowed for settling is $T_{\text{settle}}/T_{\text{hop}}$, the increase in hop rate for use of two synthesizers is $T_{\text{hop}}/T_{\text{settle}}$. A further increase in the number of synthesizers (beyond two) increases the hopping rate by a further factor of $r - 1$, where r is the total number of synthesizers. Multiple (r) synthesizers, therefore, allow for a

Figure 4.26   Read-only memory as a frequency hopping synthesizer.

hopping rate increase of up to $(r - 1)T_{hop}/T_{settle}$. For example, a single synthesizer capable of settling in 10 μsec would be useful for frequency hopping at up to 10,000 hops per second by itself. When four such synthesizers are used, however, the hop rate increases to

$$\frac{(4 - 1)\ 100\ \mu\text{sec}}{10\ \mu\text{sec}} \times 10,000 = 300,000 \text{ hops/sec.}$$

Those who wish to explore the phase-lock field further will find that it is exceptionally rich and well documented (see the references in Appendix 2). The bibliography in this book will serve to point those who are interested in the right direction.

Synthesis from Digital Words

One unique method of frequency synthesis that provides fast switching from frequency to frequency is that of digital storage of words corresponding to frequencies. For example, a 01010101 ··· pattern and a 001100110011 ··· pattern each represent a fundamental frequency that corresponds to $1/T(01)$ and $1/T(0011)$, respectively. If a word generator can be caused to output words in increments of bit times, then frequencies separated by $1/T$ can be generated.

One way to accomplish this is to store a set of patterns in a read-only memory, or ROM. Then if a particular output frequency is desired, the digital input (code word in our context here) corresponding to that frequency is repeatedly input to the ROM as long as the output frequency is needed. This is illustrated in Figure 4.26 in block diagram form.

Figure 4.27 shows typical spectra and waveforms for signal generation by a ROM synthesizer. A similar approach is that taken in the so-called numerically controlled oscillator (NCO) synthesizers, where sine function information is stored in a random access memory and output under control of a

Frequency Spectra

Wave Forms

Figure 4.27 ROM frequency synthesizer spectra and waveforms.

microprocessor. The series of digital words is then digital-to-analog converted, low pass filtered, and output as a sinusoidal frequency.

4.3   SENDING THE INFORMATION

Information modulation in spread spectrum systems is possible in most of the conventional ways; amplitude and angle modulation are both satisfactory. In addition to these more normal methods, information may be sent by various ways of modifying the code sequences used. Of course, if the code sequence is viewed as a subcarrier, which it is, the methods of modulation are revealed as the usual techniques in only a slightly new guise.

So far we have spoken little about information transmission in spread spectrum systems. Any communication system, however, has no *raison d'être* without a method of transferring information. (On-off keying is satisfactory in some cases and could be used. To be exact, on-off keying is a method of transferring information that requires no other operation but conveys only presence or absence of the signal.) In the following pages we discuss some of the techniques especially adapted to spread spectrum systems. Before proceeding to specific spread spectrum oriented modulation methods, let us discuss the effects of carrier modulation separately.

Carrier Modulation

Amplitude modulation of spread spectrum signals is not normally used because it tends to be at cross purposes with the intent of these systems. Instead of a constant envelope, low detectability signal, one with large peak power and easy demodulation by simple receivers is produced. Carrier-frequency modulation is more useful because it is a constant envelope signal, but the information sent is still readily observed. In both carrier AM and FM no knowledge of the code is required to receive and make use of the transmitted information.

In those examples in which the purpose of employing spread spectrum modulation is only to reject interference, carrier FM in combination with direct sequence modulation is quite useful. Figure 4.28 illustrates such a system. The output spectrum of a carrier-FM/direct sequence modulator is the convolution of the FM and spread spectrum code modulation. The entire $[(\sin x)/x]^2$ code spectrum is deviated as the information frequency modulates the carrier. If the composite carrier-FM/direct sequence signal is squared, however, a two-times carrier term is produced which bears all the frequency modulation information. For this reason, when some degree of privacy is required, carrier FM is not recommended.

Using carrier FM combined with frequency hopping (Figure 4.29) is also possible but requires even more care than the carrier-FM/direct sequence approach because the frequency hopping modulation is itself a form of

Figure 4.28 Simplified carrier-FM/direct sequence block diagram.

frequency modulation. Therefore, it is extremely difficult to prevent crosstalk due to frequency hopping the carrier from disturbing the information. Any phase change associated with a hop from one frequency to another is interpreted by an FM demodulator as information and is usually output as pulsed noise.

One aspect of combined-FM/frequency hopping demodulation is that noise pulses produced at frequency transitions are difficult to suppress. Bandpass filtering ahead of the demodulator tends to accentuate the problem. At a phase transition any signal traversing a bandpass filter is limited in its rate of change of phase by the phase rate of change of the filter (which is governed by the filter's bandwidth). Therefore, passing a signal with rapid phase shifts through a bandpass filter only stretches the transition time and increases the percentage of time of a disturbance. Postdetection filtering does reduce the effects of frequency hopping on an FM signal, but only to the extent that the hopping rate can be removed from the information band.

FM or AM frequency hop-modulated systems are recommended only when the hop rate is so low that it is unobjectionable. Otherwise transmit-and-receive frequency hopping signals must be highly coherent so that no phase transition will occur in the dehopped carrier signal.

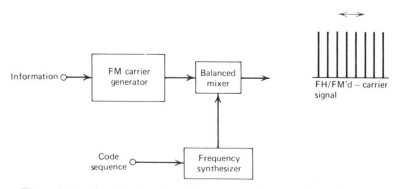

Figure 4.29 Simplified carrier-FM/frequency hopping block diagram.

For most purposes we have eliminated the usual forms of modulation from consideration for transmitting information in spread spectrum systems. Two other ways of transmitting the desired information remain: (1) clock rate modulation, and (2) code modification.

Clock Rate Modulation

Clock rate modulation is actually frequency modulation of the code clock. Demodulation is accomplished by the clock tracking circuit in the receiver, which must follow the deviation produced by both modulation and frequency offsets. For most purposes clock modulation is not used, because the loss in correlation due to phase slippage* between received and local clocks could cause degraded performance. Information sent by clock rate FM is not detectable by simple squaring, as is carrier FM, because it produces sidebands that are symmetrical about the carrier frequency, and the squaring process cancels all symmetrical signals.

For direct sequence systems an FM demodulator tuned to the RF carrier plus or minus the clock could recover the modulation if code imbalance allowed any clock component to be transmitted. As discussed in an earlier section of this chapter, code imbalance causes a clock frequency component to appear in a direct sequence spectrum, and this clock component could be tracked to recover any clock rate modulation.

For frequency hopping systems, the code clock may be at or near the data rate; therefore, clock rate modulation is not useful.

One further difficulty in implementing clock frequency modulation is that of constructing sufficiently stable oscillators that can be deviated over a wide range. Two solutions to this difficulty are given in the block diagram in Figure 4.30. In Figure 4.30*a* a crystal discriminator is used as the stabilizing element —a VCO (not crystal stabilized) is modulated by the information to be sent. Its output is fed to a crystal discriminator whose average output is zero when the VCO signal is at its center frequency. VCO drift causes a correction voltage to be generated.

A similar technique, but one that depends for stability on a reference oscillator, is shown in Figure 4.30*b*. This configuration is more adaptable to high-frequency output signals because the addition of a divider in its feedback loop can cause the VCO to lock to a multiple of the reference. The

---

*In a phase-lock demodulator the reference VCO is normally phase shifted by 90° with respect to the incoming signal to which it is locked. This condition occurs when the incoming signal is at the same frequency as the open-loop (rest) frequency of the VCO. As the input frequency deviates away from the VCO rest frequency, more or less phase shift occurs (depending on whether the deviation is higher or lower than the VCO), and this phase shift increases until it is near 180° or zero, at which point the loop will break lock. At any given frequency offset the sine of the phase offset in radians is $2\pi\Delta f/\text{loop gain} + \omega_{\text{mod}}/\omega_n^2$. The second term of this equation contributes little when the loop is the wideband loop required for information demodulation. Therefore, for our purposes $\sin\Delta\phi = 2\pi\Delta f/K$.

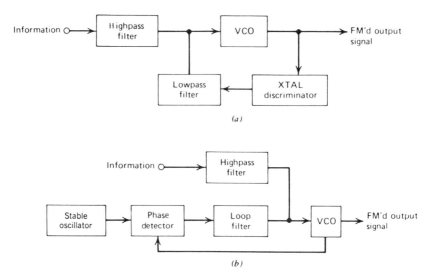

Figure 4.30   Stabilized FM modulator block diagram: (*a*) crystal discriminator stabilized; (*b*) phase locked to stable reference.

discriminator stabilized circuit is limited to the frequency range of available crystal discriminators.

Code Modification

Code modification (also called phase inversion modulation or code inversion modulation) as a modulation form is the most useful technique for general use in spread spectrum systems. It works equally well for direct sequence and frequency hopping systems, although the effect on RF output is not exactly the same (as one might expect because the outputs themselves are not the same).

What do we mean by code modification? Here we refer to the basic code sequence used in the direct sequence system to phase shift modulate the carrier or the code sequence used in a frequency hopper to choose the pattern of hopping. Modification means that we must change the code in such a way that the information signal is embedded in it and is uniquely decipherable only by someone knowing the original (unmodified) code. In addition, the desired code properties—good autocorrelation and low cross-correlation, which we discussed at some length and are so important in spread spectrum systems—must be preserved.

Satisfying all these conditions at one time would seem to be no mean trick. The simplicity of the circuits used to solve the code modification problem belies their importance. Although many different approaches have been used, all reduce to some technique for digitizing the information to be sent, and

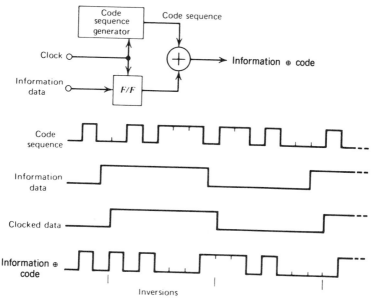

Figure 4.31 Basic code modification (modulation) using modulo-2 addition.

modulo-2 adding it to the code sequence. This process has the effect of inverting (ones become zeros and zeros become ones) the code each time a transition occurs in the information data stream. Figure 4.31 illustrates the operation in its simplest form.

Note that the input data are in digital format and reclocked by the code clock before being added to the code. This is done to ensure that inversions of the code due to information are present only at times when a code transition would normally occur. To an observer who does not know the original code sequence, the new product of information and the original code is of no more use than the original. A receiver who does know the original code needs only once again to modulo-2 add the composite (information $\oplus$ code) sequence* to its reference code, and then the information is recovered.

Because the modified code is not actually transmitted in an RF communications system as binary data, but is sent as phase transitions on an RF carrier, the receiver uses another RF carrier which is also phase shift keyed but with the unmodified code. These two phase-shift-keyed carriers are multiplied together and the effect is the same as that described for binary signals. Here we consider the transmission process as completely binary for purposes of simplicity.

---

*Here we assume synchronization between the transmitter's and receiver's code sequences. Also, because $A \oplus B \oplus A = B$, then data $\oplus$ code $\oplus$ code = data.

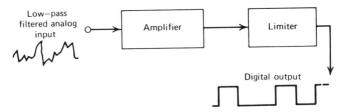

Figure 4.32   Simple limiting digitizer.

Digitizing Methods

Several ways to digitize analog signals (principally voice) for transmission in spread spectrum systems have evolved. The simplest is equivalent to clipped speech. It consists of applying the analog input signal to an amplifier and limiter whose output is a squared version of the analog signal. Voice processing of this kind has been used in both AM and FM to increase average modulation.

Figure 4.32 is a block diagram of this digitizer. After clocking, the binary signal generated by the limiter is modulo-2 added with a code sequence just as any other binary data stream would be.

In practice, digitizing in this way produces usable, though distorted, signals. When voice recognition is valuable, this technique is not recommended, although from the standpoint of getting a message across, word intelligibility is actually increased.[911] When simplicity is important and voice quality must be sacrificed, the limiting digitizer is worthy of consideration.

Pulse code modulation (PCM) is also of interest for signal digitization if only because of its wide use. Here we consider a simple form of PCM although in principle it is the same as more complex forms. Figure 4.33 shows a three-bit PCM generator capable of quantizing an analog signal into eight levels. For most applications in voice transmission systems, three-bit PCM offers satisfactory quality.

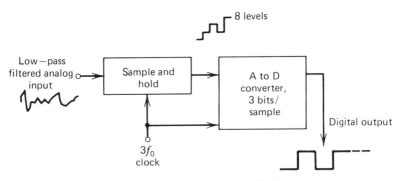

Figure 4.33   Three-bit PCM digitizer.

Figure 4.34 Delta modulator block diagram.

Higher-fidelity requirements may be met by increasing the number of quantization levels and thus the number of bits transmitted per level. Remembering that for a spread spectrum system, the process gain available is a function of the ratio of the RF bandwidth to the information rate, we see that increasing the number of bits transmitted to represent a given signal is not a highly thought of practice. Therefore, PCM is not often used in spread spectrum systems for information transfer, except as in the following.

A technique that could readily be considered to be a form of PCM is "delta" modulation, except that only a single information bit is sent for each sample. This is accomplished by comparing a given sample with the previous sample and transmitting a one if the current sample is larger and a zero if the current sample is smaller than the previous sample. Thus only representations of changes are sent as information, and the transmission rate required is greatly reduced. (For example, the transmission rate is equal to the sample rate, with delta modulation, where the transmission rate would be three times the sample rate for three-bit, or eight-level, PCM.)*

A basic delta modulator is shown in Figure 4.34. This simple digitizer consists of a sampled comparator (i.e., whose output changes only at sample times) and an integrator. The integrator is used to integrate the comparator's output, which output is the digitized representation of the signal prior to the present input. The integrated signal then is compared to the input signal, and the result is a one or zero output at each sample time.

An advantage of the delta modulator's design is that, in addition to a minimum transmission rate, the same circuit is useful for demodulating the data stream. That is, the delta-modulating analog-to-digital converter also serves as the digital-to-analog converter that regenerates a representation of the original analog information, in the receiver.

In the receive mode, a received delta-modulated data stream is input to the same data modulator, and the analog output is taken from its integrator (refer to Figure 4.34). In a simple delta modulator such as is shown, the fixed

---

*The most common form of PCM in use today employs eight bits per sample, and is used for voice transmission.

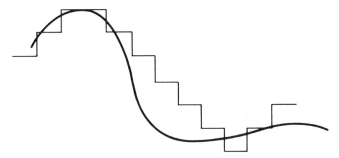

Figure 4.35 Delta-modulated representation of an analog signal showing distortion due to integrator step size.

integration time is a disadvantage, however, if the signal to be transmitted has significant dynamic range. That is, if a larger and/or rapid change occurs, then the maximum step* taken does not effectively follow the intended waveform, and distortion is introduced. On the other hand, for small and slow changes a smaller step size would be better.

Figure 4.35 illustrates the integrator's response to an analog signal.

The solution to providing for varying signal rates of change is to sense the variation and modify the integration time to compensate for the changes. Such a technique has been elegantly implemented in the "companded variable slope" or "continuously variable slope" delta modulators, both of which are typically called "CVSD" modulators. These modulators adapt to signal-level changes by counting the number of Ones and Zeros that have occurred, and when some specific number (usually three or four) have occurred in a row, by adjusting the integrator's time constant. Thus, if the requisite number of Ones occurs in a row, which signifies that the signal level is tending to increase, then the integration time decreases and this enlarges step size. Conversely, if the same number of Zeros occurs in a row, step size decreases.

Diagrams showing typical CVSD modulator/demodulators are shown in Figure 4.36. These IC units are simple to use and inexpensive, and dominate

---

*The response of an integrator to a pulse is to charge to a given level and remain at that level. Since the integrators in use here usually consist of a single RC integrator, the input pulses, together with the integrator's resistor form a current source that charges or discharges the capacitor by a voltage step

$$\Delta V_{\text{oc}} = \frac{1}{C} \int \frac{E_p}{R} \, dt$$

where E_p is the pulse level. If a series of ones is sent, then the integrator's output is a positive-going stepped (staircase) function, for zeros a negative-going stepped function, and for a steady-state signal an alternating one-zero pattern is sent (which integrates to zero charge).

Figure 4.36   Typical CVSD integrated circuits: (*a*) Motorola MC 3417/18; (*b*) Harris Corp. HC-55516/55532.

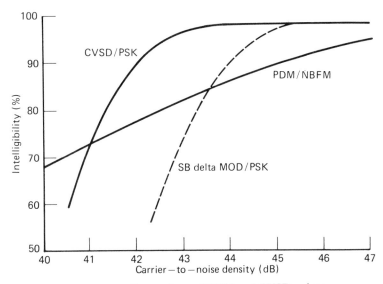

Figure 4.37   Comparison of PDM and CVSD voices.

many military applications for contemporary digital voice, including spread spectrum systems.

Performance of CVSD modulator/demodulators is good, with the capability to operate at error rates as low as 1×10^{-2}. They also provide good intelligibility (see Figure 4.37). The prime drawback to their use for spread spectrum applications is that they tend to require operation at 16 Kbps or higher for voice applications, and sound noisy even though their intelligibility scores are good. Figure 4.38 shows the signal-to-noise performance that may be expected for CVSD modulation at 16 and 32 Kbps as the input signal level varies.

As an aside, we note that vocoding techniques are also used to convert analog signals to digital (usually PCM) format. In general, these techniques break the voice band down into subbands. Information in regard to which subbands contain the power in the analog signal is then transmitted. Periodic sampling of the subbands and transmission of a digital representation of the significant subband is an analog-to-digital conversion method. Vocoding as a technique has been known for many years, but has not come into wide use largely because of its implementation cost.

Another method is suppressed-clock PDM (pulse-duration modulation). This approach is shown in Figure 4.39. An input analog signal is input to a sample-and-hold circuit operating at p times the highest frequency in the input signal. A sawtooth generator, synchronous with the data samples, is compared with each sample-and-hold output, and when the sawtooth ramp exceeds the stored level, the comparator output changes state. If the comparator output is set to the "One" state at the beginning of each sample and

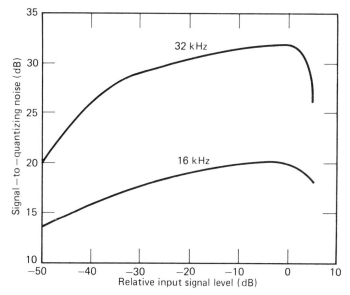

Figure 4.38   SNR using 1-kHz sine wave for CVSD.

it goes to Zero when the ramp exceeds the stored sample level, then a series of variable length Ones is generated. The length of these Ones depends on the stored level because the time required for the ramp to reach the stored level depends on that level. Noting that each PDM One varies between zero length and $1/R_s$ (the clock period) and is synchronous with the clock, we modulo-2 add a square wave generated by dividing the clock by two. This modulo-2 addition subtracts the clock from the data stream and gives the same information but with an effective $p/2$ sample rate and therefore, smaller information bandwidth. Replacing the clock information at the receiver is as simple as modulo-2 adding the half-clock rate square wave to the suppressed clock signal. This restores the original data stream.

Suppressed-clock PDM works extremely well in spread spectrum systems, for it meets the requirement of being quantizable at the code clock rate. Curves that measure performance of SCPDM and FM for various FM deviation ratios, as carrier-to-noise ratio varies, are shown in Figure 5.25. The improvement derived from SCPDM over FM is in the range of 4–7 dB, which must be considered a significant gain when we note that to get a 4-dB improvement in the process gain by increasing the code rate the RF transmission bandwidth must be increased by a factor of 2.5.

We now come to the subcarrier-FM code-modification method of modulation. In this technique, as in the others we have discussed, code modification actually consists of sequence inversion keying by information in digital form. The subcarrier-FM method employs an oscillator, frequency modulated by the information to be transmitted, whose output is shaped, reclocked, and

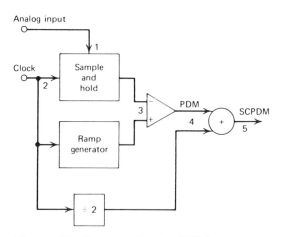

Figure 4.39 Suppressed-carrier PDM generator.

modulo-2 added to the code sequence. Subcarrier frequencies well below the code clock rate are chosen but must be high enough to be above the highest modulating frequency. Subcarriers of 50–500 kHz with code chip rates in the 1-to-100-Mbps range, are typical. Keeping the separation large between information, subcarrier, and code rate helps to reduce noise introduced in the digitizing process.

   A voltage-controlled astable multivibrator is a good choice as a subcarrier oscillator for this use. Not only is its modulation linearity good but it also has a square wave output. Thus only reclocking at the code rate is necessary before addition to a code sequence.

Several of the information-transfer techniques discussed are more useful in direct sequence approaches than in frequency hopping. Subcarrier FM, for instance, or any other code modifier that increases the required hopping rate, is undesirable in a frequency hopping system.

Remembering that data rate reduction produces as much processing gain improvement as does a bandspread increase, then we see that a lower rate voice digitization approach would be desirable. One such technique that is promising is linear predictive coding (LPC). Although not yet developed to the point of being implemented on a single IC chip as the CVSD units have, the LPC approach offers a process gain (and jamming margin) increase of 10 log $(16,000/2400) = 8$ dB over CVSD, since LPC can provide a 2400-bps digitized voice rate. This may not prove to be quite as great an improvement as it initially appears, however, since the CVSD approach works to bit error rates as poor as 1×10^{-2}, while it is doubtful that LPC techniques can tolerate such high error rates and therefore they require higher signal-to-noise ratios.*

Frequency Hopping Code Modifiers

Figure 4.40 illustrates a code modifier adapted to frequency hopping. It uses only a modulo-2 addition of the code and input data. Reclocking at the code rate is not shown; however, this would be accomplished by transferring frequency-select information to the synthesizer only at clock times.

This simple frequency hopping modulator is made up of three parts: (a) a code generator, (b) a code sequence modifier (information modulator), and (c) frequency synthesizer. In this scheme, in which P frequency choices are required from the synthesizer, the number of parallel chips of information from the code generator would be $\log_2 P$. The code generator itself would be made up of n stages, each feeding an output line. Of course, more than n stages could be used, but n outputs would still be satisfactory.

Each code generator input line connects to a data modifier whose other input is the data to be transmitted. Figure 4.41 is a diagram of a modifier for a frequency synthesizer with 2^5 frequencies or more to choose from.

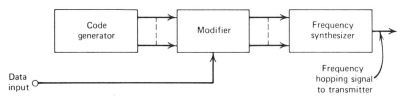

Figure 4.40   Frequency hopping code modifier (modulator).

*LPC encoding requires BER less than 1×10^{-3} for some applications.

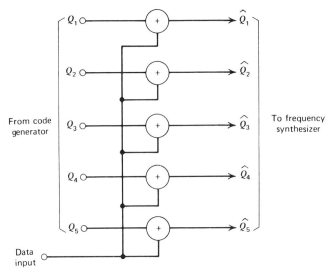

Figure 4.41   Simple code modifier.

The effect of the code modifier is to pass on a word that is the complement of a one in the code generator output, whenever the data input is a one. When the input data is a zero, however, the code generator output word is uncomplemented. Thus one frequency command is sent for a data one and another is sent for a data zero; for instance, if

$$\{Q_1, Q_2, Q_3, Q_4, Q_5\} = \{1, 0, 1, 1, 0\},$$

then

$$\{Q_1, Q_2, Q_3, Q_4, Q_5\} = \{0, 1, 0, 0, 1\}.$$

Data-modified n-tuple $\{Q_1, Q_2, \ldots, Q_n\}$ is then used to command the frequency synthesizer to give a corresponding output frequency.

There are, of course, many other ways to modify the frequency command code. Simple inversion of the command word as illustrated could cause the mark-space frequency choices to be symmetrically distributed, which is not desirable in most systems; that is, if the code word were $1\,0\,1\,0$, then a mark would be sent at frequency $1\,0\,1\,0 \doteq 11$ and a space would be sent at frequency $0\,1\,0\,1 \doteq 6$ (see Figure 4.42). The correspondence between the frequency command word and the output frequency may be modified to alter this pattern, but any fixed command-output correspondence would still give a fixed mark-space relationship; that is, if f_m were chosen for a mark, then f_s would always signify a space (or vice versa). A would-be interferer, knowing the correspondence between f_m and f_s, could render the system inoperative

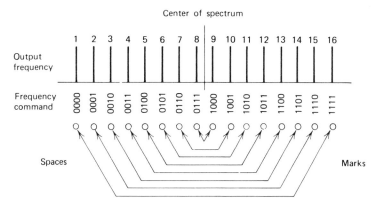

Figure 4.42   Illustration of space/mark symmetry for simple code modifier.

by transmitting the complementary frequency whenever any in the set were received. The only protection from such action by an interferer, then, is to modify the code-frequency correspondence periodically or change the code modifier so that a nontrivial relationship exists between frequency comple- ments. One simple method, useful for changing the one-zero relationship, is illustrated in Figure 4.43 (only one line is shown).

Here, instead of simply inverting the coded frequency to go from a zero command to a one command, a different code input is taken for each of the possibilities. Each of these coded frequency commands is taken from a different source (i.e., from two separate code streams). The two code inputs could come from separate code generators or from different taps on the same generator, depending on system requirements. In either case the one-zero relationship would be nonconstant and difficult to determine for an outside observer.

It is worthy of note that a frequency hopping system is inherently capable of protecting its internal code sequences to a higher degree than can the direct sequence systems. This is because the direct sequence systems modu- late their carriers with their code, and it is therefore directly observable. In a

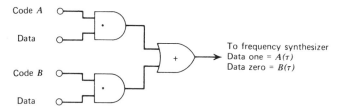

Figure 4.43   Gating for modifying code/frequency correspondence (showing one line).

frequency hopping system, the code itself is not transmitted. Instead, a frequency representing a code vector is sent. Therefore, an observer must know the translation between frequency and code vector in order to determine the code that is in use.

PROBLEMS

1. For a given data rate, compare the RF bandwidth of quadriphase modulation with biphase modulation.

2. How many mix-and-divide modules would be required to develop 8000 or more frequencies from a set of 16 base frequencies in a direct frequency synthesizer?

3. If the synthesizer is the simple phase-lock (indirect) type, draw a block diagram to produce 8000 frequencies separated by 100-Hz spacing.

4. What minimum data rate would be usable for transmitting voice by a three-bit PCM digitizing method? (Let voice bandwidth be 3 kHz.) How much RF bandwidth would be consumed in providing 40-dB process gain for this digitized voice signal?

5. Compare the frequency hop-rate limitations of direct and indirect synthesizers.

6. Determine the mean and base frequencies for a direct, 64-frequency add-and-divide synthesizer with output at 70-MHz and 10-kHz output spacing.

7. Show that carrier modulation is not effective for information protection in a direct sequence system.

8. How many chips of code are required to provide unambiguous frequency selection for a synthesizer with 2000 frequencies available?

9. Compare the relationships of base frequencies and divisors possible for a 256-frequency direct synthesizer. What is the optimum number of bases to minimize the synthesizer's complexity?

10. What is the effect of dividing the PDM signal, before clock subtraction, by 2 (toggling a flip-flop)?

5

CORRELATION
AND DEMODULATION

Spread spectrum signals are usually demodulated in two steps: first, the spectrum spreading (direct sequence, frequency hopping) modulation is removed, and second, the remaining information-bearing signal is demodulated. In either the direct sequence or frequency hopping systems removal or demodulation of the spectrum spreading modulation is accomplished by multiplication with a local reference identical in structure and synchronized in time with the received signal. (Again, we reserve the problem of synchronization for Chapter 6 and assume that received and local reference signals are synchronous.) Baseband recovery is a matter of demodulating the conventional kinds of signals—FM, FSK, PSK—in the face of noise. The noise of most interest seen by the baseband demodulator, however, is generated by cross-correlation of a local coded reference with an interfering signal.

This chapter describes the correlation (despreading or bandwidth collapsing) and baseband demodulation processes. In the area of spread spectrum communications there is no more important pair of operations, for it is here that process gain and code discrimination are effective. The other circuits used and other functions performed are peripheral to them; that is, the special properties of spread spectrum systems are realized in the correlation and baseband demodulation functions. It can be said that all other operations are performed just to make them work properly.

5.1 REMAPPING THE SPREAD SPECTRUM

The first of the operations performed by a spread spectrum receiver (neglecting amplification, downconversion) is that of despreading or removing the spread spectrum modulation. Any received spread spectrum signal is mapped from a large set of signal components, spread over a wide band, into a small or even single-valued set occupying ideally just enough bandwidth to pass any baseband modulation components. In this operation, called correlation, a received signal is multiplied with a locally stored model of the signal, and the resulting output is integrated. Demodulation then follows.

Quoting R. L. Rex and G. T. Roberts,[226] "As it applies to waveforms, correlation is a method of time domain analysis that is particularly useful for detecting periodic signals buried in noise, for establishing coherence between random signals, and for establishing the sources of signals and their transmission times." Here we are most concerned with establishing coherence between (pseudo-) random signals. Detection of the periodic information signal buried in (pseudo-) noise is left to the baseband demodulator.

For frequency hopping and direct sequence systems the correlation process is the same. The main difference is that the local reference is a frequency hopping or direct sequence signal, depending on which is to be received; that is, the local reference is matched to the desired signal. In Chapter 2 we discussed the way in which frequency hopping and direct sequence receivers realize their processing gains by transforming desired wideband signals into narrowband signals and undesired signals into wideband signals. The remainder of the process, as we also stated in Chapter 2, consists of separating out the desired signal.

Our purpose is to describe the implementation of typical correlators, which have much in common with the modulators discussed in the preceding chapter. Two types of spread spectrum correlators are discussed here. We have labeled them "in-line" and "heterodyne" correlators for obvious reasons.

In-Line Correlation

All correlators for direct sequence systems shown up to this point have been the in-line type, not because they are used more often or are superior to others but because their operation is exactly the same as a direct sequence modulator (the circuits can be identical) and can perhaps be a little more readily understood on that basis.

In the in-line correlator in Figure 5.1*b*, in which a phase-shift-modulated received signal is one input, the signal is multiplied by a code identical to that used in generating the phase-shift-keyed input signal. The effect of this is complementary to the code modulation in the transmitter: the input signal is phase inverted each time the local correlating code sequence has a One-Zero or Zero-One transition. If the transmitter's code is identical to the local code and the two codes are time synchronous,* then at each phase shift of the

---

*Actually it is important to realize that the receiver must synchronize its code sequence to the code modulation on the signal received at its input. In practice the code sequence, as seen at the receiver, is quite different from that being sent by the transmitter at any given instant. This difference in code phase is, of course, due to the propagation time between transmitter and receiver and is the basis of spread spectrum ranging.

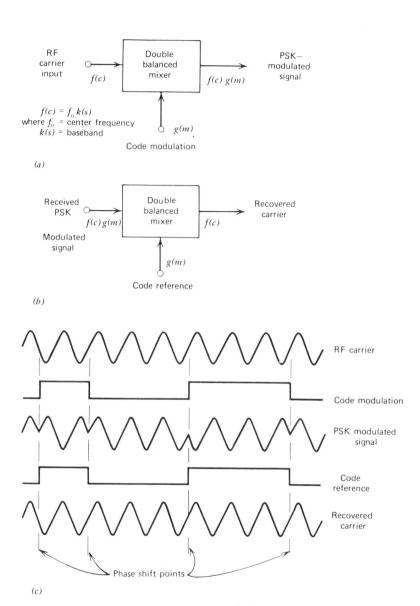

Figure 5.1 In line correlator carrier recovery: (*a*) modulating multiplier (PSK modulation); (*b*) in-line correlator (complementary PSK modulator); (*c*) waveforms in PSK (biphase) carrier recovery.

transmitted signal the receiver phase shifts it again. These two complementary phase shifts combine to cancel one another and to restore the original carrier. The spectrum spreading-despreading operation is a "transparent" process in that carrier modulation, such as AM or FM, is passed through without change.

In-line correlation, though extremely simple, is not used in the majority of spread spectrum systems. The reason for this lies in the area of practical implementation rather than being a theoretical block. First note that the input and output frequencies are the same (see Figure 5.1); that is, if the correlator's PSK-modulated input signal is centered at f_o, the recovered carrier will also be at f_o. This, in itself, would be of no consequence were it not for one of the main purposes of the correlator—to be specific, rejection of interfering signals.

The correlator in a spread spectrum receiver is that point beyond which interfering signals cannot be permitted to pass, at least in their original form. Otherwise a narrowband interfering signal could enter the post-correlation circuits and masquerade effectively as a desired, properly despread signal. Remembering, then, that interfering signal inputs are likely to be extremely large compared with the desired signal, we must consider the likelihood of direct leakage of an interfering signal around the correlator. When leakage occurs, the carrier suppression capability of the correlator's balanced mixer is of no avail because the interfering carrier does not pass through the correlator but around it. For this reason in-line correlation is not often used.

Heterodyne Correlation

A heterodyne correlator, as implied by its name and the preceding discussion, is one in which the correlated signal output is at a different center frequency from the input signal. In the process of despreading or removing the code modulation the information-bearing signal is translated to a new center frequency. This avoids the possibility of direct feedthrough and in some instances simplifies receiver design, because the circuitry following a heterodyne correlator can be at a lower frequency. Frequency hopping correlators are almost always of the heterodyne type (to be more exact, I do not know of any frequency hopping correlator that is not). Direct sequence correlators, in general, also employ heterodyne correlation.

Heterodyne correlators are shown in block diagram form in Figure 5.2 for direct sequence and frequency hopping receivers so that a direct comparison between the two can be made. (Filters that would normally be employed between the local reference modulator and the correlating mixer are not shown.) The direct sequence correlator in Figure 5.2*a* generates a local reference signal that is a frequency-offset replica of the expected direct sequence signal (with the exception that the local reference is not information-modulated). If we were to observe the signals generated in a direct sequence heterodyne correlator, we would see the spectra in Figure 5.3.

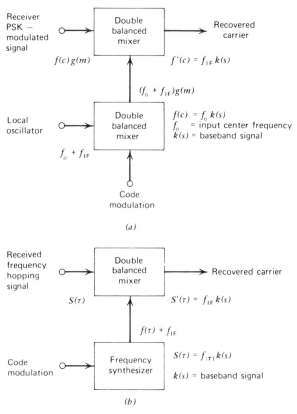

Figure 5.2   Comparison of heterodyne correlators for direct sequence and frequency hopping: (*a*) direct sequence heterodyne correlator; (*b*) frequency hopping heterodyne correlator.

The local reference signal in Figure 5.3 is generated in exactly the same way as the transmitted signal. In fact, when we have a receiver and transmitter than do not operate simultaneously, the same circuit could be shared between the transmit modulator and receive local reference generator.

In both direct sequence and frequency hopping systems the correlation process is a frequency-by-frequency multiplication (corresponding one-to-one with a chip-by-chip code comparison in the time domain) of the local reference with incoming signals.

The frequency-by-frequency correlation process is obvious when considering frequency hopping, but is not quite so obvious in direct sequence systems. If, however, one considers that each run of Ones or Zeros is a half period of a square wave related to the code chip rate (i.e., $1/RC, 2/RC, 3/RC, \ldots,$ n/RC), then we see that the modulated carrier has sidebands that are related to these runs. Thus, when the local reference code is synchronized

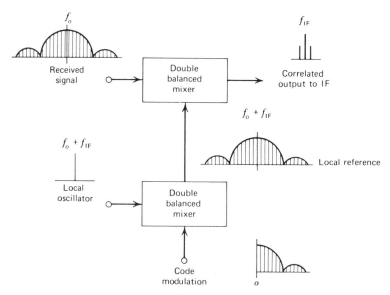

Figure 5.3 Signal spectra in direct sequence heterodyne correlator.

with the incoming signal code, the local reference sidebands correspond frequency by frequency to the input signal sidebands. This correspondence in turn allows the two signals (local reference and received signal) to correlate and produce a despread output signal.

Typical correlator output signals are shown in Figures 5.4*a* and 5.4*b*. Figure 5.4*a* shows the unmodulated despread carrier, and Figure 5.4*b* shows the suppressed carrier spectrum typical of PSK modulation by a square wave. In this case the modulation is a 1200-Hz square wave, simulating a One-Zero series at 2400 bps. No interference is present in either case. Note that in Figure 5.4*b*, the modulation sidebands are symmetrical around the suppressed carrier, and that only odd harmonics are present. (It is also noted that the two primary sidebands are not equal in amplitude, as they ideally should be. This was found to be due to amplitude nonlinearity in the receiver, but did not result in a detectable loss in system performance.)

Code Timing Offset Effects On Correlation

The prime purpose of the correlator, whether direct sequence or frequency hopping, is to match a local reference signal to a desired incoming signal and thereby reproduce the embedded information-bearing carrier as an output. So far we have assumed that a perfect match exists between our receiver's local code sequence and the received signal code modulation, both in the pattern of the codes and in their timing. Such a happy circumstance,

Figure 5.4   Typical correlator output signals: (*a*) no modulation, no interference; (*b*) PSK modulation, no interference.

however, seldom exists in a real world system; therefore, it is of some interest to examine the effect of less-than-perfect code synchronization on correlator output.

In Chapter 3 we showed that the autocorrelation function for a maximal linear sequence is the triangle in Figure 5.5. For signals encoded by such a sequence the maximum signal output from the correlator occurs when the two codes are in the zero-shift position. (For that matter maximum autocorrelation output occurs for any code type in the zero-shift position. The only reason for speaking specifically about maximal linear sequences here is that

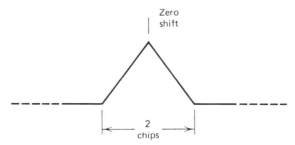

Figure 5.5   Autocorrelation for binary code sequency in a region of perfect synchronization.

their autocorrelation is well behaved in all shift positions; this, in turn, saves a great deal of qualifying explanation—such as this statement.) The optimum signal-to-noise ratio occurs when the local and received codes are exactly aligned, because this is the point of maximum signal (and signal-to-noise ratio) output. Where a receiver employs AGC to control the signal level, the autocorrelation curve is not one of signal level, but instead is a plot of signal-to-noise ratio.

Noise seen at a correlator's output derives from several sources:

Environment* and circuit noise × code modulation.

Undesired signals × code modulation.

Desired signals × code modulation.

Noise output due to environment and circuit noise is overshadowed in those circumstances in which undesired signals (jamming or other interference) are present. This is due to the ability of spread spectrum signals to operate in environments in which undesired signal power is much larger than desired signal power. Environment and circuit noise are of real concern only in maximum-range, interference-free conditions.

When the incoming desired signal is exactly matched with the receiver's reference, there is no noise generated by the desired signal. Before synchronization, or during imperfect synchronization, a part of the desired signal is output as code-related (pseudo-) noise. The amount of output noise depends on the degree of synchronization; when there is no synchronization (the local and received signals are more than one code chip apart), the output produced is all noise.

Considering the diagram in Figure 5.6 in which a pair of identical codes is not quite synchronous, we see that the corresponding frequencies produced

---

*Noise from the environment consists of radio-frequency signals generated by lightning, radio stars including the sun, and man-originated sources such as autos and neon signs.

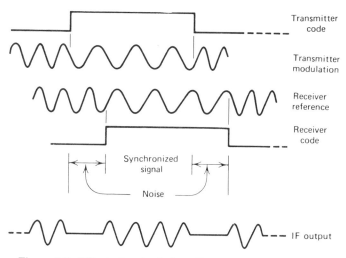

Figure 5.6   Effect of code timing offset on correlator output.

by code modulation for direct sequence, * or the frequency transmitted for a frequency hopper, overlap and that for a part of each code run the signal is properly translated into the IF. During the nonoverlapping chip time, however, the product of the incoming and reference signals lies somewhere in a region bounded by the covariance of the two signals. This part of the output is noise, some of which will fall within the IF bandwidth.

It is interesting to note that as the two codes slip from exact synchronization the correlator output code noise increases an amount corresponding to three times the offset. Observing Figure 5.6, we see that for each r reduction in the synchronized signal, due to code timing offset, there is a $2r$ increase in the noise contribution which then results in a signal-to-noise function

$$f(\tau) = A \frac{T - \tau}{2\tau},$$

---

* Here we recognize that direct sequence modulation and frequency hopping are actually similar. A frequency hopper generates a new frequency for each new code (generator) state by commanding a frequency synthesizer to output a particular frequency corresponding to that state. A direct sequence modulator does nearly the same thing by directly modulating with a code that has varying One and Zero durations. These variable-duration pulses then produce variable-frequency sideband spectra in the modulator output, which change each time the modulating code has a One/Zero transition. Both frequency hopping and direct sequence signals are then seen to be frequencies or sets of frequencies which change at a code determined rate, and frequency hopping can be said to be a single-sideband version of direct sequence modulation. In the heterodyne correlator the two work in the same way, translating the desired signal, frequency-by-frequency, to a single output frequency while translating everything else to a (pseudo-) randomly determined frequency.

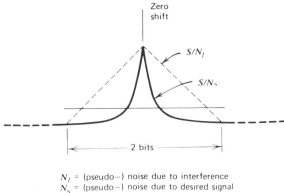

N_I = (pseudo–) noise due to interference
N_s = (pseudo–) noise due to desired signal

Figure 5.7 Comparison of signal-to-interference noise and signal-to-self-noise for code timing offset of ± 1 chip.

where A = maximum output,
        T = period of one chip,
        τ = timing offset.

This function is plotted in Figure 5.7 when no interfering signal is present and the desired signal is above threshold. If a significant amount of interference occurs, however, the signal-to-noise ratio increases as a linear function of synchronization accuracy. Why? Because the code $\times$ noise interference produced is not so much a function of code timing accuracy and is large enough in any case of interest to an antijamming system to swamp out any self-noise effects completely. Figure 5.7 compares signal-to-noise functions for jammed and unjammed input signals. (Note that the signal-to-noise function for the high noise condition is the same as the code correlation function.)

The difference in the two curves in Figure 5.7 from the standpoint of their width would seem to imply that more accurate range measurements are possible under high signal-to-noise (i.e., no interference, high-signal) conditions. This is certainly true and should be expected. After all, any signal-to-noise sensitive device (and what device is not) used to detect synchronization would be forced to recognize the synchronized signal within, for example, approximately ± 0.2 chip of the zero shift under high-signal conditions in which it would resolve to only ± 0.8 chip for the same threshold point when being jammed.

5.2   EFFECT OF NONSYNCHRONOUS INPUT SIGNALS

The performance of a spread spectrum system when faced with nonsynchronous input signals is as important as its handling of desired signals. Remember that we said that the correlator is the point beyond which an

Figure 5.8   Balanced mixer leakage at correlator output, 8191 code. No modulation.

undesired signal must not pass? The problems illustrated in the preceding chapter for balanced modulators also apply to the balanced modulators that are part of a correlator—and apply more stringently. Referring to Chapter 4 and the various possible balanced-mixer problems illustrated there (see Figures 4.4 through 4.10), let us examine their effect on a spread spectrum receiver. First, what effect has carrier suppression (or lack of it) on a correlator and its output?

For the in-line type of correlator it should be obvious that insufficient carrier suppression would allow a jamming signal to leak directly through and into the detection circuits. When an interfering signal is multiplied by a code-modulated signal, or directly by a code as in an in-line correlator, the signal is suppressed by the multiplier's carrier-suppression factor, whereas the modulation products output are equivalent to any other multiplier. Therefore, an interfering input signal is (or should be) reduced to a level well below that of the pseudonoise produced by multiplication with a code.

The spectra shown in Figures 5.8 and 5.9 illustrate the effect of jamming signal leakage in a direct sequence system, due to lack of carrier suppression. Note that the leakage-related signal appears as a narrowband carrier. Proper manipulation of the jammer's frequency can in turn cause this unsuppressed carrier to capture the receiver's carrier tracking loop, making the jammer very effective.

Figure 5.10 shows a direct sequence spectrum in which both carrier and code-balance suppression are unsatisfactory. Such a signal, if applied to a narrowband detector, such as a phase-lock loop, would cause the phase-lock loop to lock on the unsuppressed carrier. Proper carrier suppression, how-

Figure 5.9   Balanced mixer leakage at correlator output, 8191 code. No modulation.

Spurious carrier component due to
poor carrier balance

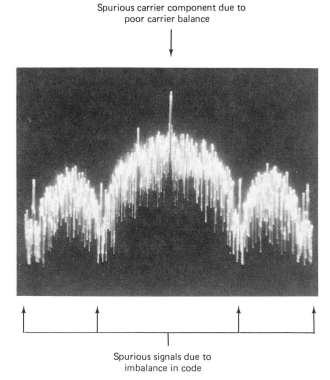

Spurious signals due to
imbalance in code

Figure 5.10   Typical direct sequence correlator output with poor suppression of carrier and code-unbalance components.

ever, would depress the signal in the noise and prevent the detector from locking to this spurious signal.

The signal outputs seen at the nulls in Figure 5.10, caused by poor code balance, can also cause the detector to lock up and falsely register signal detection. Worse yet, these narrowband signals lie at a point where they are not well masked by code-related noise. Even when good carrier suppression exists poor code balance can allow false detector lockup due to interference. (The interference, rather than occurring at center frequency, would necessarily be offset by an amount equal to the clock rate.)

Does the same thing happen in a heterodyne correlator? Yes, it does, but not quite in the same way. In the heterodyne correlator the local reference is the one that must show good suppression of spurious signals. Consider that the spectrum of Figure 5.10 is a local reference signal to a correlator in which the received signal is at f_o, the local reference center frequency is $f_o + f_{IF}$, and the signal nulls are at $f_o + f_{IF} \pm f_{clock}$. Again, interference translated into the receiver IF can cause the same problem (i.e., recognition of interference as a desired signal) as in a poorly operating in-line correlator.

Several methods (and some empirically-generated rules) have been employed successfully in efforts to overcome these problems. One is the intuitively reasonable rule of thumb that states that the carrier suppression of a balanced mixer used in a direct sequence system should be at least equal to the process gain desired in the overall system; that is, if the system RF bandwidth is 20 MHz and its detection bandwidth is 10 kHz, the process gain expected is 33 dB. The correlator then should have 33-dB or more carrier suppression.

When very high processing gain is required, beyond the capability of a simple balanced modulator, tandem arrangements of more than one have been used successfully in correlators to attain the required carrier suppression. Otherwise, many practical correlators are limited to about 30-dB carrier suppression if environmental conditions are considered.

A tandem correlator consists of a pair of mixers arranged as shown in Figure 5.11. An auxiliary code (code 2) is modulo-2 added to the primary code (code 1, which matches the received signal's code). When code 1 and the received code are synchronous, the output from the first multiplier is still modulated by the auxiliary code. Thus any leakage through the first multiplier (mixer) is masked by a broadband signal and does not appear as a coherent input to the second multiplier. At the second multiplier, the auxiliary code is multiplied with the output of the first multiplier, and, since it is synchronous with itself, remaps the desired signal into the desired narrow bandwidth. At the same time, it spreads the residue of the first multiplication due to the jamming signal. Thus the effect to the demodulator is that of a correlator whose balanced mixer has the carrier suppression capability of the first multiplier plus that of the second. In this way, two balanced mixers with nominal 30-dB carrier suppression can produce an overall carrier suppression of 60 dB—a significant result.

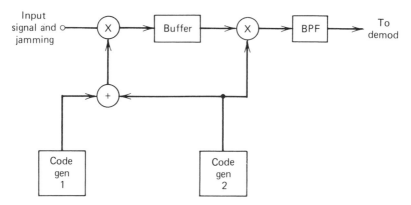

Figure 5.11 Tandem in-line correlator.

Heterodyne correlation can also be applied in the tandem correlator. Either the first multiplication or the second, or both, may be implemented with a code-modulated local carrier. These approaches provide the same protection from interference-leakage as in a single heterodyne correlator.

To overcome the possibility of false receiver-lock on code-unbalance-related signals (which can occur at all multiples of the code clock around the signal carrier), some form of filter is required ahead of the receiver's correlator. This filter may take the form of a simple fixed-frequency bandpass filter in an IF preceding the correlator or it may be a more complex tuned filter in a receiver preselector. At any rate, the bandpass characteristic of the filter must be such that spectrum roll off beyond the main lobe prevents locking due to code imbalance from signal input to the correlator at center frequency plus or minus the code clock and its multiples. Filter requirements vary from system to system because of code rates, gain and AGC values, and other parameters. In general, however, the code-unbalance lockup-prevention filter should be approximately as wide in bandwidth as the code rate (BW ≈ code rate) and roll off sharply thereafter.

It is also important that a filter having bandwidth characteristics similar to those of the IF filter be inserted at the output of the local reference. This has two effects:

1. It reduces the level of any stationary signals at the nulls of the local reference signal.
2. It provides a match between the level of the various components in the local reference spectrum and the input signal spectrum. (This is necessary to prevent generation of undesired crossproducts in the cancellation process.)

Figure 5.12 illustrates the correlator with these two filters in place.

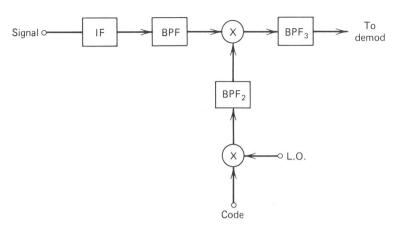

Figure 5.12 Correlator showing positioning of bandpass filters. (BPF passband same as BPF$_2$ passband, but offset in frequency. BPF may be integral to IF.)

We have spoken about the effects of carrier suppression and other effects of an improper direct sequence correlator operation. There is much less in a frequency hopping correlator to concern us in that neither carrier suppression nor code imbalance have any meaning to the frequency hopper. The nearest equivalent is a stationary spurious signal in a frequency synthesizer output, fed to the receiver correlator as a spurious local reference. This signal could cause undesired signals to be translated into the receiver IF. Frequency hopping synthesizers, however, do not typically exhibit stationary spurious outputs.* Spurious output signals from such synthesizers normally vary from frequency to frequency at the same rate as the desired output because they are generated as a function of the component signals and synthesizer configurations necessary to generate the desired signal output, and those change as the output frequency changes. For this reason frequency hopping systems are not so susceptible to "ticklish" spots, in which an interfering signal is likely to cause reception problems, as direct sequence systems.

It should be noted that the very process of frequency hopping inherently produces spurious signal sidebands. The frequency hopping signal is, of course, a series of pulses at the hopping rate, and these have spectra that are a function of the pulse shape. Typical frequency hoppers with rectangular pulse output can be expected to have $(\sin x/x)$ distributed sidebands. Work has been carried out in the area of sideband reduction through pulse shaping, but such systems are currently the exception. The key is to remem-

---

*Here the word "stationary" must be emphasized. Frequency hopping synthesizers have spurious outputs. It would be miraculous if they had none. The spurious outputs change as rapidly as the desired outputs, however (and it is hoped are at lower levels). Therefore, it is at least as hard for a would-be interferer to find them as the intended outputout.

Figure 5.13 Output of frequency hopping correlator with both desired and interfering signal inputs (exposure for approximately five chip intervals).

ber that a frequency synthesizer may have extremely low spurious output when at a fixed frequency but high spurious levels as soon as it begins to hop, because of the pulsed signal characteristic when hopping.

Correlator Output With Interference

The output of a spread spectrum receiver's correlator (before filtering) is interesting to observe. When both interfering and desired signals are present, a great deal of insight into system operation is available to the viewer. Figures 5.13, 5.14, and 5.15 show actual correlator outputs under interference conditions for frequency hopping and direct sequence receivers.

The first of these photographs shows a frequency hopping correlator's output for an exposure time covering a few frequency chips.* Inputs to the correlator are a synchronous desired signal and an equal-power CW interfering signal. The recovered carrier, which has been remapped to a single frequency, is at the center; the CW interfering signal (a single frequency input to the correlator) appears to be at several different frequencies—one for each new local reference frequency. An increase in the interference level by adding more jamming signals only causes an increase in the number of signals seen at each chip interval. (Figure 5.13 would also be identical to a correlator with five CW jamming inputs and a desired signal but with exposure for the duration of a single chip. For this reason only one photograph is shown.)

Figure 5.14 shows the output of a direct sequence receiver's correlator with various levels of CW jamming. This series of photographs is more

---

*Hop times. One Hop time = one code chip time = 1/code clock rate.

0 dB interference
(signal power equal to interfering power)

+5 dB interference
(interference 5 dB above signal)

+10 dB interference

+15 dB interference

Figure 5.14 Typical direct sequence correlator output with various interference levels.

interesting in that it shows the increase in the level of the (pseudo-) noise output with respect to a constant-level desired signal. As jamming level is increased, we can see that the signal-to-(pseudo-)-noise ratio out of the correlator decreases. The demodulator which follows must pick the desired signal out of the noise generated by the correlator in the same way that any other demodulator picks its desired input from noise. The only difference is that the noise from a correlator is not truly random but is structured by the particular code sequence that generates it.

At the 15-dB interference-to-signal level, the desired signal apparently has almost disappeared into the noise. Operation is still satisfactory, however,

15 dB

20 dB

25 dB

29 dB

(Long code. CW interference.)

Figure 5.15 Long code, CW interference. Demodulator input signal at various J/S ratios. Unmodulated signal input.

and the demodulator still produces a usable output signal. With equal signal and interference (O-dB) levels, the correlator output appears to be completely free of interference.

This particular system's process gain (the ratio of RF bandwidth to information rate) was over 30 dB. Therefore, when jamming is at $+15$ dB, the output signal-to-noise ratio in the information bandwidth is still near 15 dB.

Figure 5.14 shows correlator output where the carrier is modulated, using the subcarrier-FM technique. Figure 5.15, on the other hand, shows correlator output where the carrier is unmodulated.

Two characteristics of the spectra shown in Figure 5.14 and 5.15 should be noted. One is that the noise shown is that produced by cross-correlation of the CW jamming signal with a long $(2^{31} - 1)$ chip code. The other characteristic is that the desired signal level decreases as the jamming signal becomes large.

The first characteristic is important to us in that the noise produced is approximately "white." The second characteristic is due to the receiver's AGC action. We will refer to both characteristics in our later discussions.

Direct Sequence Reaction to Interference

One of the major properties advertised as a feature of all spread spectrum systems is their ability to reject interference. Differences in signal structures and system implementations, however, lead to variations in reaction to the same kind of interference. Single-frequency CW interference is theoretically most effective in disrupting a direct sequence receiver, for instance, although another type of receiver may well be unaffected.

Let us inspect the action of a direct sequence receiver when operating in a hostile environment. Figure 5.16 is a model we can use to examine this operation. Correlator input in Figure 5.16 is $S + I + n_{sys}$, and in the case of interest to us here $I \gg N_{sys}$. Ignoring the system noise contribution, we see that correlator output as related to its input is

$$\left(\frac{S+I}{I}\right)_{out} = G_p \left(\frac{S+I}{I}\right)_{in},$$

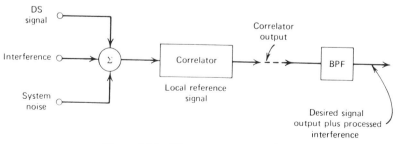

Figure 5.16   Direct sequence model.

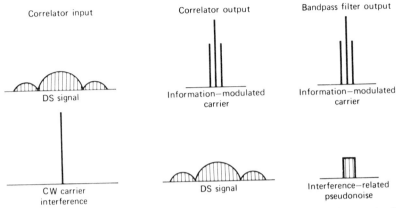

Figure 5.17   Signal transformation due to multiplication with local reference in direct sequence correlator.

except that the two inputs are transformed as in Figure 5.17. The bandpass filter at the correlator output completes the picture by accepting the narrow-banded desired signal and rejecting all of the interference lying outside the filter band. Because the interference is transformed into a wideband signal by multiplication with the local reference, most of its power lies outside the information band. The ratio of the bandwidth of the local reference to the bandwidth of the information filter is then the receiver's true process gain; that is,

$$G_p = \frac{\text{BW}_{\text{RF}}}{R_{\text{info}}} \approx \frac{\text{BW}_{\text{local reference}}}{\text{BW}_{\text{info filter}}},$$

and the filter output due to interference is I/G_p.

Relating the information filter output to correlator input, we then see that

$$\frac{S + I/G_p}{I/G_p} = \frac{SG_p + I}{I},$$

where S = signal power,
$\quad\quad\quad I$ = interference power,
$\quad\quad\quad G_p$ = process gain.

Thus, if process gain is defined, filter output S/N is approximately

$$\frac{SG_p + I}{I},$$

which is the signal-to-noise ratio delivered out of the correlator output filter to the receiver demodulator and synchronization detector. The demodulator and synch detector following the correlation process operate as if they were not included in a spread spectrum system at all. In fact, the signals presented to them are not wideband but consist of the reconstituted, modulated carrier signal and of interference decomposed into pseudonoise. (We continue to ignore the small contribution of system internal noise compared with interference. System noise effects are considered in Chapter 7.) Differences in the effect of various kinds of interference depend on the type of interference and the specific demodulator and synchronization circuits used, plus the cross-correlation of the code used with the interfering signal.

At first glance we might consider that a spread spectrum system would be most affected by spread spectrum interference. Except under very restricted conditions, however, in which the interfering signal has high correlation with the desired signal (implying high code cross-correlation), this is not true. Any signal (including the desired signal when not in exact code synchronism) is spread in bandwidth by multiplication with the local receiver reference. Furthermore, the bandwidth of a signal at the correlator output is equal to its own bandwidth before multiplication plus the bandwidth of the local reference signal. An interfering signal with 2-MHz bandwidth being input to a direct sequence receiver whose expected signal (and local reference) is 10 MHz wide would produce a correlator output signal into its bandpass filter that is 12 MHz wide. The wider the bandwidth of the input interference, the wider the output signal. Thus the wider the input signal, the less is its effect on the system because the power density of the signal after processing is lower, and less power falls in the bandpass filter.

Conversely, then, we might guess that the most effective interference to a direct sequence receiver is that which has the narrowest bandwidth—a CW carrier. In fact, CW interference is the most effective generally, against a direct sequence system. This has been proved in tests of numerous direct sequence systems, because the power density in the correlator output due to a narrowband signal is higher than that due to wider band signals.

How big is the difference in signal output due to wideband and narrowband interference? If we consider that a CW carrier is the narrowest interference to be seen and that the widest interference bandwidth is equal to the local reference bandwidth (a reasonable assumption because a direct sequence receiver is likely to filter all input signals to equal its reference bandwidth), then the correlator's output varies between one and two times the reference bandwidth. Noise in the correlator output filter bandwidth then varies over a range of two (3 dB) for equal power but varying-bandwidth interference. Figure 5.18 illustrates correlator input and output signals and bandpass filter output signals for the cases of CW and wideband interference.

It would be derelict to continue without mentioning the role that the code sequence plays in determining interference effects. We have mentioned code cross-correlation and discuss it in Chapter 3 in some detail. Here, we must

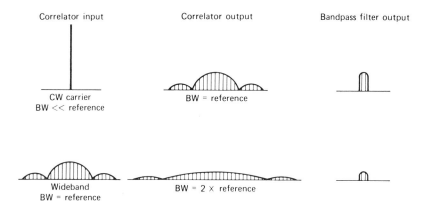

Figure 5.18 Correlator input and output and bandpass filter output for CW and wideband interference.

point out that any significant degree of cross-correlation between a spread spectrum receiver's local reference and an interfering signal can detract from performance. A common point of failure in the design of direct sequence systems lies in using codes that are too short (too few chips between repetitions) for the projected interference level. Worse yet, the shorter codes when multiplied with interference (especially narrowband interference) tend to produce correlations that are not at all noiselike. Therefore, a synchronization detector or demodulator in such a system is likely to give surprisingly poor performance when it has been assumed (as is the usual case) that correlator output products due to interference are characteristically Gaussian.

Noting again that cross-correlation is expressed by

$$\int_0^\infty f(t)\,g(t-r)\,dt = \psi_c,$$

we see that the implementation is a multiplier with two signal inputs and an integrator, which is usually a bandpass or lowpass filter. For $f(t)$, an interfering signal, and $g(t-\tau)$, a coded local reference, ideally the resultant ψ_c should be zero, where the product is pure noise. A demodulator or synch detector would then register no difference between the condition of having interference and having none. Unfortunately this happy circumstance is not even approached unless the codes used are satisfactorily long.

For short sequences code/interference cross-correlations often produce a repetitious correlation envelope like that shown in Figure 5.19. This signal obviously does not integrate to zero and can therefore cause false synch recognition or offsets in signal demodulation.

Figure 5.19   Typical short sequence/CW interference cross-correlation pattern (note repetition).

It is of some interest to note that when observing cross-correlation patterns like those in Figure 5.19, variation of the interference frequency causes a variation in the cross-correlation pattern. With a CW interfering signal, for example, each time the interference frequency is changed by an amount approximately equal to $BW_{IF}/2$ a whole new cross-correlation pattern appears. Because some of these patterns have higher peaks than others, it is normal for a direct sequence receiver to be more sensitive to CW interference at some frequencies than at others. The shorter the code sequence (in general), the more pronounced the frequency-sensitivity property in a direct sequence receiver.

A good rule of thumb for selecting minimum code length in a direct sequence system is to choose the sequence used such that code chip rate/code length $< F_{low}$, where F_{low} is below the lowest frequency of interest in the information being demodulated. (What this really says is that the code repetition rate should not fall in the information passband.) Otherwise, code/interference crossproducts may fall into the demodulated signal band and thereby reduce receiver output S/N. Such crossproducts can also greatly increase the incidence of false synch recognition. For this reason the response of the initial synch detector in a direct sequence receiver is often rolled off at low frequencies—usually by simple AC coupling (a discussion of synch detection will be found in Chapter 6).

Figure 5.20 shows the frequency spectrum of the signal shown in Figure 5.19. This figure features a line spectrum, with lines separated at the code repetition rate. It shows beyond question that the cross-correlation of the short code used here (8191 chips long) with a CW jamming signal is *not* Gaussian.

Figure 5.20 Spectrum of the signal produced by cross-correlation of an 8191-chip linear maximal code with a CW jamming signal.

At any rate, it is simple to say that a direct sequence system has a given amount of process gain. We must exercise much more care when specifying useful jamming margins, because it is necessary to know the type of interference to be seen, the interference frequency, and the amount of cross-correlation between that particular interference and the code sequence being used.

A series of photographs showing spectra internal to a direct sequence receiver is shown in Figures 5.21 through 5.28. These figures show the signal at the input to the demodulator in a typical system. Figure 5.21 is a typical noise spectrum produced by cross-correlation of a jamming signal with the receiver's local (long code) reference. It is noteworthy that there is little difference in this spectrum, whether there is no signal present, the desired signal is present, or a jamming signal is present. The demodulator sees a noiselike input (depending on the code) under all conditions.

When synchronization occurs, the demodulator's input changes from a noiselike signal to a coherent one as illustrated in Figure 5.22. In this figure, the spread spectrum signal has been collapsed to a coherent carrier, which is unmodulated by data.

In the next photographs, Figures 5.23a and 5.23b, the carrier is phase shift keyed at a 2400-bit-per-second rate, with a one-zero-one-zero repeating pattern (a 1200-Hz square wave). Figure 5.22a shows the two-sided spectrum that results, with sidebands at ± 2400 Hz, ± 7200, $\pm 12,000$, etc. (all odd harmonics). With interference, in Figure 5.23b, only the major sidebands are visible above the noise. Figure 5.23a shows this receiver with no jamming and

Figure 5.21   Demodulator input (correlator output, band limited by IF) prior to code synch.

Figure 5.22   Demodulator input (band-limited correlator output) at code synch, 0-dB interference. No data modulation.

Figure 5.23   Correlator output signals. PSK data 8191-chip code: (*a*) no interference; (*b*) 29-dB interference.

5.23*b* shows the same receiver with a CW jamming signal 27 dB above the desired signal at the RF input. In both photographs in Figure 5.23, the code being used is 8191-chip linear maximal. Figures 5.24 and 5.25 show the same system at slightly lower jamming levels.

For comparison, Figures 5.26 and 5.27 show the same system, with similar jamming-to-signal ratios, but with a long $(2^{31} - 1)$ chip code and the 2400 bps modulation removed. Note the striking difference in the level of the noise with respect to the desired signal level in these photographs. Also, note that as in previous photographs, the absolute signal level decreases as the

Figure 5.24   8191-chip code. PSK modulation. 25-dB interference.

Figure 5.25   8191-chip code. PSK modulation. 27-dB interference.

Figure 5.26   Long Code. No modulation. 25-dB interference.

Figure 5.27   Long code. No modulation. 29-dB interference.

Figure 5.28 Correlator output signals showing more detail: (*a*) 2400-bps modulation with 0-dB interference (8191-chip code); (*b*) 2400-bps modulation with 15-dB interference (8191-chip code).

receiver approaches its jamming threshold. This is due to AGC action, and is a desirable characteristic of the design.

More-detailed spectra are shown in Figure 5.28, which is an expansion of the desired, correlated signal spectrum under two conditions with 2400 bps modulation present. In Figure 5.28*a*, with 0-dB jamming, only the modulation-produced sidebands are apparent, although a jammer-produced carrier is just visible at the center of the spectrum. With the CW jamming signal increased to 15 dB above the desired signal, jammer-produced lines are visible between the desired signal sidebands. This illustrates a serious possible design flaw: In this system, the data rate is 2400 bps, while the 8191-chip code repetition is 2500 Hz. Thus the lines produced by cross-correlating a CW jamming signal with the 8191-chip code are almost indistinguishable from the desired signal. The demodulator, under the correct circumstances, cannot tell the difference, and may synchronize to the jamming-produced spectrum.

These photographs show typical signals seen by a direct sequence system's demodulator. They illustrate the possibly difficult task of differentiating between desired and undesired signals that is left to the demodulator under high jamming conditions.

Frequency Hopping Reaction to Interference

Frequency hopping systems are often termed "avoidance" systems from the standpoint of rejecting interference or jamming; direct sequence systems, by comparison, employ "averaging." The direct sequence approach spreads an interfering signal's power over a wide band of frequencies, and this power is thereby said to average over the band of frequencies (even though the power is not evenly distributed).

A frequency hopping receiver also spreads any received interfering signals over a broad band,* but the process is different. While code modulation causes an incoming CW signal to a direct sequence receiver to occupy many frequencies, the frequency hopping receiver simply translates it (a CW signal) to a new frequency, and each time the frequency hopper goes to a different frequency the CW signal is translated differently. From a modern algebraic standpoint the operation of correlation in a direct sequence system is homomorphic,* *whereas in a frequency hopping system it is isomorphic.*‡

We have already shown the direct sequence approach in which the local reference phase modulates incoming signals at a code-determined rate. Now let us examine the frequency hopping approach.

The frequency hopping receiver's local reference signal, which it multiplies with all incoming signals, is a replica (neglecting modulation) of the signal it expects to receive from a desired transmitter; that is, the local reference hops from frequency to frequency on command from a code generator (again identical to that contained in the transmitter). There is one difference—the local reference is offset by an amount equal to the center frequency of the correlation filter following the local/received signal multiplier. When the received signal is f_1, the local receiver must generate $f_1 + f_{IF}$, etc., and the local reference must continue to generate new frequencies at the same

---

* In both direct sequence and frequency hopping receivers an interfering signal is spread over a band equal to the covariance of the interference and the local reference; i.e., $BW_{spread} = BW_{interference} + BW_{reference}$.
† A one-to-many mappings from a set G_1 to a set G_2.
‡ A one-to-one mapping from G_1 to G_2.

rate and frequency (plus IF) as the transmitter to remain synchronized. The difference frequency (IF) is output through the correlation filter.

What, then, is the effect of a single frequency (CW) interfering signal? First, we point out that it is the least effective interference for a frequency hopping receiver—exactly the opposite of its effect on a direct sequence receiver. Next we define a CW interferer as one whose power is contained completely within one frequency hopping channel and need be only ϵ^* greater than that of the desired signal to cause an error. Here ϵ is considered to be the decision threshold of the receiver's bit-decision device; for example, if the desired signal transmitted is in a channel corresponding to a One, the decision device could be forced to make an error (decide that a Zero was sent) by an interfering signal with $S + \epsilon$ power in the Zero channel. For a single interfering signal, however, the frequency hopping receiver could be expected to average one error in every n transmissions (n is the number of channels available).

If only one frequency (or chip) is sent per bit, a single CW interferer would cause an error rate of $1/n$. For this reason frequency hopping systems are forced to use some form of redundancy such as sending multiple chips per bit of information. The particular frequency or power (once the $S + \epsilon$ criteria is exceeded) level of a single CW interferer is of no consequence to a properly designed frequency hopper, for the probability of using one channel is the same as for any other channel.

To be effective against a frequency hopping receiver, a would-be jammer must meet at least one of the following criteria:

1. Place $S + \epsilon$ power in a large number of the possible channels.
2. Be able to predict the next frequency the hopper will use (involves breaking the code and the command pattern).
3. Be able to receive the hopped signal, determine its frequency, and retransmit a signal calculated to change the effect of the desired signal at the receiver (a "look-through" jammer). All this must be done in significantly less than the hop time to be effective.

Here we are speaking of binary frequency hopping systems in which the choice of a One or Zero frequency is made on a chip-by-chip basis; that is, each chip transmitted has only two possible frequencies for transmission: f_1 to signify a One and f_2 to signify a Zero.

Effective methods are available for a frequency hopping system to combat both the second and third jamming approaches: the codes used can be made so difficult to break that it is not worth the effort, and by hopping at high rates and/or avoiding disadvantageous geographical situations the look-through threat can also be circumvented.

---

*This is the case for unmodulated frequency hopping carriers. Where the frequency hopped carrier is modulated, interference is effective at much lower levels.

It is also possible to negate the look-through jamming threat by employing simple, noncoherent pulsed transmission techniques. Let us suppose that we employ a frequency hopped carrier that is turned on to send a one, and off when a zero is intended. (In other words, we employ on-off keyed frequency hopping.) At the receiver we make the bit decision as a function of whether a signal is present at the frequency of interest. Therefore, if we intend to send a One, we transmit a carrier at f_{one}, and if we intend a Zero, we do not transmit. The look-through (or frequency-following) jammer then can:

1. When we send a One, transmit at the frequency we are using, which only increases the probability of our making the correct decision, since any energy at f_{one} signifies a One.
2. When we "send" a Zero, transmit randomly, as he has no transmission to follow.

This means that the look-through frequency-following jammer is no more effective against such a system than the simple jammer who spreads his or her power over as many frequencies as possible. In the case of a One being intended, the probability of a jammer-related error is the probability that the jamming signal will be equal to and opposite in phase with the desired carrier (to cancel it completely). Where a Zero is intended, the desired transmitter does not transmit, so the jammer cannot follow, and his probability of causing an error is just

$$\frac{J}{n} = \frac{\text{the number of frequencies jammed}}{\text{the number of frequencies available}}.$$

The most effective general threat to any frequency hopping system, then, is that of placing $S + \epsilon$ power in a large number of the possible channels.

So far, the effect of varying the relative phase of an interfering signal with respect to the phase of the desired signal in a channel has been ignored. Unfortunately, this does not cause it to go away. We have concerned ourselves with jammers that transmit $S + \epsilon$ in J/n percent of the channels available to a frequency hopping receiver (J = number of channels jammed; n = number of channels available) and said somewhat naively that the probability of chip error (remember that a chip is a single-frequency transmission signifying a one or zero) is simply J/n. This is not quite true because the probability of interference hitting an intended channel is just as good as that of hitting a complementary channel. Thus interference can cause errors in several ways:

1. By hitting the complementary channel.
2. By hitting both the complementary channel and the intended channel.
3. By hitting the intended channel with a signal that cancels the intended signal.

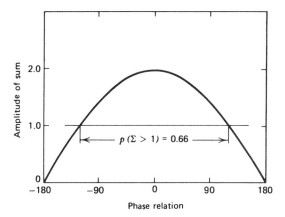

Figure 5.29 Plot of $A \sin \omega\tau + A \sin(\omega\tau + \phi)$ showing pdf for phase addition of random sine waves.

We have already considered the effect of interference hitting the complementary channel only. The $p(e)$ for this is just J/n.

The probability of hitting both complementary and intended channels is $(J/n)^2$, and the probability that the intended signal and a jammer hitting in the intended channel will add in phase is 0.66* (see Figure 5.15). Therefore, the probability that an error will occur when both complementary and intended channels are hit is $(1 - 0.66)(J/n)^2 = 0.34(J/n)^2$.

Finally, we must consider the possibility that interference will hit in the intended channel without hitting the complementary channel. Again, the probability of hitting the intended channel is J/n. The probability that the interference is in phase with the intended signal so that they add is shown in Figure 5.29. This figure is the probability distribution function (pdf) for all phase positions of two signals $A \sin \omega\tau$ and $A \sin(\omega\tau + \phi)$ and is such that $p(\Sigma > 1) = 0.66$.

Considering that the effect of phase addition must be to give a sum not greater than ϵ more than zero, we see that the probability of an interfering signal causing complete loss of an intended signal due to phase cancellation is

$$1 - p[A \sin \omega\tau + A \sin(\omega\tau + \phi)] > \epsilon.$$

For small epsilon,[†] which is the desired case, this probability should be negligible when compared with the other two sources of error.

To give an example, let us assume that we have a frequency hopping receiver and an interfering signal that is effective in 10% of the receiver's

---

* Optimistically, from the jammer's viewpoint.
† Remember that ϵ is the bit-decision threshold in the receiver and that the designer's goal is (or should be) to make ϵ extremely small to avoid bias errors.

channels; that is, J/N is 0.1. Total chip errors caused by the interference would then be

$$p(e) < \frac{J}{n} + 0.34\left(\frac{J}{n}\right)^2 + f(\epsilon)\left(\frac{J}{n}\right)$$

$$= 0.1 + 0.0051 + 0.001 \; [\text{giving } f(\epsilon) \text{ a value of } 0.01]$$

$$= 0.1061.$$

(With nothing else considered, this in itself is a horrible error rate. A five-chip-per-bit redundancy, however, with simple majority coding, reduces this error rate to 0.008, which is much more reasonable and could be usable.) At any rate, it has been shown that the contribution of phase effects is small but must be considered when designing low-error-rate frequency hopping systems for data transmission under jamming.

We must also reiterate here that frequency hopping systems should not be designed without some form of redundancy and/or error correction coding. We also point out that while such error rates may be acceptable in a do-or-die military system, they would not be tolerated by a consumer.

5.3   BASEBAND RECOVERY

The baseband recovery process in a spread spectrum receiver is identical to the baseband recovery process in any other receiver employing the same type of baseband modulation.* Frequency-modulation, frequency-shift-keying, phase-shift-keying, and even amplitude-modulation forms are quite compatible with direct sequence or frequency hopping (admittedly some combinations display more compatibility than others). Whenever they are employed, the conventional modulation forms work well with their usual types of demodulators, but space has not been taken here to give such demodulators more than honorable mention.

In the remainder of this chapter we discuss some techniques that have proved to be well adapted to demodulating the recovered carrier in a spread spectrum receiver. (These same techniques are just as useful in nonspread spectrum applications.) In general, all types of threshold extension demodulators are useful: phase-lock loops, FM-feedback demodulators, Costas demodulators, others are best suited to the task at hand; operating well at the lowest possible signal-to-noise ratio.

Frequency hopping systems, unlike direct sequence systems, are difficult to implement with coherent transmit/receiver functions. Coherence to a fre-

---

*Remember that we said that spread spectrum systems are transparent—that removal of the spectrum spreading modulation restores whatever signal existed before it was overlaid with a coded spread spectrum signal.

quency hopper means that every new frequency generated must be not only precisely defined in frequency but must also bear an exact phase relation to the preceding frequency. Because coherence in the signal itself is difficult to achieve or maintain, few frequency hopping systems employ coherent detectors such as phase-lock loops. The reason is that a phase-lock loop would be forced to reacquire lock each time the signal changes frequency. This could take a significant amount of the overall chip time, especially when fast frequency hopping or high process gain is a goal.

Frequency hopping demodulators often employ simple envelope detection in conjunction with integrate-and-dump matched filtering to make the most of the chip timing available.

Baseband recovery in a direct sequence receiver is inherently a coherent process in that at least the receiver's local code reference must be an accurate estimate of the received code modulation. Also, because of the superior threshold performance of coherent detectors over alternative types, they are almost always used. Phase-lock loops, feedback-FM loops, Costas demodulators, and their variations are often found in direct sequence receivers. Another type of demodulator that takes advantage of the fact that squaring any double-sidebanded signal produces a carrier (see Appendix 4) at twice the signal frequency may also be used. It is called a squaring loop[1220] and is a phase-lock loop used to track the twice-frequency carrier; the loop VCO is divided by two and used as a coherent reference for demodulating the signal. Performance of the squaring loop has been shown to be equivalent to that of the Costas loop, although the phase ambiguity introduced by having a twice-frequency (though divided down) coherent reference must be resolved.

Phase-Lock Loops

A phase-lock loop is a complex device to which literally thousands of pages of scientific material[1201–1223] have been dedicated, and deservedly so. Yet its function can be described in terms so deceptively simple that we tend to lack appreciation of its power. A word of caution: do not let the simple terms of the simplified design process lull you. Optimal design often requires intimate knowledge of a phase-lock loop's internal workings, including third-order reactions to step-function frequency changes. Because the simple phase-lock loop contains at least one nonlinearity, an exact prediction of its reactions can be a rather difficult proposition. Fortunately phase-lock loops have been submitted to thorough examination by both analytical and empirical media, as an examination of the references will reveal.

The integrated-circuit art, for example, has reached a point such that a complete (but simple) phase-lock loop is available in a single small package to handle many applications.

A phase-lock loop is a feedback loop that contains an oscillator, a phase detector, and a filter, connected in such a way that the oscillator "locks on"

Figure 5.30 Simple phase-lock loop.

to an input signal. (Figure 5.30 shows a basic phase-lock loop for reference.) Some of its outstanding characteristics are:

1. A phase-lock loop tracks the frequency of an input signal and not its phase. This makes it a unique frequency-modulation detector or, with some extension, a coherent AM detector. Demodulated-FM information is taken from the line that controls the loop's (voltage-controlled) tracking oscillator, keeping it at the same frequency as the input signal.

2. The loop oscillator tends to track in phase quadrature* with the input signal to which it is locked. This phase angle is not maintained, however, as the input frequency departs from the rest frequency of the loop oscillator. At any given input frequency the phase error between the loop oscillator and input carrier is approximately

$$\sin \phi = \frac{2\pi |f_0 - f_i|}{k},$$

where f_0 = loop oscillator rest frequency (open-loop frequency),
         f_i = input carrier frequency,
         k = open-loop gain for the phase-lock loop.

3. The loop acts as a tracking filter such that its demodulated output has a loop-filter-determined bandwidth over the range in which the oscillator can be caused to track the input signal. A phase-lock loop is the equivalent of a tunable bandpass filter which centers itself at the frequency of a desired signal, accepting only the information (and noise) contained in a two-sided (mirror image) replica of its low-pass response.

4. Best of all phase-lock loop is a low-threshold demodulator. Because it offers a coherent reference carrier (the loop VCO input to the phase detector), a phase-lock demodulator can operate at a signal-to-noise ratio approaching 0 dB.

---

*This is true of analog phase detectors and some digital phase detectors. Some digital phase detectors (used primarily in frequency synthesizer) track with zero or 180 degree phase shift.

Why does a phase-lock demodulator offer so much in a spread spectrum receiver? One reason is its low threshold. A second reason is that the loop VCO, in tracking the incoming signal, generates a local reference carrier that can be used in the receiver for code clock tracking, carrier tracking, Doppler compensation, or any other receiver-frequency optimization requirement. Third, the tracking-filter characteristic allows a single basebandwidth filter to pass the baseband modulation, even though the carrier signal frequency varies over a wide range.

From the black box point of view, a phase-lock loop does exactly the same thing as any other FM demodulator. Its input signal is a frequency-modulated carrier and its output is the carrier-modulating waveform. The magnitude of this output signal depends on two things—the amount of carrier-FM deviation and the loop VCO gain K_2; for example, a phase-lock loop containing a VCO whose sensitivity, or gain, is 10 kHz/V would output a 1-V (peak-to-peak) signal when its input signal deviates ± 5 kHz.

Output frequency response is determined by the loop filter, which limits the VCO tracking rate. The proper loop filter is a lag-RC type, and typical loop frequency response due to such a filter is shown in Figure 5.31. Often it is advantageous to modify the filter by adding a second capacitor in parallel with C and R_2 (see Figure 5.32) to give a third breakpoint, beyond which the loop response again rolls off at a 6-dB/octave rate. This reduces the high-frequency response of the loop without significantly affecting the loop operation in the bandpass region. (The high-frequency rolloff capacitor C_2 is usually chosen to be no greater than one-fifth to one-tenth of the main filter capacitor C_1.) Output signal-to-noise ratio can be significantly improved by this simple addition, which also decreases loop jitter with noisy input signals.

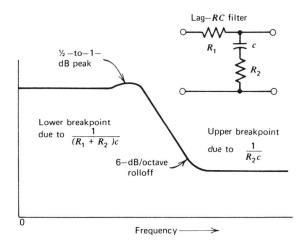

Figure 5.31   Typical low-pass frequency response for phase-lock loop with lag-RC loop filter.

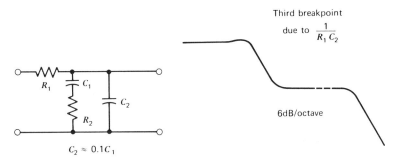

Figure 5.32   Modified loop filter for additional high-frequency rolloff.

A phase-lock loop can easily be augmented to act as a coherent amplitude demodulator, and this makes it possible to generate coherent AGC or to generate an "in-lock" or "signal present" output. Figure 5.33 is a block diagram of a phase-lock loop, which is augmented to include both AM and FM demodulation. The only addition to the simple loop that is necessary is a phase shifter, a second phase detector, and a low-filter, with the phase-lock portion of the overall demodulator itself unmodified.

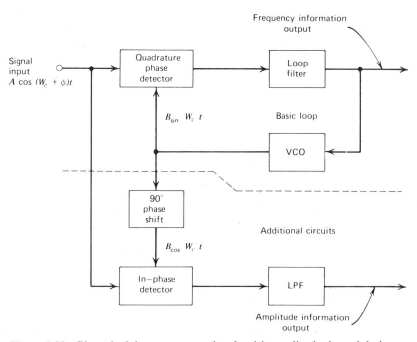

Figure 5.33   Phase-lock loop augmented to furnish amplitude demodulation.

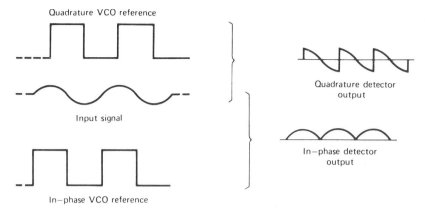

Figure 5.34 Phase-detector output waveforms.

The phase-lock loop of an augmented loop demodulator such as that in Figure 5.33 continues to operate in its usual manner so that an input signal, when locked to the loop VCO, is in phase quadrature with the VCO. In the additional circuitry a 90° phase shifter supplies a VCO signal to a second phase detector, which is in phase with the incoming signal. The result of multiplication (in the second phase detector) of the two in-phase signals is a signal that, when filtered, is the amplitude information we desire. The signal waveforms of Figure 5.34 illustrate why this is true. The output signal from the quadrature detector, after integration, is zero when the signal is symmetrical around the zero axis; that is,

$$A \int_{\pi/2}^{0} \sin \omega_c t \, dt + A \int_{0}^{-\pi/2} \sin \omega_c t \, dt = 0,$$

if the phase-lock loop is locked and the VCO rest frequency is the same as the input signal frequency.

The in-phase detector output, on the other hand, is similar to the output of a rectifier and when integrated gives a value

$$\frac{2}{T} A \int_{0}^{T/2} \sin \omega_c t \, dt = 0.63 A.$$

In one case, then, we have an output that varies as the phase of the input signal varies. In the other case the output varies as A, the signal amplitude varies.

Phase-lock loops are used primarily in direct sequence systems, for unless a frequency hopper is highly coherent a phase-lock loop must reacquire each time the frequency changes.

Coherent frequency hopping systems have not been demonstrated success-fully at this time. It is possible to construct hardware that provides coherent frequency hopping signals, and transmitters and receivers can be built that are coherent when operating side by side. When these same systems are separated, however, and the signal must pass through the atmosphere, it is no longer coherent when it reaches the receiver. This is due to differences in delay from one frequency to another. (The atmosphere is frequency disper-sive.)

Squaring Loops

We have not previously discussed squaring-loop demodulators here, for they differ primarily from phase-lock loop in that a square-law device is placed ahead of the phase-lock loop that tracks the twice-frequency carrier. The tracked carrier is then divided by two, phase shifted by 90°, and multiplied with the input signal to demodulate it. Squaring loops, of course, are used for demodulating double-sideband suppressed-carrier signals because the phase-lock loop could be used directly at the carrier frequency without the necessity of squaring if the carrier were not suppressed.

Performance of a squaring loop is a function of the loop bandwidth, the squaring device's performance, and any presquaring filter bandwidth. A squaring loop has performance exactly equivalent to that of a Costas loop.

Figure 5.35 is a block diagram of a squaring loop demodulator. The input biphase signal (PSK) is squared,* which produces a twice-frequency unmodu-lated carrier. A phase-lock loop is then locked at twice the input frequency to the squared signal. After dividing the VCO output by two, a carrier is produced that may be multiplied with the input PSK modulated signal to produce the desired data. (Multipliers shown can be double-balanced mixers.) Coherent-AM demodulation for AGC, signal detection, and other purposes is also easily accomplished with a squaring loop. All that is necessary is to phase shift the VCO signal by 90° and multiply it by the phase-lock-loop input signal, as in the simple phase-lock-loop approach.

Costas Loop Demodulators

A special type of compound phase-lock loop, called an *I-Q* or Costas[109] loop (after J. P. Costas), is often used for demodulating double-sideband sup-pressed-carrier signals, and this, of course, is exactly the type of signal produced by biphase or quadriphase phase-shift keying. (See phase-lock references, Appendix 3). Also, if code-inversion modulation is used, Costas demodulation is a good choice. Like the other phase-lock loops, Costas loops have their primary application in direct sequence reception.

---

*In practice, a frequency doubler can be used instead of the squaring device. These are readily available from a number of sources.

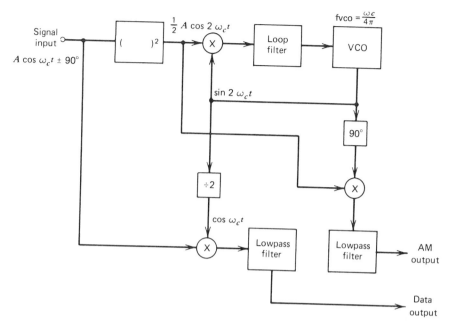

Figure 5.35 Squaring loop demodulator.

The basic configuration of a Costas loop is shown in Figure 5.36. At first glance it appears to be the same as the augmented phase-lock loop in Figure 5.33, and it is the same in some ways. Again a VCO is used as the carrier reference, which is multiplied in-phase and in quadrature with the input signal, and the multiplier outputs pass through low-pass filters. The difference is in the addition of a third multiplier to which the filter outputs are applied and whose output is used to control the loop VCO.

How does it work? An input signal $\pm A \cos(\omega\tau + \phi)$ (biphase modulated) is applied to the I and Q multipliers, where it is multiplied by signals $B \cos \omega\tau$ and by $B \sin \omega\tau$, respectively. Outputs from the multipliers are then the following:

For the I multiplier,

$$\pm \frac{A}{2} [\cos \phi + \cos(2\omega\tau + \phi)],$$

For the Q multiplier,

$$\pm \frac{A}{2} [\sin \phi + \sin(2\omega\tau + \phi)],$$

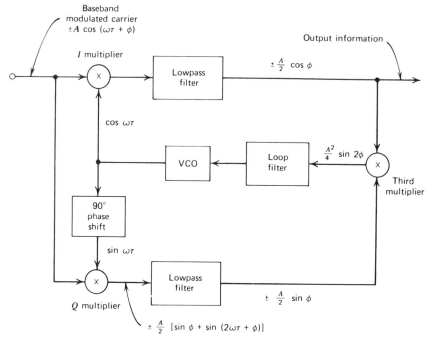

Figure 5.36 Costas loop demodulator for double-sideband suppressed-carrier signals.

which, when passed through the I and Q low-pass filters, become $\pm (A/2)\cos \phi$ and $\pm (A/2)\sin \phi$.

These two signals, which contain both the phase-shift-keyed information and carrier phase, are then multiplied together to obtain $(A^2/4)\sin 2\phi$. After filtering, this signal is used to correct the loop VCO and cause it to track the incoming (virtual) carrier. (Of course, we remember that there actually is no incoming carrier signal. That is the purpose of a Costa demodulator—to demodulate double-sideband suppressed-carrier signals.)

Information is available from two points, depending on the modulation method used. Phase-shift-keyed information that results from the data transmission is available at the output of the I multiplier output filter. Frequency-modulation information is available as in the simple phase-lock loop—at the loop filter output. The Costas loop has no amplitude output and cannot in this form furnish either "signal present," "in-lock," or coherent AGC signals. Its main advantage over a simple phase-lock loop is its ability to demodulate phase-shift-keyed* and/or suppressed-carrier signals.

---

*Phase-lock loops cannot demodulate PSK information. The abrupt phase shift causes only a transient change in a phase-lock loop's VCO control voltage, after which it settles to its previous level. Furthermore the transient change may be positive or negative, depending on whether the loop VCO is advanced or retarded to get back into proper phase lock with the incoming signal.

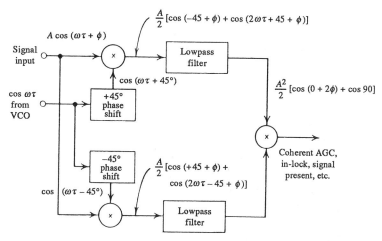

Figure 5.37 Multiplier configuration for coherent amplitude output (added to Costas demodulator).

The I filter output $\pm(A/2)\cos\phi$ is just $\pm A/2$ for small ϕ and $\pm A/2$ is the desired binary data. $\cos\phi$ is approximately one when the phase approaches $0°$. It should be noted that the loop does not know, and has no way of knowing, which is one and which is zero. Therefore, we must use a form of data transmission that in itself is unambiguous. Such techniques as differential phase-shift keying or transmission of polarity-determining bits are well enough known not to be discussed further here.

When a signal-presence, in-lock, or coherent AGC signal is required, the Costas loop must be augmented. Three more multipliers, connected as in Figure 5.37, may be used. The signal input applied to the Costas loop, $\pm A\cos(\omega\tau + \phi)$, is also applied to the multipliers. Their other inputs are $+45°$ and $-45°$ shifts of the VCO.

The signals that result are shown in Figure 5.33. After filtering, which removes the twice-frequency components ($\cos 2\omega\tau + \cdots$), a third multiplier multiplies the resultants and gives an output

$$\frac{A^2}{4}\left[\cos(0° + 2\phi) + \cos 90°\right] = \frac{A^2}{4}\cos 2\phi.$$

For ϕ small, the in-lock condition, $\cos 2\phi$, is 1 and the output is $A^2/4$. If the Costas loop is not locked to the incoming signal, however, $\cos 2\phi$ rapidly approaches zero. Therefore, the configuration in Figure 5.37 is a true coherent amplitude demodulator and can be employed to give coherent AGC, etc.

Filters used in the amplitude detector would usually be the same as those employed in the Costas demodulator itself, although they would not necessarily be the same. The low-pass filters used in the arms of the Costas demodulator would be designed to pass the information band, with cutoff at the highest desired frequency or at some other frequency chosen to maximize the information signal-to-noise ratio while offering minimum effect to the desired signal. The loop filter used in the Costas loop is the same type used in phase-lock loops but may be quite narrow because it need pass only carrier frequency changes, and unless information is sent on the carrier only Doppler shifts need to be tracked.

Of course, the exact filter bandwidths used depend on the use of the Costas detector or the coherent amplitude detector. Gain constants have the same effect on a Costas loop as on a simple phase-lock loop. It should be noted, however, any imbalance in the I and Q paths can cause offset in the third multiplier output and possible tracking problems as a consequence.

Other approaches to providing a coherent-AM output from the Costas loop involve squaring the inputs to the loop's third multiplier, which gives I^2 and Q^2 signals (that is, the in phase and quadrature multiplier output signals, after low-pass filtering). These two signals are then added or subtracted to give either $I^2 - Q^2$ or $I^2 + Q^2$ signals, which can be used as representations of the carrier amplitude.

Both Costas and squaring-loop demodulators have degraded performance with respect to the simple phase-lock loop. That is, the phase-lock loop's threshold is 3 dB better than either the Costas or squaring loop.

Frequency Modulation Feedback Demodulators

Frequency modulation feedback (FMFB) demodulators are akin to phase-lock-loop demodulators in that a controlled oscillator is caused to vary as a function of the input signal frequency, and the control input to the oscillator is taken as the demodulated signal. A typical demodulator is illustrated in Figure 5.38. The components making up this FMFB loop are nearly the same as might be expected in a common FM receiver, especially one containing an AFC (automatic frequency control) loop. The demodulated output signal in a

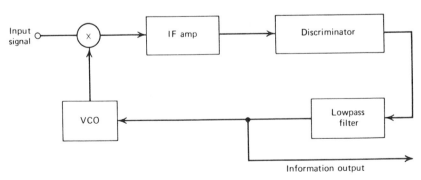

Figure 5.38 FM feedback-demodulator block diagram.

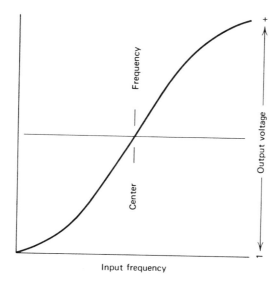

Figure 5.39   Typical discriminator output diagram.

conventional receiver would come from the discriminator, however, and not from the VCO control line as it does in an FMFB loop.

The loop filter in an FMFB demodulator is different from an AFC loop filter in that the AFC loop is designed to track only carrier drift; the discriminator outputs the desired signal. For the FMFB loop the filter is designed to pass the information modulation, which causes the VCO to be varied at the information rate. The VCO signal, mixing with the incoming signal, then tends to track the incoming signal and hold the frequency input to the discriminator at the discriminator's center frequency. Any change in the input frequency causes the discriminator to put out a positive or negative DC voltage (depending on the direction of frequency offset) which in turn corrects the loop VCO to zero the discriminator output. (A typical discriminator frequency-to-voltage curve is shown in Figure 5.39.) We will not discuss the design of discriminators for FMFB loops here because excellent, highly stable crystal discriminators that can perform far better than any conventionally constructed unit for this type of system are available.

FMFB loops are not used in spread spectrum systems as often as phase-lock loops, though their performance from the standpoint of threshold is nearly the same for deviation ratios greater than 3, as shown by J. T. Frankle.[1013] The phase-lock loop does offer better performance for low-deviation ratios, but another reason exists for the prominence of phase-lock usage; that is, a phase-lock loop's internal oscillator is locked in frequency to its input frequency, with only a phase shift between the two which is never greater than $\pm 90°$. The FMFB loop's oscillator does exhibit frequency error, which is a function of loop gain and frequency offset from the discriminator center frequency.

Since spread spectrum systems (and especially direct sequence systems) require frequency tracking as close as possible, phase-lock detection often is more advantageous. Even so, FMFB demodulation is useful and can often be combined with phase-lock methods to the advantage of both.

PDM Demodulation

A demodulator for the SCPDM (suppressed clock pulse-duration modulation) technique described in Chapter 4 is shown in Figure 5.40. We note here that in spread spectrum systems in which PDM modulation is employed the PDM demodulator commonly follows another demodulator such as a Costas loop because the information is not in PDM form as transmitted and must be converted to it.

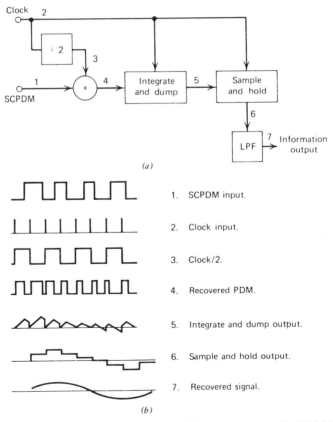

Figure 5.40 SCPDM demodulation process: (*a*) suppressed carrier PDM demodulator; (*b*) waveforms in SCPDM form.

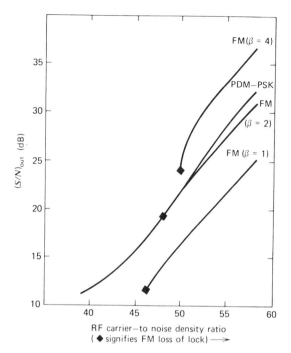

Figure 5.41 Performance of PDM demodulator compared with FM.

The SCPDM demodulator shown[241] has exhibited superior threshold characteristics to FM, and curves demonstrating this performance are given in Figure 5.41.

Frequency Hopping Demodulation

The demodulators discussed so far have been unsuited to frequency hopping systems. Few (if any) frequency hopping links are able to support coherence,* so that the signal into the demodulator changes phase each time the system hops to a new frequency. In addition, the signal is likely to be pulsed due to sampling mark and space frequencies. Therefore, phase-lock and FMFB loops are not usually employed in frequency hopping demodulation. Instead, simple envelope detection, which does not care about input phase shifts and can respond rapidly to pulsed signals, is most often employed.

---

*Here and throughout this book the term "coherent" means carrier coherence when referring to frequency hopping systems. Frequency hoppers, like their direct sequence counterparts, must maintain code and code clock coherence (synchronization) in order to communicate. Carrier coherence is seldom found in frequency hopping systems because of the difficulty (note that we did not say impossibility) of maintaining it in an RF link. Direct sequence systems, on the other hand, are almost always code, code clock, and carrier coherent.

The most common method of transmitting information in a frequency hopping system is to send a frequency f_1 if the datum is of One, and a frequency f_2 if the datum is a Zero. Usually some transmission redundancy is employed so that one code-determined set of frequencies is used to convey a One, and another set to convey a Zero. If more than one frequency is employed for redundancy, the transmitter has the choice of transmitting all of them at once or one at a time. This choice has a significant impact on the receiver and demodulator.

Before proceeding, let us digress for a moment to list the demodulators most suitable for spread spectrum reception, together with the modulation format for which they are usually employed.

One or more of the demodulators we have discussed in this chapter are suitable for demodulating the forms of modulation typically used in spread spectrum systems. Matching modulation types with demodulators, we have the following:

Modulation Type	Detector
Code Inversion	Costas or squaring-loop detector
Subcarrier FM/FSK	Phase-lock detectors
Carrier FM/FSK	Phase-lock detector
Clock FM/FSK	Costas or squaring-loop detector
PDM/PSK	Costas or squaring plus PDM demodulator
Frequency hopping (coherent)	Phase-lock detector
Frequency hopping (noncoherent)	Envelope and integrate-and-dump detectors

Of course, we can demodulate clock or carrier FM signals with a discriminator or double-sideband suppressed carrier by doubling and multiplying the result with the input signal, but for spread spectrum use these noncoherent methods offer little because of their threshold characteristics. Only the noncoherent frequency hopping systems can benefit from the use of demodulation of other than the phase-lock variety.

Figure 5.42 shows a typical frequency hopping demodulator in a noncoherent receiver. The block diagram is for a multiple-chip-per-bit receiver, in which a decision is made on the basis of each sequential set of chips. This demodulator is designed for compatibility with sequential sampling of One and Zero frequency channels; that is, the frequency synthesizer inserts into the receiver a frequency corresponding to a one being transmitted, and follows it with the frequency corresponding to a zero transmission. Each of these frequency alternates is sampled for one-half the chip period. (We pay for the reduction in receiver complexity by doubling the effective bandwidth of the postcorrelation signal.)

In the demodulator in Figure 5.42 an input signal, a series of pulses, is passed through an envelope detector to an integrate-and-dump filter, which

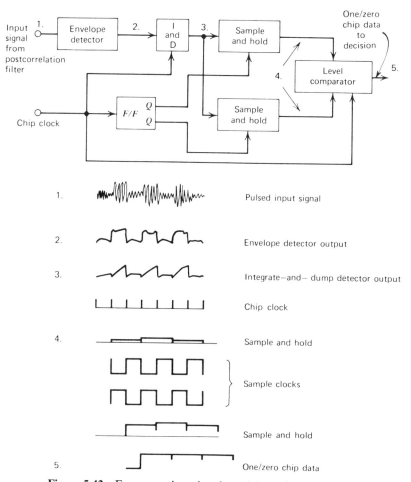

Figure 5.42 Frequency hopping demodulator (noncoherent).

is dumped at the chip rate. Alternate chips are sampled and compared, and the largest of each pair is assumed to be the intended signal. Therefore, if a One is intended, then one-half a chip (say the first) would contain a pulse of RF, the second half would not. Comparison of the signal level in the two half chips would then allow the One/Zero decision to be made for that chip.

We might wonder why this seemingly overcomplex chip decision process should be used when a level detector such as a Schmitt trigger placed at the envelope detector output could give the same information. The difference is that when interference is added to the input signal, random hits by the jammer in complementary channels cause both half chips to contain an RF burst (again on a random basis). An integrate-and-dump operation, with subsequent comparison of the signal level in the two half chips, allows us to decide which channel contains the largest signal rather than basing the decision on whether a signal exceeding threshold is there (which the simple level detector would be forced to do).

The output signal to the data decision circuit consists of n-tuples of chip data in which the number of chip-related bits is the same as the number of chips sent for each bit. For a two of three majority decision system each n-tuple consists of three chip-related bits. The decision circuit then determines on the basis of these bits whether a One or a Zero was intended by the transmitter.

When an interfering signal is received with the desired signal, occasional hits in the unintended channel will occur (i.e., the interference will appear in the One channel when a Zero is being sent and vice versa). When this occurs (and the probability is precisely the same as the ratio of the number of interfering carriers to the number of frequency hopping channels available), the chip comparator must decide whether the real signal or the interferer is that sent by the transmitter. This decision is based on a simple comparison of the power in the two alternate channels sampled by the receiver; that is, if the interfering signal has more power in its channel than the desired signal has in the intended channel, the wrong decision will be made.

The threshold of the chip decision is just the decision threshold ϵ of the level comparator (Figure 5.42). Whenever either signal power or interfering power exceeds the other by ϵ, the chip decision will be made in favor of that

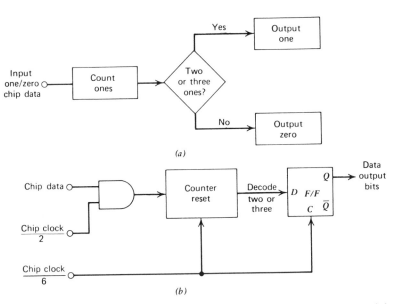

Figure 5.43 One/Zero chip-to-data decision for two of three decision; (a) flow diagram for a two of three majority decision based on three chips per bit; (b) block diagram for a two of three majority decision.

channel. *For this reason a jammer does not increase his or her effectiveness by increasing his or her power at any one frequency once it exceeds the signal by ϵ.*

If an interfering signal does cause a chip error, the bit decision input gets as a signal a single Zero in place of a One, or a single One in place of a Zero. When the bit decision is based on two of three, as in our Figure 5.42, this single chip error will not cause a bit error. If two or more chip errors occur in the same set of chips, however, a bit error will follow. The chip clock/6 input to the two of three decision circuit in Figure 5.43 is used for output data because the actual data rate in a frequency hopping system, with respect to the receiver, is slower than the receiver's chip rate (which is twice the transmitter's chip rate because it samples both One and Zero channels) by a factor of two for chip sampling and a factor of three because there are three chips per bit.

Neither the three-chip-per-bit format nor majority decision as advanced here is the optimum. They are given only as representative of typical techniques.

Integrate-and-Dump Filters

An important type of filter, called "integrate and dump" (I and D), is shown in Figure 5.42 as part of the frequency hopping demodulator. It is matched to the period of a bit at the chip rate. Figure 5.44 shows two implementations of this filter—one for operation at RF or IF and the other for baseband (as in

RF/IF I and D filter using high Q parallel—tuned circuit for integration.

Baseband I and D filter using
RC section for integration

Figure 5.44   I & D filters.

our frequency hopping illustration). The two are equivalent from the standpoint of their function, but here we emphasize the operation of the baseband
filter.

Two conditions must be satisfied for any filter to be "matched":

1. Its impulse response must be a time-reversed replica of the signal to
   which it is matched.
2. Any particular unit of information must be independent of that preceding and succeeding it.

That integrate-and-dump filters meet these conditions will be evident as our
description progresses.

The signal input to the baseband integrate-and-dump filter, which would
be most likely to be used in a frequency hopping system, is just the
envelope-detector output. Thus the input signal is a series of what is
intended to be rectangular pulses but in the real world is distorted by noise
and only approach rectangularity. At any rate, the series of pulses is applied
to an RC integrator whose period is long compared with the pulse period
$(T_{RC} = R \times C > T_{\text{signal}})$; that is, the integration performed is restricted to the
more linear part of the integrator's charge characteristic, where V_c closely
approaches $V_t/RC = I_{pk}t/C$.

The integrator is allowed to charge for the period of an input pulse and is
then sequentially sampled and dumped (discharged). If the signal existing at
the input is a square wave, the integrator during Δt will charge to $V\Delta t/RC$.
For any other signal shape, however (assuming that the peak voltage is the
same), the charge at the end of Δt is lower and a decision threshold can be
set on the basis of a desired signal being present in a given signal space. For
the integrate-and-dump detector the process of discharging the signal stored
is a matter of shorting across the integrator, preparatory to the start of
accumulating energy during the following chip period. Figure 5.44 shows
shorting and sampling switches connected to the integrator. (Of course, in
practice, electronic switches are used.) To accomplish the purpose of signal
detection the switches shown are closed (1) to sample the integrated signal
and (2) to dump the integrator's charge.

Independence of signals is ensured by the dumping process, which isolates
the energy in one chip from that in the others; it accomplishes one of the
requirements of a matched filter. The second requirement—that its impulse
response be a time-reversed replica of the desired signal—is also met.

Consider the RC integrator, with both its switches open. An impulse at its
input charges the capacitor to a level $(l_{pk}/dt)/C$, which remains until
dumped. This is exactly the same as the desired flat-topped square pulse. For
the integrator that uses a high-Q tuned circuit, the input impulse causes the
circuit to ring at the tuned frequency—again forming the duplicate of a
desired input signal. In both types of integrator, time inversion has no

meaning, for square wave and sine wave pulses are the same when their leading and trailing edges are reversed.

Both types of integrate-and-dump detector, then, are matched filters, and the advantage of a matched filter is demonstrated for signal detection. When the baseband integrator charges to its highest level for a desired input and to a lower level for any other signal, the detection process is optimized.

The Dehopped Signal

A frequency hopping receiver must perform the same operations as its companion transmitter, except that it does them in reverse order. First, the spread spectrum modulation is removed by a chip-by-chip remapping of each received frequency into a narrowband filter and demodulator. The "de-hopped" signal must then be processed to demodulate the information sent.

Frequency hopping receiver performance depends on how well the band-pass filter following the dehopping multiplier is able to select the desired signal from among other signals (and noise). In many frequency hopping receives frequency shifts caused by instability, Doppler shift, or even nonopti-mum signaling may force the filter bandwidth to be wider than the minimum signaling bandwidth, which in turn results in a loss in performance.

In a binary* frequency hopping transmitter, data transmission (in this section only digital information transmission is considered) is performed by sending one frequency to signify a space or an alternate frequency for a mark. Whether only one or multiple chips are sent for each bit of informa-tion, one of two frequencies is sent for each chip. The receiver must then be able to observe both frequencies and decide which was intended. (It is often jammer strategy to try to send a signal in the complementary channel so that the receiver will make an incorrect decision.)

The receiver must be doubled (essentially two receivers) to observe both alternate channels at once.

Ignoring for the moment that if two receivers were used for data reception the system complexity and cost would be increased, let us look at the received signals. Figure 5.45 shows the output signals from a pair of mark/space data receivers. We see that the input to the mark or space band-pass filters is a series of pulses at the data rate (if multiple chips are sent per bit, the pulses are at the chip rate). These RF pulses are envelope detected and used to make a bit decision.

Observing that the mark/space-pulsed-RF streams are complements im-mediately suggests that an alternate receiver without this redundancy might

---

*There are also "*m*-ary" frequency hopping systems (of which one is TATS).[111] These transmit-ters choose from a set of frequencies corresponding to data inputs; for example, three binary bits could be sent for each 8-ary signal chip, with each of the eight levels corresponding to one of eight possible binary 3-tuples $000, 001, 010, 011, \ldots$.

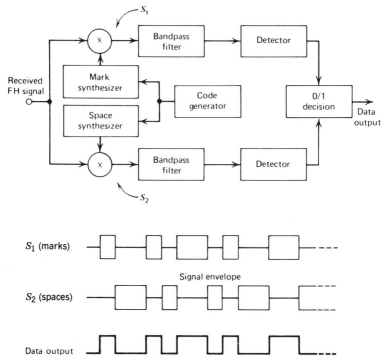

Figure 5.45   Dual mark/space receiver diagram.

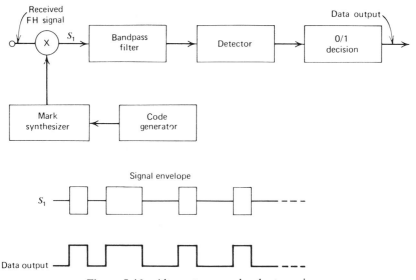

Figure 5.46   Alternate nonredundant receiver.

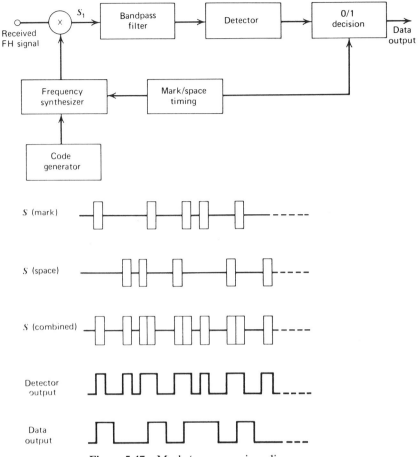

Figure 5.47 Mark/space receiver diagram.

be possible. If we consider, for instance, that the absence of an RF pulse is a space, the mark channel receiver above could be used and only half the receiver, or Figure 5.46 would be required.

Alternatively, if the synthesizer in the receiver were made to hop twice as fast as the transmit synthesizer—first to the mark frequency and then to the space frequency—it could sample both. Its block diagram and the waveforms generated are shown in Figure 5.47. Note that even though the output data are the same as in the preceding two cases, the signal input to the band-pass filter in Figure 5.47's mark/space sampling receiver consists of many pulses of half the duration of the transmitted frequency hopping signal. Because the pulsed-RF input to the postcorrelation bandpass filter is of different duration, depending on the specific receiver implementation, we see that the filter must have a different bandwidth for proper reception. We note, however,

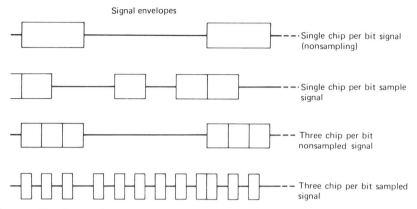

Figure 5.48   Comparison of RF envelopes of postcorrelation signals in a frequency hopping receiver.

that the data rate is the same for all three and therefore there must be a difference in receiver performance because of the different filter bandwidths.

Before considering this difference and the effect of it, we must point out one more factor—the coherence* or noncoherence of the frequency hopping synthesizers used. If the entire transmit and receive system is phase coherent, the postcorrelation bandpass filter output signal is phase continuous; that is, the pulsed-RF signal appears to be simply a pulse-modulated RF carrier. On the other hand, if the entire system is not coherent, the pulses are a series of RF bursts, each at a different phase, and the signal appears to be both amplitude and phase modulated.

What difference does this make? A significant difference. The bandwidth of the coherent signal is less than that of the random-phased noncoherent signal; thus the postcorrelation filter bandwidth may be decreased for a net gain in threshold and/or processing gain. When receiver sampling and multiple chips are used, the difference becomes even more pronounced. Figure 5.48, when the various chip envelopes are compared, shows that a three-bit-per-chip sampled signal, for instance, is very nearly a square wave. If that square-wave RF envelope were considered to be 100% AM modulated by a square wave, it would readily be sen that the spectrum was that of an

---

*Coherence in a frequency hopping synthesizer requires that, when jumping from one frequency to another, the synthesizer output signal phase be well defined; for instance, one frequency burst might end at a zero crossing and the next start at a zero crossing. In a frequency hopping system coherence implies that *both* the transmit and receive frequency synthesizers maintain constant phase relationships so that the remapped single frequency correlator output is phase continuous. Such coherence, by the way, is no mean undertaking, for it requires not only a precise transmitter but uniform link (propagation) characteristics and identical receivers. There are those who speculate that overall system coherence in an actual operating over-the-air wideband frequency hopper is impossible to achieve.

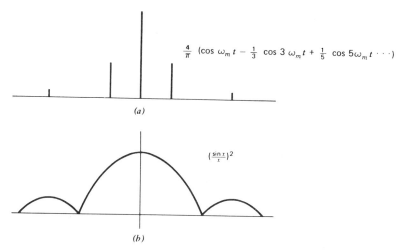

$$\frac{4}{\pi} \left(\cos \omega_m t - \frac{1}{3} \cos 3 \omega_m t + \frac{1}{5} \cos 5\omega_m t \cdots \right)$$

(a)

$$\left(\frac{\sin x}{x}\right)^2$$

(b)

Figure 5.49 Comparison of coherent and noncoherent frequency hopping signal spectra after dehopping; (*a*) coherent square-wave-pulse AM spectrum; (*b*) noncoherent square-wave-pulse AM spectrum.

AM transmitter with modulation sidebands distributed according to the distribution of harmonic energy in the square wave. This distribution is

$$\frac{4}{\pi} \left(\cos \omega_m t - \frac{1}{3} \cos 3\omega_m t + \frac{1}{5} \cos 5\omega_m t + \cdots \right)$$

according to the Fourier expansion. The most significant power falls within the first and third harmonics; therefore, a filter that passed only these harmonics would output a reasonable RF pulse. If only the first pair of sidebands were passed by the filter, the output would be near-sinusoidal.

For noncoherent reception the phase of each RF pulse is different and the effective modulation rate is twice that of a coherent pulsed signal. This leads to a signal spectrum at least twice as wide as that of the coherent signal, thus requiring a wider filter and causing a loss in receiver performance. What is the expected frequency distribution for a randomly phase-shift-keyed signal? What else but $[(\sin x)/x]^2$? Figure 5.49 compares the spectra of coherent and noncoherent frequency hopping signals after dehopping. Note that the noncoherent signal main lobe occupies twice as much bandwidth as the coherent signal and its first sidebands.

Figure 5.50 is a photograph of an actual noncoherent dehopped signal's spectrum, which confirms our expected result.

Of course, there is nothing special about the $[(\sin x)/x]^2$ noncoherent signal that would prevent filtering it to the same $2\omega_m$ bandwidth as the coherent signal. This would result in a similar RF envelope from the

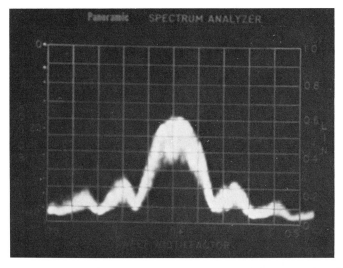

Figure 5.50 Photo showing actual noncoherent dehopped frequency hopping signal.

postcorrelation bandpass filter but would give a decrease in signal power because of the different distribution of power. To be more exact, filtering to the ω_m bandwidth would result in a loss of 25% of the power in a $[(\sin x)/x]^2$, spectrum but only 10% of the coherent signal's power. This small difference (about 1 dB overall) could be significant in determining system operation or failure.

m-ary Detection

Detectors for m-ary signals are similar in principle to the alternate-chip-comparing detector already described, except that they simultaneously compare all possible signal channels and make a decision based on the largest signal detected. An m-ary frequency hopping signal is one in which the transmitter has a choice of m frequencies in each chip time, rather than two as in our previous discussion. The transmission can then represent more than one bit of information; that is, an 8-ary transmission with one of eight frequency choices being transmitted at any given time would signify three binary bits by each chip. A 16-ary transmission would convey four binary bits per chip, and so on.

The m-ary transmitter can transmit on any of the frequencies made available by its synthesizer, of course, but allowing any set of m frequencies from a large set can pose a difficult demodulation problem for the receiver. To be exact, if any m of the synthesizer frequencies are used, the receiver must monitor all m, which forces the receiver to employ as many as m

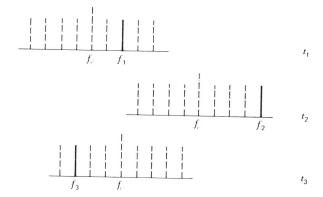

Figure 5.51 Typical transmission format for *m*-ary frequency hopping.

synthesizers or even an array of *m* receivers. The approach taken in TATS (tactical transmission system)[111, 124] is to use a frequency frame like that shown in Figure 5.51, in which each possible frequency has a definite relation to *m* − 1 others.

In Figure 5.51 we see a set of frequencies shown for three different chip times, t_1 through t_3. Each set of frequencies consists of eight possible transmission frequencies clustered around a center frequency. The center frequency f_c changes at each chip time, so each *m*-ary set is hopped around and becomes a frequency hopping transmission in itself. In addition, however, the information to be transmitted offsets the center frequency to $f_1, f_2, f_3, \ldots$, depending on the data being sent. Therefore, except for synchronization, the center frequency is never actually transmitted. Only the frequencies offset from the center by the input data requirements are transmitted.

In addition to the more obvious advantage that is afforded by *m*-ary transmissions over binary, that is, more data per chip, a simple receiver format is permitted. Only one receiver with separate detectors operating at baseband or IF, rather than a separate receiver for each possible frequency, is satisfactory. The demodulator in Figure 5.52 is satisfactory for receiving such transmissions. This demodulator is a simple extension of those shown for frequency hopping systems except that all possible frequency inputs are sampled at once, and the channel with the largest signal in a given chip interval is chosen as the correct one. If, for instance, the channel judged to have the largest signal corresponds to a 110 input, then 110 would be the demodulator output. Similarly, any of the other seven channels (referring to Figure 5.52) containing the largest signal would cause an output 000 through 111, depending on the 8-ary to binary correspondence.

One problem with the *m*-ary transmission format does exist; that is, a system's jamming margin may be reduced below the level that would be expected for a given number of available hopping channels. This is due to the

Figure 5.52 Block diagram of m-ary demodulator.

extra vulnerability of m-ary transmissions to interference. If any one of the m channels other than that intended is hit by an interfering signal, an error is possible. In an 8-ary receiver, for instance, an interfering signal in any one of seven unused channels could cause an error and that error could represent three bits. To overcome this problem error correction coding can be used. The TATS system employs a Reed-Soloman encoding technique for this purpose. A determined jammer, however, could note frequency, retransmit that frequency offset by the channel spacing, and cause an error almost every time. Only the hopping rate of the desired signal and the geometry of the particular situation preventing the jamming signal from reaching the friendly receiver during the correct chip could avert this type of vulnerability.

When employing the m-ary format, it is common to employ one of two techniques to avoid the offset jamming signal vulnerability mentioned above. The first of these is to choose the m frequencies used from the complete set available, rather than restricting them to a contiguous set. The second is to increase the number of frequencies available by a factor of 7, which reduces the probability of a hit by the jammer.

PROBLEMS

1. Correlation filter output signal-to-noise for a 20-dB correlator interference-to-signal input and 30-dB process gain would be how much (neglect implementation loss)?

2. Draw a block diagram implementation of a correlator designed to represent the function

$$\int f_{(t)} g(t - \tau)\, dt.$$

3. What is the overall probability of an error in a frequency hop chip if 30% of the available frequency channels contain an effective interfering signal?

4. Draw impulse response waveforms for the integrate-and-dump detectors in Figure 5.44.

5. Compare the binary and m-ary frequency hopping approaches.

6. What, in a direct sequence system, is the equivalent of m-ary information transmission?

7. What bandwidth would be expected of a common channel spread spectrum signal (i.e., one similar to the desired but employing a different code sequence) at the output of a correlating multiplier?

8. Given 30-dB process gain, what would be the output $(S + I)/I$ ratio in a correlator when input interference is 10 dB above the desired signal?

9. What repetition rate, or line spacing, would be expect to see at the output of a correlator's multiplier for a 4095-chip code at a 1-Mcps rate when multiplied with a CW carrier.

10. What is the phase angle between the input signal and reference in a phase-lock loop when the input is offset by 10 kHz from the loop internal oscillator, and loop gain is 10^5?

11. Compare phase lock and Costas receivers for subcarrier FM or FSK demodulation.

12. What would be the effect of employing simultaneous carrier FSK and subcarrier FSK in a direct sequence system?

6

SYNCHRONIZATION

Now we must talk about the hardest part. Throughout this book we have assumed good synchronization (on the part of the code) between transmitters and receivers. By "good" synchronization we mean that the coded signal arriving at a receiver is accurately timed in both its code pattern position and its rate of chip generation (with respect to the receiver's reference code). So far we have assumed that the codes in the system we have talked about were already synchronized and would remain so.

What an assumption! More time, effort, and money has been spent developing and improving synchronizing techniques than in any other area of spread spectrum systems. There is no reason to suspect that this will not continue to be true in the future.

Code synchronization is necessary in all spread spectrum systems because the code is the key to despreading the desired information and to spreading any undesired signals. The overall requirements of frequency hopping and direct sequence systems with regard to synchronization are similar, but somewhat more latitude is possible in the frequency hoppers. Here we discuss the synch requirements of both types of system, describe ways of attaining synchronization from a cold start, and discuss ways to maintain synch once it has been acquired. Chirp sytems are not considered because they do not require synchronization in the same sense as other types of spread spectrum systems. To be sure, chirp systems do synchronize, but synch is on a pulse-by-pulse basis because the most important property of a chirp matched filter is rapid total-pulse correlation.

Sources of Synchronization Uncertainty

In spread spectrum systems two general regions of uncertainty exist, with respect to synchronization. These are code-phase and carrier-frequency uncertainties, and both require resolution before a spread spectrum receiver can operate. The code phase must be resolved to better than one chip, and the center frequency, as seen at the receiver, must be resolved to the degree that the despread signal is within the aperture of the postcorrelation filter. Furthermore, the carrier frequency is often constrained to be accurate enough to work well with a demodulator.

The code clock rate must also be resolved, but because the clock rate can be related to the carrier frequency, if a system is so designed, it is not necessarily a separate source of synchronization uncertainty.

Most of this code-rate, phase, and carrier-frequency uncertainty may be eliminated simply by providing accurate frequency and time sources in both the transmitter and receiver. Not all, however, can be overcome in this way. Doppler-related frequency errors often cannot be predicted and will affect both carrier frequency and code rate. In addition, mobile transmitters and/or receivers see a relative code-phase change with every change in their relative positions. Even fixed-position transmitting and receiving stations may see changes in code phase, center frequency, and/or code rate due to propagation-path-length changes. Such effects are especially pronounced in tropospheric scatter communications but may also be seen in all but the most stable direct line of sight links.

The amount of Doppler-frequency uncertainty applied to a received signal is a function of the relative velocity (range rate) of the receiver and transmitter and of the frequency transmitted. Consider a sinusoidal traveling electromagnetic wave. If that wave is observed passing a given point, it will repeat at a rate of $f = c//\lambda$, where f is the frequency of repetition, λ is the wavelength, and c is the speed of light in the propagation medium* being considered. Thus, if neither the transmitter that sends the wave nor the receiver that observes it is moving, the frequency of the received wave is the same as that sent by the transmitter. Let the point of observation move with respect to the transmitting point, however, and the number of repetitions in a given period is changed; for instance, if the receiving point moves toward the transmitter, the rate of reptition of the electromagnetic wave at the receiver increases because it reaches the receiver at a rate higher by an amount exactly proportional to the receiver's approach velocity.

The "Dopper-frequency shift" is seen to be equal to the number of wavelengths at the frequency of interest that the receiver covers in a 1 sec period. The amount of Doppler-frequency shift then is just $D = V/\lambda = fV/C$ Hz, where V is the relative velocity of transmitting and receiving terminals, and f is the frequency transmitted. When a given frequency f_t is sent, the received frequency is just

$$f_R = f_t \pm \frac{f_t V}{C}.$$

---

*2.997925 $\times$ 10^8 m/sec in vacuum, per NBS Applied Mathematics Series 55, Handbook of Mathematical Functions, February 1971.

The plus sign for increased frequency is applicable to terminals approaching one another, and the minus sign for decreased frequency is valid with departing terminals.

A rule of thumb for determining Doppler shift when frequency and velocity are known is that

$$D = 3.336 \text{ Hz/MHz} \cdot \text{km} \cdot \text{sec}$$

$$= 0.000926 \text{ Hz/MHz} \cdot \text{km} \cdot \text{hr}$$

$$= 0.00148 \text{ Hz/MHz} \cdot \text{mi} \cdot \text{hr}.$$

Thus, the Doppler effect must be considered in optimizing a communication system, for it is often the major contributor to frequency uncertainty when stable transmitting and receiving frequency elements are employed.

Often the frequency sources used in a communication system are not so stable as we might wish, and their contribution to frequency uncertainty cannot be ignored. In spread spectrum systems another consequence of frequency uncertainty is also apparent—any clock rate offset is cumulative in code-phase offset; that is, a 10-Hz offset in clock rate for a code generator would translate to a 10-chip-per-second cumulative code error, which would cause a code-phase drift of 36,000 chips.

Factors that contribute to oscillator-frequency instability are temperature, attitude of the crystal (i.e., vertical, horizontal, or otherwise), amplitude stability, noise contributions from the amplifier stage, and others. Oscillator instability has been characterized in detail by Lindsey and Lewis.[1729]

The prime sources of uncertainty for synchronization, then, are those that are time* or frequency dependent. If the receiver could know range and transmitter time accurately enough, there would be no synchronization problem, or if both transmitter and receiver possessed oscillators accurate enough, they could derive the timing needed. Furthermore, that same accurate oscillator could act as a source of precise frequencies to remove all frequency uncertainty except for that produced by the Doppler effect. A highly stable frequency source is to be desired for any spread spectrum system. Figure 6.1 illustrates the amount of code drift that occurs per day between a pair of systems with various relative code rate differences.

6.1 INITIAL SYNCHRONIZATION

The initial synchronization problem is the most difficult of all. When synchronization has already occurred, it is often possible to base a subsequent

---

*Time uncertainty here includes any propagation time uncertainty due to unknown range. Uncertainty due to time offset, expressed in code chips is Δt times the code rate. That is offset in code chips $= \Delta t \, Rc$.

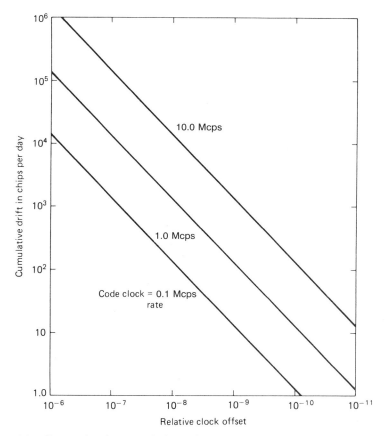

Figure 6.1 Curves showing cumulative daily drift for various code rates and relative code-rate errors.

synchronization on the knowledge of timing gained. However, a cold start without prior knowledge of timing, or at best very minimal knowledge, is the usual rule rather than knowing the proper time frame for achieving synchronization with a desired transmitter or receiver. Many techniques for achieving synchronization have evolved, some with simple requirements, others with complex implementation implications. Several are described in this section. Which is best depends on the application, the amount of time allowed, and the amount of uncertainty involved.

Certainly, if the time of day (and in some cases the date) is known to an accuracy of sufficient degree by both transmitting and receiving stations, the code generators used to determine the order of system functions could be set to coincide, and the only uncertainty would be that caused by unknown range; for example, if two identical code sequences were started at precisely the same time and the clocks used to generate their chip rates continued from that time to give identical code rates, then at any point in time after the original start the two sequences would be at the same state. A transmitter and receiver employing this pair of sequences could be sychronized in real time, and would remain so forever if no outside influence disturbed them. They still could not talk together, however, unless the distance between them were less than $\frac{1}{R} \cdot c = \frac{c}{R}$ (R = code chip rate, c = speed of light), the distance corresponding to one code chip in range. Otherwise, the propagation time delay for a signal modulated with the transmitter's code sequence to reach the receiver would be such that the receiver's sequence could have reached a new state (and so would the transmitter), whereas the signal sent would be in transit and the coded signal as received would not be in synch with the receiver's code.

The obvious cure for this small difficulty is to move the transmitter's sequence ahead in time by an amount equal to the signal propagation delay, and therein lies a new problem; that is, the distance from transmitter to receiver may not be known well enough to time the code sequences satisfactorily. Thus we see that even perfect knowledge of time and perfect timekeeping does not completely solve the sychronization problem. This is especially true when one or both of the terminals communicating are mobile, for a poorly defined distance uncertainty is the result.

Even for fixed terminals with known range and precisely known time, it is possible for some phenomenon such as multipath to require a special sychronization procedure. Bearing in mind, therefore, that even the stablest of code sequence generators does not completely overcome either initial sychronization or tracking problems, we now launch into the subject of initial sychronization methods.

The "Sliding" Correlator

The simplest of all correlation techniques uses a "sliding" correlator, so called because the receiving system, in searching for sychronization, operates its code-sequence generator at a rate different from the transmitter's code generator. The effect is that the two code sequences slip in phase with respect to each other, and if viewed simultaneously (say with an oscilloscope), as in Figure 6.2 they seem to slide past each other, stopping only when the point of coincidence is reached. The flow diagram of Figure 6.3 illustrates this process.

Figure 6.4 shows spectra typical of those seen at a direct-sequence demodulator's input. Figure 6.4a shows the noise spectrum, band limited by the predemodulator IF, generated by the cross-correlation of the receiver's internal code generator with any unsynchronized input signal. This is the typical spectrum seen any time the receiver is not sychronized.

At the instant that the receiver's code sequence comes into synch with an incoming signal, a coherent output like that in Figure 6.4b appears. The demodulator then has the task of acquiring carrier lock on the signal. The signal shown in Figure 6.4b is double-sideband, suppressed-carrier modu-

Transmitted code
reference

Phase—slipping
receiver code

Transmitted code
reference

Receiver code
reference (synchronized)

Figure 6.2   Sliding correlator action: (*a*) sliding process in operation; (*b*) synchronized codes after acquisition.

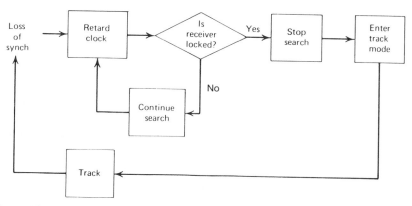

Figure 6.3   Flow diagram for sliding correlator synchronizer.

Figure 6.4   Signals seen at input to direct sequence demodulator: (*a*) noise input to demodulator, unsynchronized condition; (*b*) DSB, 2500-bps signal after synchronization, 25-dB J/S ratio, CW jamming.

lated by a 2500-bps square wave. Also, the input jamming to signal ratio for the receiver in Figure 6.4*b* is 25 dB.

The advantage of the sliding correlator is its simplicity in that nothing more is required than some way, on command, of shifting the code clock of the receiver to a different rate. The difficulty of using a simple sliding correlator for synchronization, however, is that when a large degree of uncertainty is encountered, examination of all possible code-phase positions is impractical because of the time involved. Recognition of sychronization, which must occur to stop the sliding or "search" process at or near the point of sychronization, is limited in response time by the bandwidth of the system's postcorrelation receiver. Thus the rate of searching, or sliding through all of the possible code-phase offset positions, is also limited by that bandwidth. Therefore, to give an example, a system with a postcorrelation (baseband) receiver bandwidth of say 1 kHz could recognize synchronization in approximately 0.35 msec (applying the general rise time-to-bandwidth relation $T_\tau = 0.35/\text{BW}$), and the time used to slide through the point of correlation (remember that the correlation function is two chips in width) should be a minimum of this amount. The maximum search rate then would be approximately $2/T_\tau$ or 5.7 kcps. Because it is possible under some circumstances for a receiving code-sequence generator to be offset by trillions of chips, we can easily see that hours, days, or even years might be consumed in the initial synch process if the sliding correlator alone were used. As a rule of thumb, one can expect to search at a rate approximately equal to the data rate for which the receiver has been designated.

In practice, sliding correlation is almost always employed, though the technique is often augmented by other methods for restricting the area of search. Most of the remainder of this chapter (at least that part having to do with synchronizing) discusses methods of initial sychronization that, in general, employ techniques to supplement sliding correlation; that is, these other methods bring the two code streams of interest into a range in which sliding correlation can reasonably be used.

Synchronization Preambles

One of the most effective techniques for making use of a sliding correlator employs special code sequences, short enough to allow a search through all possible code positions in some reasonable time but limited in how short they can be by correlation requirements. A well-chosen code sequence (called a preamble when used for synchronization) is a good solution to almost all sychronization problems. Range estimates are unnecessary when a preamble is used, and acquisition time is set by the length of the preamble. Consider an avionics system, for example, in which no knowledge of direction, speed, or distance can be derived before receipt of a signal. A sychronization preamble sent at the beginning of each transmission would allow normal push-to-talk system operation without having prior knowledge of relative positions. Of course, any time required for the synch preamble acquisition process would be added at the beginning of each transmission. A reasonable alternate might be to transmit a preamble code for acquisition only at the beginning of a series of transmissions, which would synchronize all link users. Thereafter synchronization would be maintained by the system clocks, broken only by transmission of an "out" signal.

Typical synchronization preambles range in length from several hundred chips to several thousand, depending on the specific sytem's requirements. When the information being sent lies within a specific band, it is advisable to select the preamble length such that its reptition rate does not fall within that band. Spectral components from code repetition × interference occur at intervals of R_c/L Hz, where R_c is the code chip rate and L is the code length in chips. If the preamble repetition rate is within the information bandwidth of the system, two or more of these spectral components will lie in the receiver's IF bandwidth and could appear to the demodulator as a pair of modulation sidebands, thereby interfering effectively with the desired signal. Though the preamble code repetition rate can act as an interference-enhancing element, it cannot be chosen to produce less than one spectral component in the information band. No matter how far apart the spectral components may be, a variation in interference frequency can cause one or more components to fall into the signal band.

At the opposite extreme, we may choose the preamble length and code rate such that the repetition rate causes multiple spectral components to lie in the information band (this implies longer codes and/or lower code rates). If the code-produced frequency lies below the information band, the interference produced can be acceptable.

Figure 3.39 shows the line spectrum of an 8191-chip code when cross correlated with a CW jamming signal.

Preamble synchronization methods have one significant weakness that comes about as a result of the very code property that makes them work well; that is, the relatively short sequence length which allows rapid synch acquisition tends to be more vulnerable to false correlations and to possible reproduction by a would-be interferer. With the exception of the possible vulnerability problem, however, preamble synchronization is by far the least critical, easiest to implement, least complex, and best for all around use.

It is possible to employ short synch preambles if they are hidden by transmission well below the level of a masking signal. For example, suppose that a long code and a short code modulate a pair of orthogonal carriers, but the short-code-modulated signal is attenuated by at least 10 dB. Then, the two signals are summed together. The preamble-modulated signal is then 10 dB or more below the longer code level, and cannot be demodulated unless the code is already known.

Bounds on the preamble's length are generally set by four criteria: the minimum preamble code-sequence length is bounded by cross-correlation and interference rejection requirements, whereas the maximum code sequence length is set by maximum available acquisition time. The fourth criterion, that of code repetition rate, may tend to bound the preamble code length in either direction.

Frequency Hop Synchronization

The comparatively slow code rates employed in frequency hopping systems give rise to greater ease of synchronization than those of direct sequence systems, because at the slow code rates used a given clock error accumulates much more slowly; that is, given a clock error of one part in 10^6, a 50-Mcps direct-sequence generator would accumulate a one-chip offset in 20 msec, whereas a typical frequency hopping code generator at 5 kcps would drift only one chip in every 200 sec. Thus the area of code-phase uncertainty is significantly different for frequency hopping systems than for direct sequence systems. For this reason direct sequence systems may employ a frequency hopping preamble to facilitate more rapid synchronization.

Frequency hop synchronization in direct sequence systems is implemented by allowing for two code sequences which are generated at different rates, one for the direct sequence mode and one for frequency hopping during synch acquisition. The codes must be related, in that frequency hop code synch allows for decreasing the direct sequence code-phase unceraity; for example, assume that the direct sequence code to be synchronized is 10^6 chips long* and its chip rate is 10 Mcps. We then can take a slower 1000-chip code sequence at say 10 kcps, sychronize the two codes so that their starting points coincide, and arrive at a pair of code sequences for which each chip in the lower code corresponds to 1000 chips in the higher-speed code.

If the slow code is then used for synch acquisitions, the synch uncertainty which would exist could be no worse than 1000 chips instead of the 10^6 chip uncertainty possible with the higher-rate code. Therefore, an initial synchronization operation can be carried out in one-thousandth the time.

Initial acquisition of the slow code, however, does not complete the acquisition process. Because synchronization to a slow code chip resolves the high-speed code only within 1000 chips, serial code search must next be carried out to sychronize the high-speed code within 1 chip. So we see that by searching over a 1000-chip uncertainty twice we can resolve a 10^6 chip ambiguity, while searching for a total of 2000 chips.

A slow code, unlike a high-speed code, is not adapted to direct sequence transmission. A 10-kcps code could hardly be expected to give a useful amount of spread spectrum processing gain in a direct sequence modulator/demodulator. The solution, therefore, is to use the slow code to drive a frequency hopping synthesizer during an initial synchronization period, with a switch to the high-speed code and direct sequence operation following frequency hopping synch acquisition.

Frequency hop synch provides for rapid synch acquisition but does not allow for simplification when direct sequence modulation has been chosen to avoid a frequency hopping synthesizer requirement. When a flexible frequency synthesis technique already exists, frequency hopping synch methods are a good alternative to direct sequence preamble synch methods.

---

*A number chosen here for convenience; a more realistic number would be $2^{20} - 1$, or about 1.05 million.

Transmitted-Reference Methods

When a receiving system must be the simplest possible, a transmitted reference[1502, 1506] may be used for initial synch acquisition, tracking, or both. Transmitted-reference receivers employ no code sequence or other local reference generator. Instead the coded reference is generated at the transmitter and transmitted at the same time as the intended information-bearing signal. Both frequency hopping and direct sequence sytems are amenable to transmitted-reference methods. Figure 6.5 shows a transmitted-reference system.

The operation of a transmitted-reference receiver is precisely the same as that of any other receiver using an offset local reference signal.* The difference is that the local reference is generated in the transmitter and sent to the receiver along with the signal to be demodulated. The two carrier frequencies are offset by an amount equal to the first IF in the receiver, and when mixed, a correlated IF signal is produced.

Transmitting the reference is obviously advantageous when the receiver must be limited in cost, weight, or size because there is no need for a code-sequence generator, search or tracking circuits, or any of the code-related mechanisms. Of course, if the reference is transmitted only to achieve rapid synch acquisition with a subsequent switch to normal direct-sequence or frequency hopping demodulation, we must also still include these subsytems.

If the transmitted-reference technique is to be employed when interference or active jamming is a threat, some other synch provision may be necessary. Otherwise any two identical signals offset by the amount of the IF would produce false synchronization in the post-correlation circuits. One method of protecting the IF in this situation would be to frequency hop one of the signals transmitted, which would require the receiver IF to hop with the same frequency pattern, or a frequency hopping local oscillator would be needed to act as a local reference for the frequency hopping subsignal. In either case the synchronization time would be increased to account for frequency hop synch acquisition.

One other disadvantage exists, and this is that noise introduced in the reference due to being transmitted through the link (rather than locally generated) degrades the receiving system. When interference appears at both desired and reference-signal frequencies, correlator output products are produced which can fall into the IF band, thus also degrading signal demodulation.

A simple way to send data in a transmitted reference system of this type is to modulo-2 add the data to either of the two code streams that modulate the carriers.

---

*i.e., A heterodyne correlator.

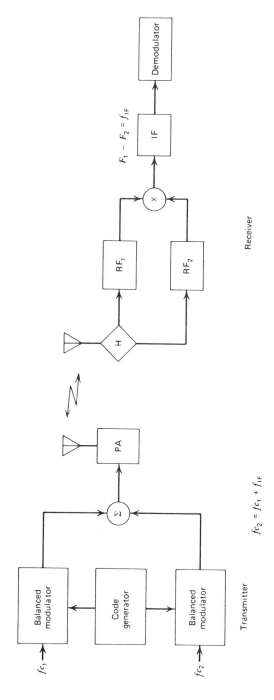

Figure 6.5 Transmitted-reference synch method (direct sequence application).

231

Universal Timing

When accurate, dependable, universally known time is available, it is possible to devise a system that never loses synchronization. (This does not necessarily mean, however, that a search procedure is not needed for synch acquisition. Even if, in real time, transmitting-sytem and receiving-system code sequence generators are in perfect alignment, a signal sent from one to the other would not be synchronous when it reached the receiving sytem because of the time required for signal propagation. For this reason, and others, some search-and-track capacity is always required, even when code chip rates are precise and distances are well known.) Only very low rate frequency hopping systems can operate without search-and-track capability.

Consider that every communication terminal knows the time of day and the date within 1-μsec. Then at prearranged times, say at each hour on the hour, all code generators could be set to a new state derived from that time and date information. Resetting to a new state could, of course, be done more often when, minimum waiting, entry to the network by new users is desired. If the code sequences in question have rates of 1 Mcps or less, universal 1-μsec timing ensures real-time synch to within one chip time. In any case, however, some search-and-track capability would still be required.

Universally timed spread spectrum systems have been used principally in satellite communications in which ephemeris data provide accurate distance information and ambiguity can be reduced to the point at which search processes may be minimized. Future spread spectrum systems for all uses will probably employ universal timing concepts for synchronization and tracking, as frequency sources for mobile use improve and as computerized position estimation methods provide more accurate information for range compensation.

Burst Synchronization

The idea behind burst synchronization is to transmit a short, high-speed message that provides the spread spectrum receiver with enough information to gain rapid access to a subsequent direct sequence or frequency hopping signal. It is, of course, hoped that any would-be interferer is to be caught napping by the sudden appearance and disappearance of the synchronizing signal. A determined interferer might leave his or her interfering signal on continuously, however, which would negate any element of surprise. On the other hand, the high peak power available to a low-duty cycle burst transmitter could cause our would-be interferer to require an inordinately large amount of continuous RF power, thus providing interference protection.

No information, other than code and/or carrier synch, is normally sent during a synch burst. Information to be transmitted then follows a switch to direct sequence or frequency hopping.

Sequential Estimation

Much akin to burst synchronization methods are those of sequential estimation, in which the receiver to be synchronized simply demodulates the
incoming coded signal, inserts the data (code) as a vector input to its own
sequence generator, and makes a trial check for synchronization. The code
being received may or may not be easily acquirable. The received signal is
demodulated in exactly the same way we would demodulate any PSK data
stream (this technique is not readily applicable in frequency hopping). Once
the local code generator is loaded with a code input taken from the received
signal, it must race ahead (search forward) by a number of bits $n + T_p/f_k$,
where n is the code generating register in length in bits, T_p is the receiver's
processing delay time, and fk is code clock rate.

Sequential estimation (see Figure 6.6) has a real advantage as a rapid
synchronization method. Ward's results[65] show a time improvement factor of
23 for replacing a sliding correlator with a sequential estimator. Sequential
estimation, however, does have the drawback that it is more vulnerable to
interfering signals than some other techniques. This is because the demodulator demodulates the code received without benefit of processing gain.

Average acquisition time recorded by Ward shows 90 msec at -6-dB, 300
msec at -10-dB, and 1 sec at -12.5-dB signal-to-noise ratios. (A white-noise
source was used for these measurements.) Although this performance against
noise may not appear to be impressive when deliberate interference is
expected, those spread spectrum applications in which the system contends
only with natural disturbances—and there are many of them—could benefit
by synch acquisition through sequential estimation techniques.

Special Coding for Synch Acquisition

Special codes, specifically designed for synch acquisition, have been employed to advantage in the quest for more-rapid spread spectrum system
lockup. Short preamble codes might be considered special codes, but in
general the only difference between them and the code sequences used after
acquisition is length. Here we are more interested in codes that have special
properties to make them more easily acquirable than the maximal sequences
or to give some implementation advantage that can be translated to a simpler
or more-rapid synch procedure. Emphasis in this section is placed on use of
these codes rather than on the codes themselves. Those who are interested in
code properties and code generation in more detail are referred to Chapter
3, and the references listed in Appendix 2.

The best-known easily acquirable code type is the JPL component
code,[219, 330] made up of shorter codes. These shorter codes, each of
different length, are combined to result in a code with a length
$(2^m - 1)(2^n - 1) \cdots (2^r - 1)$, where $m \neq n \neq r$, and with a special correla-

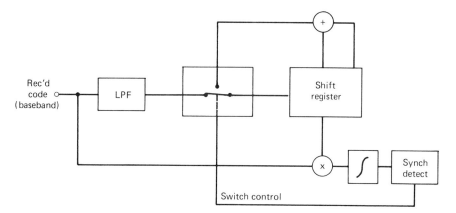

Figure 6.6   Illustration of sequential estimation synchronizer.

tion property, which is different from the maximal linear sequences in that the m sequences have only one point of autocorrelation. The component codes, on the other hand, have one more correlation possibility than there are components in the code (i.e., if a code is made up of three component subcodes, there are four autocorrelation points associated with the overall code). Moreover, all but one of these autocorrelation levels are associated only (and separately) with the individual codes making up the composite. The highest correlation level corresponds to total composite code synchronization.

Synchronization, by using the JPL component codes, is accomplished by first cross-correlating one of the component codes with the composite code. Once this component code reaches the point of synchronization with its mate, which is embedded in the composite code, a partial correlation occurs. The partial correlation is then the signal for the second component code cross-correlation to be initiated, which causes the partial correlation level to be increased. This process continues until all the component codes making up the overall composite code are individually synchronized with their counterparts in the received signal. When all are individually synchronized, the correlation is the same as if the process had simply synchronized the composite code. The advantage of this technique is that it provides for rapid synch acquisition without the use of a preamble or anything other than the composite code itself. When the component codes are, say, 200, 500, and 1000 chips in length, separate search processes over these individual lengths (a total of 1700-chips search) can be accomplished much more rapidly than a search of the composite 10^8 chips.

This advantage to lock-in time is paid for by a decrease in the signal-to-noise ratio in the correlator output when all code components are not locked. Therefore, when interference is a problem, the partial correlation levels combined with code $\times$ interference noise could reduce the correlated signal-to-noise ratio below a usable level. Also, interference could cause a correct decision regarding correlation level to be difficult to make.

The Gold[312,313] codes previously discussed are also specially formulated for synch acquisition, though they are not normally expected to reduce synch time. It may be, however, that use of a Gold code makes it possible to reduce the overall code length, which in turn reduces the search time. Gold codes are actually sets of nonmaximal linear codes whose correlation properties are uniform and well defined over the entire set. Two other code types have been advanced to serve in similar functions to the Gold codes. These are the Kasami codes and the Bent codes[382]. Both have lower cross-correlation bounds than Gold codes. That is, while Gold codes have cross-correlation bounded at $2^{(N+1/2)} + 1$ or $2^{(N+2/2)} - 1$, the Bent and Kasami code cross-correlation bound is $2^{(N/2)} + 1$. For example, a 1023-chip Gold code set would have a cross-correlation bound of 63, while either Bent or Kasami code sets would have their bound at 33, a difference of approximately 3 dB. However, the size of the Bent and Kasami sets is much smaller than that of the Gold code sets: for the Gold codes, each set has $2^N + 1$ codes, while the Bent and Kasami codes each have only $2^{N/2}$ codes apiece. Thus the Bent and Kasami codes are not as useful as the Gold codes in multiple-access applications where large numbers of users are to be accommodated. One other difference should be noted: the Bent codes are nonlinear codes.

Anderson[31] has also defined a set of codes with well-defined correlation properties that may be employed for synchronization. Although they are not really special codes, short sectors of longer codes or even very short codes (say 63 chips) are sometimes sent as synch preambles to be detected by a delay-line matched filter. This technique is used, as all the other special encoding techniques usually are, to reduce synch acquisition time. The delay-line matched-filter signal detector is discussed in a subsequent paragraph.

Is a special synchronization code generally useful? No. But can it be used for a specific purpose? Definitely. Special codes are useful for specific purposes. Most are intended to reduce acquisition time but may pay for this by providing reduced interference rejection, poor cochannel-user rejection, reduced ranging capability, or some other performance loss. As in many other areas, treatment of a special synch code is a matter of matching it to a specific limited use.

Matched-Filter Synchronizers

The matched-filter synchronizers referred to here are not the same sort of matched filters already discussed—the integrate-and-dump detector—although both are generically matched filters (Figure 6.7). A matched filter is any device that in effect generates a time-reversed replica of its desired input signal, when its input is an impulse. As Turin[1711] states, "The transfer

Figure 6.7   Integrate-and-dump filter showing impulse response.

function of a matched filter is the complex conjugate of the signal to which it is matched."

Both IF and baseband matched filters have been implemented and for comparable filter lengths give similar performance. IF matched filters tend to be the surface acoustic wave type, with lithium niobate, lithium tantalate, or quartz as base materials, depending on the length and temperature range required. Baseband matched filters have been implemented as digital integrated circuits by using normal integrated-circuit techniques. The length of acoustic matched filters is limited by the material and by the acceptable loss in the devices themselves; the longest acoustic matched filter presently known is made up of a series of short lines to reach 4000 total chips. Digital baseband matched filters (see Figure 6.8a) are limited only by the ability to sum the correlated signal properly and by the power dissipation of the circuits employed. (Digital baseband matched filters employ active circuits, as opposed to the surface wave filters, which are strictly passive.)

Matched-filter synchronizers, generally made up of delay elements, are quite different from integrate-and-dump detectors and chirp filters. They do, however, act as conjugate signal generators to the proper signals, as do all truly matched filters. Figure 6.8b illustrates the typical delay-line matched filter that one would expect to find in a matched-filter synchronizer, although in a usable filter there would undoubtedly be more delay elements.

The delay-line matched filter[*] is intended to recognize a code sequence and, in a specific configuration, recognize a particular sequence and that sequence only. Each delay element has a delay equal to the period of the expected code clock so that each element contains energy corresponding to only one code chip at any one time. Still considering Figure 6.8, let us assume an incoming signal, biphase modulated by a seven-chip code sequence, 1110010. This signal, with its phase reversals occurring at each one/zero transition in the code, travels down the delay line until chip one is in T_7, two is in T_6, and so on. When all delay elements are filled and the signal-modulating code corresponds to the filter delay elements, the signal phase in T_2 is the same as that in T_5, T_6, and T_7. Also, delay elements T_1, T_3, and T_4 all contain the same signal phase.

Summing the sets $\{T_2, T_5, T_6, T_7\}$ and $\{T_1, T_3, T_4\}$ separately, inverting the summed $\{T_1, T_3, T_4\}$ output, and summing both together, we find that the

---

[*]Often called a "phase coded" matched filter.

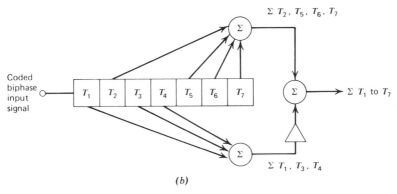

Figure 6.8   (*a*) Baseband, digital matched filter. (*b*) Delay-line matched filter.

signal energy contained in all seven delay elements adds in phase and the total output level is seven times greater than the unprocessed signal level. This phase addition property is similar to that of the chirp filters in Chapter 2, though here the signal is a code-determined discrete phase-shift-keyed signal, whereas a chirp signal exhibits continuous phase change.

The signal enhancement afforded by this technique may be increased by adding delay stages, and process gain available in this fashion is simply

$$G_p(\text{delay line matched filter}) = 10 \log n,$$

where n = number of delay elements summoned.

Practical limits to the maximum length of a delay-line matched filter exist, although recent strides have been made toward avoiding size limitations. Acoustic surface wave delay lines appear to be excellent from the size standpoint, since they have been constructed with as many as 200 delay elements in less than 1 in.[3] The prime practical limit, however, appears to be the accuracy with which the delay elements can be constructed, especially in the smaller types of lines. A second limitation that exists is that of length in time. For production units the length limit appears to be in the range of 20–40 msec, which means that the process gain is limited to approximately 40×10^{-6}/(chip rate) or 36 dB for 100-Mcps codes. Also, the necessity for matching to a particular code length means that the tap structure must be changeable if more than one code is to be accommodated. This has not been accomplished in phase-coded matched filters at lengths greater than 128 chips.

A significant point in the use of delay-line matched filter synchronizers is that they must accurately represent the clock period of the code sequence to be detected. When the delay-line elements exactly match an input signal, the signal summation is perfect. If, however, the code chip rate does not match the delay line, only partial correlation between delay line outputs occurs. Doppler shift, clock frequency drift, or any other cause for clock offset can cause the delay-element periods in the filter to be mismatched with the incoming signal. For this reason it may be necessary to employ an array of delay-line matched filters with graduated delay periods whenever a large amount of clock-rate uncertainty exists.

How does a mismatched input signal affect a matched filter of this type? A code input in which the code is correct but the code period is incorrect has only partial signal phase addition because the signal phase shifts do not occur at the proper time. This means that signals overlap into adjacent delay periods and thus do not add properly. If the mismatched input is a CW carrier, it is summed in phase for both parts of the delay line (Figure 6.8); that is, $\{T_2, T_5, T_6, T_7\}$ adds and so does $\{T_1, T_3, T_4\}$. Inversion of $\{T_1, T_3, T_4\}$ before summing with $\{T_2, T_5, T_6, T_7\}$, however, causes the overall summation of $\{T_1, T_2,, \ldots, T_7\}$ to have a net output equal to the input. The delay-line matched filter, therefore, has an advantage over a CW interfering signal of approximately N, where N is the number of delay elements. As another example of the utility of linear maximal sequences, we should note that the balance between the two sets of signals summed in our example is due only to the one/zero balance in the sequence used.

An equivalent but more flexible approach than the phase-coded matched filter is the surface acoustic wave "convolver." In such a convolver a wave is inserted into each end of a slab of surface wave material. The two waves then propagate toward one another just as the waves in a pond travel. If the two waves match, then the surface displacement is maximum, and if the energy can be integrated over the surface, a pair of matching waves generates peak energy.

In a surface wave convolver, the two waves entered from the opposite ends of the surface wave material (see Figure 6.9) are typically modulated by a code and the mirror image in time of that code. The mirror-image code is generated as the local reference in the receiver. (With linear maximal codes as a base, this is easily accomplished.) Thus the two coded signals generate a

Figure 6.9   Convolver (courtesy Anderson Labs).

maximum correlation when they are propagating in opposite directions in the surface wave device. All that remains is to integrate the energy over the length of the wave interaction. The best approach to the energy integration task is to employ a configuration called an "elastic convolver" in which a piezoelectric material (usually zinc oxide) is deposited on the surface of the surface wave device, with electrodes attached. Then, as the surface wave material vibrates, a signal is generated that is a function of the material motion. Thus, as two matching phase-coded signals travel past each other, they generate a correlation peak that is identical in shape to the autocorrelation of the code but with half the period. That is, if a code at p chips per second is being convolved with its time-reversed replica, the correlation function generated is $1/p$ sec wide (at the base). The code's autocorrelation is $2/p$ wide, however. A typical convolver output is shown in Figure 6.10.

Convolvers have been constructed with process gain as high as 40 dB. They are limited in length and dynamic range, just as the phase-coded matched filters are. This means they must be employed in receivers having either automatic gain control or some alternate means of restricting their input signal range.

Acousto-optical correlators called Bragg cells also have promise. These devices employ a surface wave device combined with a coherent light source (a laser). The laser is aimed at the surface of the surface wave device at the "Bragg angle." If the laser beam is modulated by an input signal and the surface wave device is modulated by an identical reference signal, then the reflected wave from the surface wave device is concentrated and can be detected by a sensor. If the input and reference are not identical, however, the laser beam is scattered. Thus it is possible to recognize the difference between a particular modulating code pattern and any other. Bragg cell

Figure 6.10   Typical convolver output. (Courtesy Anderson Labs).

detectors are still limited by the same thing that limits phase-coded matched filters and convolvers; that is, their internal surface wave device limits the period of the correlation practical at this time.

Synch Recognition

Recognizing synchronization is an inherent part of the synch acquisition problem in a spread spectrum system because the receiver must recognize that it is synchronized, stop the searching process, and starts its local code to tracking before the received and local codes drift apart. To complicate the process all this must be done with a minimum of misses or false starts, even in the face of deliberate interfering signals.

The search process in most spread spectrum receivers in implemented by causing the receiver's code to slip in phase with respect to the received code. This phase-slipping process may be accomplished by simply offsetting the receiver's code clock rate by a given amount and holding that offset until the proper point of synchronization is reached. This, however, is seldom the case. Instead the method of search normally used phase shifts the receiver's clock phase periodically by some small increment so that the two codes in question "slide" past each other in jumps. Figure 6.11 illustrates this effect on the correlation function. Instead of having a smooth triangular shape, the corre-

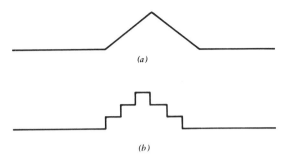

Figure 6.11 Comparative correlation functions for continuous and stepped search processes: (*a*) continuous correlation function for a frequency-offset clock; (*b*) incremental correlation function for a phase-shifted clock.

lation becomes a double staircase in which the number of steps in the stepped-search correlation function is equal to $2s - 1$ (s is defined as the number of phase increments used to shift one clock period). The exact shape of the staircase function varies, depending on the basic clock-phase relationship. Figure 6.12 shows the staircase correlation functions generated by phase stepping one sequence by $\frac{1}{10}, \frac{1}{8}, \frac{1}{5}, \frac{1}{3}, \frac{1}{2}$, and 1-chip increments with respect to a second sequence. In the first column the clocks generating the two codes are synchronous, only the codes are offset. In the second column the code clocks are offset by a constant equal to $s/2$, or one-half the phase increment taken in the search process (i.e., if the search process causes one code to move $\frac{1}{5}$ chip per shift, with respect to the other, the two-code clocks are offset, worst case, by $\frac{1}{10}$ chip). This is the worst case from the standpoint of generating a maximum correlation value, although the peak value generated does dwell for a longer period of time. The question is, "Is it better to employ a large number of phase increments (small steps) or is it better to reduce the number of steps while increasing the time spent at possibly lower peak correlation?"

The answer to this question depends on the type of signal detector used to demodulate the input signal fed to the synch recognition circuits. Where the demodulator is a phase-lock type, threshold is minimized but lock-in time is a factor. If, on the other hand, the demodulator is an envelope detector, there is minimum lock-in time but the threshold is higher.

Let us examine the searching process to see if there is an advantage to be gained in manipulating the manner of search. Figure 6.11 shows a continuously searched code correlation function with height

$$H = \int_0^T f(t)f(t - T)\, dt \quad \text{and} \quad \text{basewidth} = \frac{2}{Rs},$$

where Rs is the search rate for synchronization. The threshold point for

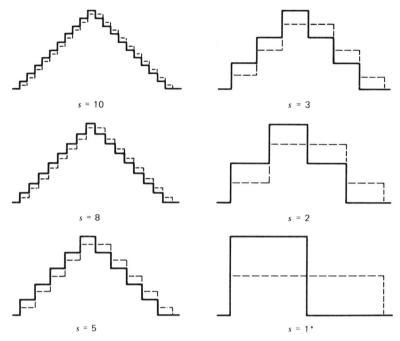

Figure 6.12 Staircase correlation function generated by incrementally stepping code phase in the code search process.

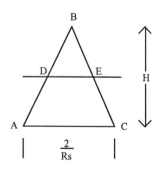

$$\text{Area of ABC} = \frac{1}{2} \times \frac{1}{Rs} \times H \times 2 = \frac{H}{Rs}$$

Figure 6.13 Continuous-search correlation function due to constant offset Rs in receiver code clock.

$$\text{Area of DBE} = \frac{1}{2} \times \frac{1}{2Rs} \times \frac{H}{2} \times 2 = \frac{H}{4Rs}$$

synchronization is taken to be the 50% point on the curve. This is the point where the signal-to-noise ratio is high enough to allow the carrier-tracking loop to lock, which in turn triggers the synch-recognition detectors.

For the correlation curve of Figure 6.13 the total area under the curve is simply

$$\frac{1}{2} \times \frac{2}{Rs} \times H = \frac{H}{Rs}.$$

The area above the threshold point is

$$\frac{1}{2} \times \frac{1}{Rs} \times \frac{H}{2} = \frac{H}{4Rs},$$

and the time at or above the 50% point is

$$\frac{1}{Rs}.$$

For stepped search, as in Figure 6.12, the correlation functions are somewhat different, having a shape developed by the number of steps taken and the step rate. Observance of Figure 6.12 will show that the highest correlation (solid lines) is reached when the initial offset of the codes is an exact multiple of the chip period. The longest dwell time at the high point occurs, however, when the initial offset is exactly half the step size (dotted lines) plus any integral number of chip periods.

The time offset, in other words, in chip periods is

$$\frac{U_t}{T} = \text{number of chip periods } p + r,$$

where r = remainder,
    U_t = total time uncertainty,
    T = (code chip rate)$^{-1}$.

The correlation function has a shape that is determined by

$$\frac{U_t}{T} = p + r,$$

and if $r = 0$, then the highest correlation occurs. If $r = $ (step size)/2, then the longest correlation at a high point occurs. Continuing, the number of steps in the correlation function with $r > $ (step size)/2 is $2s - 1$, but for $r < $ (step size/2 $\neq 0$, the number of steps is $2s + 1$.

The worst-case maximum height for the correlation function occurs when $r = 0$, and is

$$H\left(1 - \frac{1}{2s}\right) = \text{worst case height},$$

but the width at the worst case height is always $2/(\text{step rate})$, where the step rate is just the search rate times the number of steps per chip (s).

The smooth and stepped correlation functions can be compared with respect to their areas. A summary of this comparison follows:

	Smooth Curve	Stepped Curve	
Total area:	$\dfrac{H}{Rs}$	$\dfrac{H}{Rs}$	
Area at or above 50% point	$\dfrac{H}{4Rs}$	$\begin{cases} \dfrac{H}{4Rs}, & s \neq 1 \\ \dfrac{H}{4Rs}, & s = 1, \quad 0 < r < \dfrac{\text{step size}}{2} \\ 0, & s = 1, \quad r = \dfrac{\text{step size}}{2} \end{cases}$	
Time at or above 50% point:	$\dfrac{1}{Rs}$	$\begin{cases} \dfrac{1}{Rs}, & s \neq 1 \\ \dfrac{1}{Rs}, & s = 1, \quad 0 < r < \dfrac{\text{step size}}{2} \\ \dfrac{2}{\text{step rate}}, & s = 1, \quad r = \dfrac{\text{step size}}{2} \end{cases}$	

Thus we see that the stepped-search approach is equivalent to the smooth-rate search approach if the area of the correlation function is taken as a criterion.

A typical synch-recognition lineup is shown in Figure 6.14 (That shown is for a direct sequence system. Frequency hop synch-recognition arrangements are similar, however.) A received signal is multiplied in the correlator with a local reference whose code is slipping in phase with respect to the incoming signal. During this time the signal output for the correlator is the input signal convolved with the nonsynchronous local reference, which produces pseudonoise whether the input is desired or undesired. (The IF and demodulator cannot discern the difference before the occurrence of code synchronization.)

When the point of code synchronization is reached, the correlator output is transformed into a coherent narrowband signal which is passed by the IF and demodulated, and a signal-presence indication (carrier level, etc.) is fed

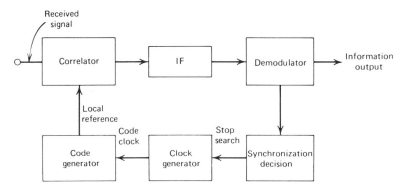

Figure 6.14 Typical direct sequence synchronization-recognition lineup.

to the synch decision block. A positive synch decision then causes the search process to stop and the clock tracking to start. All operations must be carried out, preferably while the receiver's code sequence is at or near the point of maximum correlation. For this reason careful attention must be paid to the relation between the signal detector used and the time allowed for recognizing synchronization—especially when the signal detector is the phase-lock type. Laboratory observations have shown that the lock in time for a phase-lock loop increases by as much as 8–10 times over its high signal-to-noise lock-in time when its input signal is at threshold. This increase must be taken into account in the design of a synch-recognition subsystem.

One other consideration that bears heavily on synch decision circuitry is that the noise output from a spread spectrum system's correlator is often not random, but pseudorandom. This difference, though not apparently significant, forces design provision that must compensate for this lack of randomness. No spread spectrum synch-recognition system can perform at its best without taking the true correlator output noise statistics into account. This is especially true when relatively short-code sequence lengths are used, because the shorter the code, the less it resembles real noise.

Figure 6.15a shows the lack of randomness in a short pseudorandom sequence by cross-correlating an 8191-chip sequence with a CW carrier. The short-term correlation peaks in this signal tend to cause subsequent synch-recognition circuits to register false correlations. Worse yet, each time the CW carrier or the code is changed, the pattern of short-term partial correlation changes and, therefore, it is difficult to design a correlation detector to be optimum for all the possibilities. For the purpose of comparison Figure 6.15b also shows the same cross-correlation between the code and a CW signal except that the code employed in Figure 6.15b is many millions of chips long and its cross-correlation with a CW signal appears to be random. We see from this comparison then that the code length can play an important part in the design and performance capability of a spread spectrum system,

Figure 6.15   Comparison of cross-correlation of short and long-code sequences with a CW signal: (*a*) typical cross-correlation of a short synchronizing code with a CW signal (note pattern repetition); (*b*) typical cross-correlation of a much longer code under conditions identical to *a* above.

especially when the system must be designed to reject large cross-correlating signals.

The synch-recognition function is performed by some device capable of detecting the presence of a coherent signal. The desired signal must then exceed the synch-recognition-circuit's threshold sensitivity, whether the synch recognizer is preceded by a phase-lock loop, envelope detector, or whatever. Typical synch-recognition circuits are shown in Figure 6.16. It is easy to perceive that the only physical difference in the two synch-detector configurations in Figure 6.16 is that one employs a phase-lock loop and the other an

Figure 6.16 Typical synch-detector configurations.

envelope detector. Both the phase-lock loop and the envelope detector perform the same function insofar as the threshold detector is concerned; that is, they output a signal that is a function of the input signal level. The interior or low-pass filter that follows, in combination with a schmitt trigger as a level sensor, then smooths the envelope-generated signal and, if the smoothed signal exceeds the Schmitt's triggered level, a signal is recognized. This in turn, stops the search operation.

Rise time of the synch-detection circuit is set by the integrator, except for those cases in which a phase-lock loop's lock-in time is actually the determiner of recognition time. When phase-lock time is not a limiting factor, the integrator portion of the threshold detector determines the lock-in time. In fact, this detector is most often not one but a bank of detectors, each having different integration times. The purpose of such a bank of threshold detectors, combined to make up a single synch detector, is to optimize the probability of detecting synchronization if it occurs while holding the occurrence of false alarms to a level as low as possible. Figure 6.17 is a block diagram that shows a typical multiple-integration-time synch detector. In the synch detector of Figure 6.17 a DC level corresponding to an input signal's envelope would be applied to all three integrators whose function is to determine the firing rate of the Scmitt triggers shown as threshold-detection devices. The problem is to design the filters (integrators) and threshold detectors so that the probability of detecting the desired signal, when it appears, is maximized, while minimizing the probability of recognizing the wrong signal.

The error rate, and thereby the probability of detecting a desired signal, is set by the first detector, which has the fastest rise time (or widest predecision bandwidth). The third detector sets the synch-false-alarm rate. It is the final decision maker which starts the tracking process, allows data to be demodu-

Figure 6.17   Multiple-integration synch detector.

lated or validates a range measurement. (See the flowchart and timing diagrams in Figure 6.18.) The second detector's prime purpose is to bridge the gap between first and third detectors; that is , because the third detector is delayed for some time before its decision is made, the first detector can be extinguished by noise before the third can fire. This would permit the search process to be restarted even after synchronization has occurred, thus aborting an otherwise successful acquisition. The second detector is then designed to have a probability of signal detection lower than the first detector but with an attendant lower error rate, so that its synch decision can be made and held until the third decision occurs.

Figure 6.19*a* illustrates the relationship between detector-threshold setting, noise level, and signal-plus-noise level. This relationship holds true when a signal level is to be detected by a threshold-sensitive device that might be used with the signal of Figure 6.19*b*. When the signal-and-noise density functions and the detector threshold are known, it is possible to determine probability of signal detection and false alarm for a given signal-to-noise ratio. Referring again to Figure 6.16*b*, the probability of signal detection, $P(d)$, where a detection threshold setting prevails, we can find the probability of detection by integration:

$$p(d) = \frac{\int_{\gamma}^{\infty} f(s + n) \, dV}{\int_{0}^{\infty} f(s + n) \, dV},$$

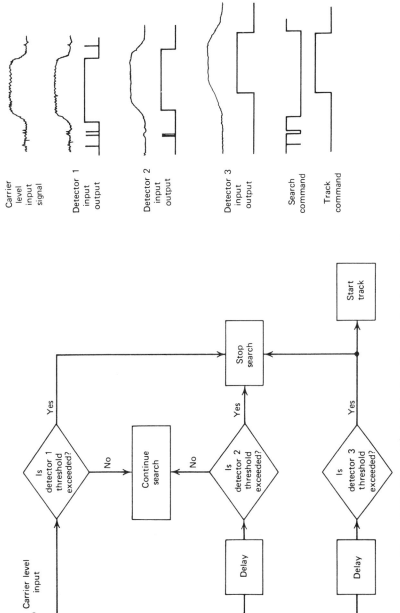

Figure 6.18 Flow chart and timing diagram showing a typical synch-decision process.

Carrier level input signal

Detector 1 input output

Detector 2 input output

Detector 3 input output

Search command

Track command

Carrier level input

Is detector 1 threshold exceeded?

Yes

No

Continue search

Stop search

Delay

Is detector 2 threshold exceeded?

Yes

No

Delay

Is detector 3 threshold exceeded?

Yes

Start track

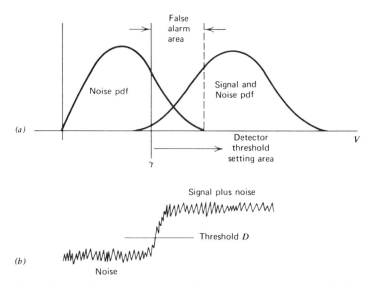

Figure 6.19   Relation of signal and noise with regard to detector-threshold setting.

and the probability of false alarm

$$
P_{(fa)} = \frac{\int_0^\infty g(noise)\, dV - \int_0^\gamma g(noise)\, dV}{\int_0^\infty g(noise)\, dV}.
$$

These derivations are obvious because any signal greater than γ (lying to the right of the threshold setting γ) will trigger the threshold detector, whereas any noise exceeding γ gives a false alarm. Determining the areas in the signal and noise density functions with relation to γ then gives a measure of the false alarm rate and probability of detection. Note that the lower the setting of γ, the more of the signal density function will lie above the threshold, thereby giving higher assurance that a signal will be recognized. Note also, however, that lowering γ increases the noise lying above the threshold, which increases the false alarm rate.

Figure 6.20 plots the probability of false alarm versus signal-to-noise ratio for various detection probabilities. This curve shows that for a 0.95 probability of detection and 10-dB signal plus noise-to-noise ratio, the probability of false alarm is 2×10^{-2}. Thus it is seen that we can determine the probability of signal detection and false-alarm rate for the synch-detection subsystem by setting synch-detector threshold(s) to the proper level.

Referring once more to the three parallel synchronization detectors shown in Figure 6.17, we find that this configuration can be used to maximize the

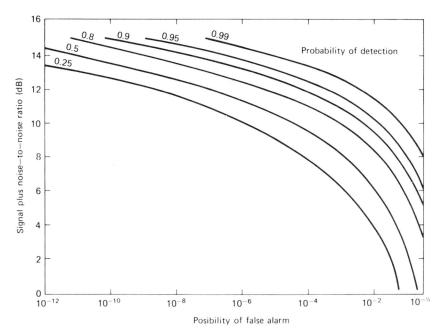

Figure 6.20 Plot of false-alarm probability and detection probability versus signal plus noise-to-noise ratio when signal and $S + N$ pdf's are Gaussian.

probability of recognizing a correct desired signal when it occurs. An alternate configuration in which the three detectors are placed in series (i.e., Schmitt trigger H must fire before J can fire and J must fire before K can fire) tends to force a more positive identification of the desired signal while minimizing any tendency to false alarms in the final code track-start decision process. It is interesting to note that this configuration does not differ significantly in the rate of search stoppages because they are determined primarily by the first decision level (H) and there is little or no difference in this decision circuit, whether the series or parallel configuration is employed.

There is, of course, nothing sacred about three synchronization detectors. Any number can be used, depending on the requirements for a particular system with respect to recognizing synchronization. The whole idea is to make sure that the search process stops when a possible desired signal is present, that the receiver looks at the signal for a long enough period to allow its demodulator lock to the desired signal if it is truly present, and that the search process goes on about its business with minimum delay when there is no desired signal present. In the proper setting, one detector would work well. In others, we might need many more than three.

One other latitude of control over synch detection is in the area of the filters preceding the threshold detectors. These filters delay the decision as to whether the signal desired is present and also as to its loss. Such delayed

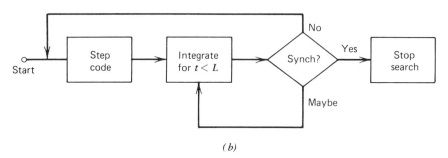

Figure 6.21   (a) Serial and (b) sequential search flow diagrams.

decisions can be advantageous in prevailing recognition of a false signal as well as in holding synch when a desired, but marginal, signal is present. If, for instance, a synch decision is delayed for a period longer than the period of the code being used, it is certain that a partial correlation cannot trigger a false indication of synch. Therefore, at least one synch-detector circuit is almost always set to require code synch for a time much longer than a synch code period. This same circuit, with a long decision delay, is effective in preventing loss of synch when short signal losses or interference bursts occur whose duration is shorter than the detector delay time.

One way of providing an adaptive integration time based on signal likelihood is an approach called "sequential" searching (having nothing to do with "sequential detection"). Figure 6.21 illustrates the difference between serial search (the simplest form of sliding correlation) and sequential search, with a pair of flow diagrams. In serial search, the code is slipped in time by an increment (usually less than one chip interval). Then, the receiver integrates for a fixed interval, usually as long a time as possible, and samples the integrator's output. If the signal is deemed to be present (the integrated output is above a set threshold), then "synch" is declared and the receiver starts to track the incoming signal.

Sequential search is different in that a second threshold level is set, lower than the "synch" threshold, which is considered to be what we will call a "maybe" threshold.* If at the sample time the signal is below the maybe

---

*The "maybe" threshold is also called a dismissal threshold, where any signal that falls below the threshold is dismissed.

threshold, the search processor takes another step. If it is above the maybe threshold, however, the same code position is held and the receiver integrates for another period. This technique allows the receiver to increase selectivity its integration time when the likelihood of a synchronized signal being present is high, and to go on to more fruitful territory when it is low. Comparing the time required for serial and sequential search:

Serial search,

$$\text{synch time} = \frac{U_t}{T \text{ chip}} \times \text{steps per chip integration time},$$

where U_t = time uncertainty, and

$$T \text{ chip} = \frac{1}{\text{code rate}}.$$

Sequential search,

$$\text{synch time} = 1 + p(f.a.) \times \frac{U_t}{T\text{chip}} \times \text{steps per chip} \times \text{integration time},$$

where $p(f.a.)$ is usually 0.1 to 0.2, and integration time (sequential search) $\leq$ [integration time (serial search)]/10. Comparing the two, then,

$$\frac{\text{sequential search time}}{\text{serial search time}} = \frac{1.2}{10} = 0.12,$$

if integration time is $\frac{1}{10}$ for sequential search and the false alarm rate is 20%. Even if the false alarm rate is 50% and the integration time is one-half that of serial search, the synch time would be 0.75 for sequential search. In typical sequential search implementations, the rate of improvement in search is a factor of 3–5 over serial searching.

Now let us discuss a typical synchronization detector design, assuming that we have a sliding correlator synchronizer consisting of a signal detector, a code-rate offset device, a coded signal multiplier (correlator), and a synchronization detector. In the first detector in the triad we choose to have a 0.95 probability of detecting the desired signal, which gives a $10^{-1/2} = 0.314$ probability of false alarm at a 6-dB$(S + N)/N$ ratio. In our second detector we allow a detection probability of 0.8, for a 0.09 probability of false alarm, and the third detector we set for a 10^{-3} false alarm rate and detection probability of 0.25. The overall probability of signal detection, because any one of the parallel detectors can register the signal's presence, is 0.995.*

---

*The probability of occurrence of one or more events from a set of independent events is ΣP_n minus the sum of all the possible combined probabilities; that is, $P_{(\text{overall})} = P_{(a)} + P_{(b)} - P_{(a)}$ $P_{(b)} = P_{(a \text{ and/or } b)}, \ P(a) + P(b) + P(c) - [P(a) \ P(b) + P(a) \ P(c) + P(b) \ P(c) + P(a)(b)(c)]$ $= P(a \text{ and/or } b \text{ and/or } c)$.

Considering the time elements now, the first detector would have no filtering other than that provided by the demodulator preceding it. The second detector's filter would delay its decision for a period such that the dropout probability of the first detector does not prevent the third decision from occurring before synch is lost, and the third detector delays until enough time has passed to ensure that a large portion of the receive code has been synchronized with the incoming signal before making its decision.

The multiple parallel decision network approach is useful in many spread spectrum systems, providing for synch recognition and rejection of spurious cross-correlated signals under all conditions.

6.2 TRACKING

The second part of the synchronization problem follows initial acquisition. Once a spread spectrum receiver has synchronized with a received signal it must continue to operate in such a way that it remains locked with its code reference exactly tracking the coded incoming signal; that is, once the code sequences are matched in time, the receiver must cause its own code chip rate to match the incoming code chip rate as precisely as possible (usually so that the codes remain matched within a small fraction of a clock period).

Without some active method to force the receiver's code clock rate to track an incoming signal's rate, the two will not maintain synchronization. The two most common techniques for doing this, actually very similar, are called "tau-dither" (or "tau-jitter") and "delay-lock." A third technique, is called "carrier lock." For frequency hopping, a method called split-bit tracking is also feasible.

Tau-Dither Tracking

Tau-dither code tracking is an error-signal-generating method similar to those used in some servo sytems for position sensing and correction. Those familiar with servo systems will recognize the word "dither" as describing a small amplitude, usually high frequency, vibrating signal, applied to such things as hydraulic actuators to reduce friction and provide positioning information.

Tau-dither clock tracking in spread spectrum systems makes use of the triangular code correlation function possessed by binary codes. The idea is to cause the code phase in the receiver to remain as close as possible to a received code phase so that the timing of the two codes will remain essentially at the peak of correlation. The way this is done is to degrade the correlation deliberately by a known amount, observe the effect, and employ the information gained to improve the degree of correlation.

A typical tau-dither clock-tracking loop is illustrated in Figure 6.22. In a tau-dither loop a phase modulator capable of shifting the phase of an input

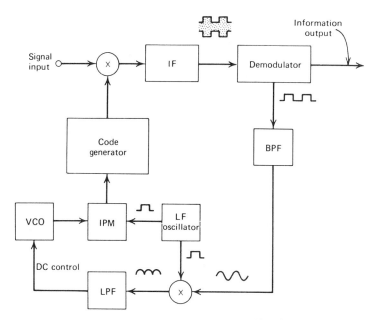

Figure 6.22   Tau-dither clock-tracking loop.

signal by some increment—say one-tenth clock period—is used alternately to set the clock that drives the receiver's code reference back and forth between shifted and unshifted positions. When this occurs, the degree of correlation between the received and local codes also changes, which causes a shift in the amplitude of the signal input to the demodulator. (The information demodulator, usually for PSK or some other form of angle modulation, is not seriously affected by a small amount of amplitude change, because of the orthogonality of angle-modulated and amplitude-modulated signals.) As the clock phase is shifted back and forth, the signal is amplitude modulated (see Figure 6.23) at the phase-shifting rate.

By observing Figure 6.24 it is easy to envision the tau-dither operation. Consider that we have a pair of identical code sequences, which we can vary (with respect to their phase relationship) in any way we wish. Now, as we approach the point of perfect synchronization and pass it, we generate the familiar triangular function. Let us, however, set the code-phase relationship so that we are at point 1*a* on the curve. This means that our two codes are shifted approximately one-half chip apart in a direction we call negative. The signal output from a correlator in which these two codes are inputs would be about half its maximum value.

Now assume that the code phase is moved (closer to maximum correlation) to a more positive point 1*b*. The correlation between our two codes is now greater. Similarly, for point 2*a* on the positive side of the correlation function, the correlation between the two codes is greater than at point 2*b*.

Modulation phase shifts due to data

Tau—dither
— envelope
modulation

Figure 6.23 Correlator output signal showing envelope modulation due to code phase shifts.

The correlator output signal is also greater when code phase is at points 1*b* and 2*a* than at 1*a* or 2*b*. We have, then, the basis of a coherent error generation and detection loop.

Let us postulate a square wave, which, when true, causes a code sequence to be advanced by a fraction of a chip, and when false, causes it to go back to its unadvanced state. Assume that the relative code states at the beginning of our operation are such that we are at point 1*a* and that the advance in phase moves us to point 1*b*. Our square wave then causes us to go back and forth between 1*a* and 1*b*, thereby amplitude modulating the correlator output at our square wave rate. Once this amplitude-modulated signal is passed through a detector, filtered, and fed back, the amplitude modulation gives an indication of the direction in which the code pair must be moved to reach perfect synchronization. (In the region of the correlation peak the amount of amplitude modulation is also an indication of how far the two codes are offset. This effect is discussed further in a later paragraph.) It should be

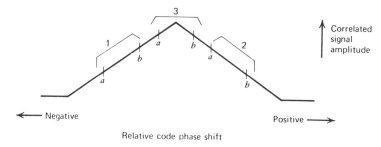

Figure 6.24 Correlation function showing phase-shift increments.

Figure 6.25 Phase-detector operation.

Figure 6.26 Dithered clock loop waveforms (*a*) no interference; (*b*) 15-dB interference.

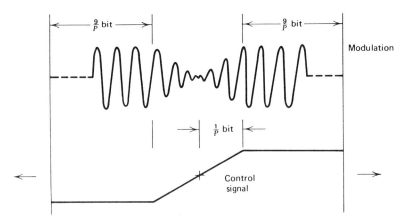

Figure 6.27 Modulation and control signal amplitude in the region of ± 1-chip synchronization for tau-dither loop.

obvious that the sense of the amplitude modulation (i.e., upward or downward modulation) produced by phase shifting between points $1a$ and $1b$ is opposite to that produced by going from point $2a$ to $2b$. Detection of this amplitude modulation in a phase detector then allows us to generate a DC signal which has the proper polarity (and amplitude) to set the VCO of Figure 6.22 to run faster or slower, so that the code sequences in which we are interested speed up or slow down, as required for correct synchronization. Phase detector operation is illustrated in Figure 6.25, and Figure 6.26 shows signals typical of an actual operating tau-dither tracking loop.

Referring again to Figure 6.24, let us examine area 3, in which the code pair being examined has relative phase shifts ahead of, and behind, perfect synchronization; that is, if the overall increment phase shift over which the codes are shifted is $\frac{1}{10}$ chip, the shift in area 3 of Figure 6.24 is $\frac{1}{20}$ chip negative and $\frac{1}{20}$ chip positive, and we see that the tau-dither operation centers on the peak of correlation so that it is never more than $\frac{1}{20}$ chip from perfect synchronization.

When the code loop is tracking so that the phase shift does straddle the correlation peak, as in area 3, the phase jumps produce no modulation* of the correlation signal. As the code drift apart, the modulation increases from zero at perfect phase synch to a maximum that is a function of the total phase shift. If the total incremental phase shift used is $\frac{1}{10}$ chip, the maximum modulation will be limited to that produced by a $\frac{1}{10}$-chip phase shift. Figure 6.27 plots the level of modulation amplitude as a function of code phase offset over the region of synchronization, where the phase increment used for tau-dither error generation is $1/P$ chip.

---

*In practical systems some residual modulation remains.

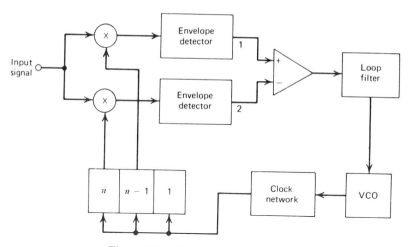

Figure 6.28   Delay-lock tracking loop.

Considering that our two codes are synchronized and that the tau-dither phase shift is centered on the peak of correlation, what is the effect of the tau-dither operation? Remember that the two-position ahead-and-behind phase shift operation always holds the codes at a position slightly offset from the optimum, so the correlated signal is never at its best. The amount of correlation loss, however, is only $1/2P$th of the maximum, so that when the tau-dither phase jump is $\frac{1}{5}$ chip, the maximum correlation loss is $\frac{1}{10}$ or 10% (about 1 dB). Thus we see that a receiving system is degraded by no more than $100/2P\%$ by a tau-dither phase shift of $1/P$ chips.

Delay-Lock Tracking

The second common type of tracking loop is called a "delay-lock" loop, because two separate correlators are used which are driven by two local code reference signals—identical except that one is delayed in time with respect to the other. The amount of delay between the local codes is usually one or two chips, either of which is readily available, because adjacent shift register stages (if properly chosen) have outputs delayed by one chip. Figure 6.28 is a block diagram of a delay-lock loop. Except for the details of error generation, a delay-lock loop is identified to a tau-dither loop. (Here we will not conjecture as to which type preceded the other.)*

In a delay-lock loop two local reference signals are generated. Identical except for the time delay mentioned in the preceding paragraph, these references are used for comparison with a single incoming signal in two separate correlators. Just as before, the correlation between an encoded

---

*A tau-dither loop is a sampled delay-lock loop.

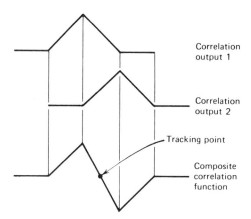

Figure 6.29 Correlation waveforms in delay-lock loop.

incoming signal and the local reference is a triangular function two chips wide, but now there are two correlated signals available (one at the output of each correlator), each with a triangular correlation function but with their peaks of correlation offset by the amount of delay between the two local reference signals. Thus the composite correlation function for a delay-lock loop has a double-peaked triangular shape (see Figure 6.29) after summing, in which one-half of the double triangle is inverted, so that the composite correlation has a linear region centered around the point halfway between the two correlation maxima.

Now, if the summed correlator outputs are filtered and used to control the reciver's clock source (the VCO shown), the receiver's code will track the incoming code at a point halfway between the maximum and minimum of the composite correlator output. (If the composite waveform is viewed as a discriminator response, the delay-lock loop is quite similar to a common AFC loop.) Because the receiver's two code references are forced to track the incoming code with an offset equal to one-half the delay between them, it may be necessary to decrease the delay to some value much smaller than one chip. Otherwise the code correlation would never be better than half its peak value. An alternative to decreasing reference code delay would be to take a separate $\frac{1}{2}$-chip-delayed code to a third correlator. This would provide an optimally correlated signal output for information demodulation. Figure 6.30 illustrates this concept in block form.

Coherent Carrier Tracking

Both RF carrier and modulating signal making up any communications signal are acted on in exactly the same way by the path from transmitter to receiver; that is, if either the transmitter or receiver is in motion, the Doppler effect

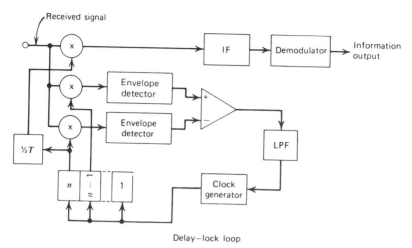

Delay–lock loop

Figure 6.30   Delay-lock loop with half-chip delayed correlator for signal demodulation channel.

will be equivalent (though of different magnitude) on a modulating signal, to whatever it may be on its carrier. Because modulating signals are, in general, at low frequencies, the Doppler shift seen is usually insignificant. With spread spectrum systems, however, and especially direct sequence systems, the high modulation rate is prone to significant Doppler-frequency shifts. Because the carrier and modulation signals suffer from identical shift mechanisms, it is possible to correct for both carrier and modulation (code) uncertainties with a single tracking loop.

In carrier-coherent systems the mechanism used is just what the name implies. In other words, the code-sequence clock rate is designed to be a submultiple of the carrier frequency or both carrier and clock are derived from the same source. Common sources must be used in the transmitter and receiver (i.e., each must have common code rate and carrier sources—not implying that both transmitter and receiver have the same source).

When the modulation rate and carrier frequencies are coherently related, tracking one will also provide for the other; for instance, consider a transmitter in which a 50-Mcps code-sequence direct sequence modulates a 500-MHz RF carrier. The carrier is derived as a 10-times multiple of the code clock. At the receiver the local oscillators are also derived from the code clock, and the postcorrelation signal is translated down to a convenient demodulation frequency by mixing with these code-rate-related oscillators. Thus the transmitter's 50-Mcps code clock rate is precisely related to the carrier signal seen at the receiver's demodulator, and a single phase-lock demodulator can be used to track code clock, and carrier-frequency shifts. It may be necessary to provide a method for resolving the phase ambiguity which can result from locking to a frequency related to code rate but different from it. A simple

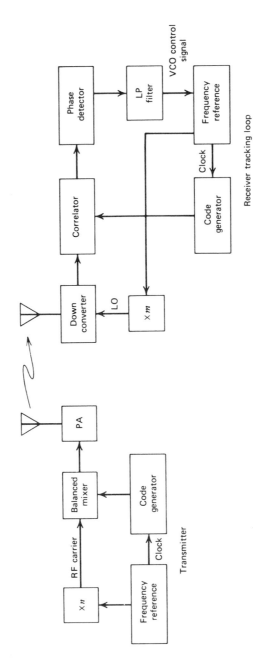

Figure 6.31 Carrier-lock tracking system.

± 1-chip-correlation search is sufficient to satisfy the difficulty, and once the proper phase is chosen only the frequency needs to be tracked.

A carrier-coherent tracking system is illustrated in Figure 6.31. It is interesting to note that whenever the frequency reference in a carrier-lock receiver is adjusted to compensate for an input signal frequency change, the frequency of all local oscillators also changes. This fact can be put to use by designing the receiver so that local oscillator changes tend to cancel input signal frequency changes. The overall effect therefore is that of combining a frequency feedback loop with a phase-lock loop. In effect, the demodulator's tracking range is reduced by the amount of frequency feedback cancellation, which often permits less complex demodulator design.

When it can be employed, the carrier-lock tracking method is quite useful. It is limited, however, in that the transmitter from which it receives signals must have code clock rate and carrier phase coherence for the receiver to operate. Otherwise unrelated frequency drifts between clock rate and RF carrier would cause the receiver to lose lock.

Split-Bit Tracking

Split-bit tracking takes advantage of the envelope shape of the signal at the output of a frequency-hopping-receiver's correlator. This envelope is constant in amplitude, though not in phase (phase shifts occur at the hopping rate) when the receiver is synchronized perfectly. If however, the receiver is ahead in time, then the correlator output is high during the first part of the hop period but low during the second part (see Figure 6.32).

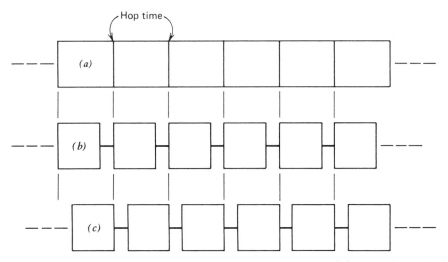

Figure 6.32 Illustration of timing for split-bit tracking loop: (*a*) perfectly synched signal envelope; (*b*) receiver ahead in time; (*c*) receiver behind in time.

When the receiver is behind the transmitter in time, the correlator output is low during the first part of the hop period and high during the second part (which is just the opposite of the case when the receiver is ahead of the transmitter). Therefore, by integrating over the two halves of the hop time and comparing the energy present, a loop can be constructed that causes the receiver to adjust its code timing so that the energy is equal in both halves of the chip time. Thus the frequency hopping receiver can be caused to track the incoming signal timing.

Correlation Loss Due to Bandpass Filtering

The correlation function generated by the delay lock and other tracking loops can be degraded by bandpass filtering of the transmitted signal. This is illustrated in Figure 6.33 (from C. R. Cahn). Figure 6.33 shows the changes in

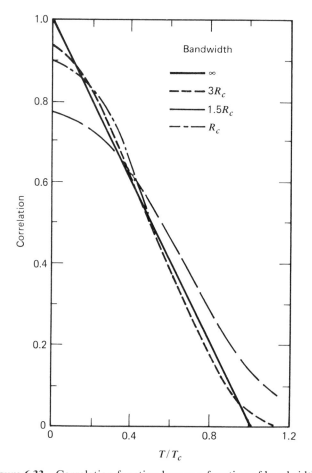

Figure 6.33   Correlation function loss as a function of bandwidth.

the shape of the correlation function as filter bandwidth is varied from the chip rate of the code to three times the chip rate, and when the filter is removed. Note that only with no filtering is the function sharply peaked.

Loss expected is as follows:

Filter Bandwidth	Loss (dB)
$2R_c$	0.45
$1.6R_c$	0.5
$1.2R_c$	0.75
R_c	1.1
$0.8R_c$	1.6

Tracking error loss as a function of phase jitter caused by jamming is

Error	Loss (dB)
$0.1T$	0.72
$0.2T$	1.51
$0.3T$	2.37
$0.4T$	3.34
$0.5T$	4.42

where

$$\text{loss} = 10 \log 1 - \sqrt{\frac{2}{\pi}} \frac{\sigma}{T},$$

where σ = phase jitter,
T = clock period.

PROBLEMS

1. What synchronization uncertainty might one expect from a 10-Mcps code generator whose chip-rate accuracy averages 1×10^{-9}, after $2\frac{1}{2}$ days?

2. What if the above system employed a 100-kcps code generator?

3. Given a system with 35-dB process gain, 20-MHz RF bandwidth, and a 4095-chip synchronization preamble, what minimum time must be allowed for synch?

4. For a probability of detection of 0.9, what probability of false alarm might be expected of a detector whose input signal plus noise-to-noise ratio is 10 dB?

5. What system loss might be expected from the use of a tau-dither loop for clock tracking, whose total time shift is $\frac{1}{2}$ chip?

6. A direct sequence system with a 10-Mcps code clock rate is traveling at a relative velocity of 980 mph away from us. What is the Doppler shift on his code clock as received by us?

7. The system turns and approaches us at 580 mph. What is the Doppler shift on the clock now?

8. The carrier frequency of the transmissions in question is 375 MHz. What is the carrier frequency that would occur due to Doppler?

9. A frequency hopping system that is 112 miles away transmits to us. Each of the terminals is synchronized in real time to within 10 nsec. The hop rate is 100 khps. How far must our receiver search to gain initial synch?

10. We reply without resetting our code timing. How far must the other terminal search to synchronize with our signal? What if we reset to our original timing before transmitting?

11. A delay-line matched filter is to be constructed to match a 1023-chip code sequence. The code rate is 5 Mcps. What are the major delay characteristics required?

12. What signal-to-noise ratio would be required to give a probability of detection of 0.99 for a probability of false alarm of less than 10^{-2} (for a single detector)?

13. Compare chirp signals, from the standpoint of synchronization, to direct sequences and frequency hop signals.

7

THE RF LINK

Spread spectrum modulators and demodulators are dependent, in general, on transmitting in a frequency band capable of supporting a wideband signal, in which enough bandwidth can be allocated. In the usual case modulation and demodulation are performed at an intermediate frequency and translated up or down, to and from, the transmission frequency. Some systems have performed the code modulation and demodulation functions at the transmission frequency. Implementation problems, however, with regard to the required signal multipliers and filters have almost invariably forced these operations to be carried out at lower, fixed intermediate frequencies.

Careful spread spectrum system design also extends to the propagation medium. Even the atmosphere can limit the band of usable frequencies. When a broad band of frequencies is to be used, both phase shift and amplitude distortion can be introduced on the signal just by passage through a supposedly benign atmosphere. This chapter outlines some of the special problems that must be addressed in designing spread spectrum receivers and transmitters—actual bandwidth requirements, effect of poor receiver noise figure, AGC, and dynamic range, and other considerations peculiar to spread spectrum systems. Its most important purpose is to show that even though spread spectrum processing is in general transparent, the task of actually constructing a workable system involves more than adding a code generator and modulator to an existing conventional system. To be sure, such appliques are possible and sometimes even practical, but their performance is always limited by the system to which they are grafted.

7.1 NOISE FIGURE AND COCHANNEL USERS

Noise and undesired cochannel signals (also noise) are just more interference, or jamming, to a spread spectrum receiver. They are added to any other undesired incoming signal and processed in the same way. Our first inclination, then, is to expect a spread spectrum receiver to have greater sensitivity than a conventional receiver. Not so.

Except for narrowly defined and even (some might say) picayune gain made possible by a combination of bandwidth manipulation applicable only in direct sequence systems and noise × code-reference spreading, little is gained. Because the effective gain can be up to about 2 dB, however, there are situations in which this gain could be useful.

Any receiver's sensitivity is limited by the amount of noise it must take in along with its desired signal, including its own internally generated noise. These sources of noise are additive, as is any other interfering signal. Noise power into the receiver is kTB watts, where k = Boltzman's constant = 1.38 $\times 10^{-23}$ $J/K°$; T = temperature of the receiver input termination in degrees kelvin; B = noise bandwidth of the receiver.

In a spread spectrum receiver the receiver bandwidth must be commensurate with the desired signal bandwidth. Even though a given baseband signal is spread over an RF band many times its own bandwidth, the receiver must necessarily accept more noise in exactly the same ratio as the increase in bandwidth; that is, if a 10-kHz (two-sided bandwidth) signal is spread over a bandwidth of 10 MHz (yielding a processing gain of 30 dB), the increase in noise intake would also be 30 dB.

For a given information bandwidth, the sensitivity of a narrowband radio is defined as that point where:

$$\text{input}\,\frac{S}{kTB_{\text{IF}}NF} \geq \frac{S'}{kTB_{\text{IF}}NF}(\text{minimum}) = \frac{S'}{B_{\text{info}}NF}(\text{minimum}),$$

where $S' = CS$,

C = conversion factor depending on modulation type and index,
NF = noise factor.

With a spread spectrum radio, sensitivity is still defined at a given output signal-to-noise ratio, but the receiver's processing gain enters the picture:

$$\text{input}\,\frac{S}{kTB_{\text{RF}}NF} \geq \text{output}\,\frac{S'G_p}{kTB_{\text{RF}}NF}(\text{minimum}),$$

but

$$G_p = \frac{kTB_{\text{RF}}}{B_{\text{info}}},$$

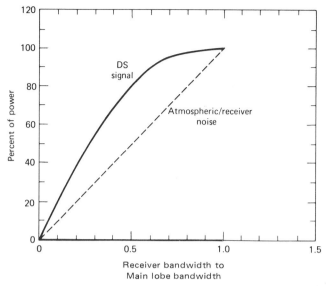

Figure 7.1 Comparison of power versus bandwidth for direct sequence and flat noise signals (normalized to direct sequence mainlobe bandwidth).

therefore,

$$\text{output} \frac{S'kTB_{\text{RF}}}{kTB_{\text{RF}}\,NFB_{\text{info}}} = \frac{S'}{B_{\text{info}}\,NF}\,(\text{minimum}),$$

which is identical to the narrowband result. Thus we see that a spread spectrum receiver should have similar sensitivity to a narrowband receiver, for the same information bandwidth.

By judicious choice of receiver bandwidth, however, we could reject more of the potential noise than signal and produce some gain. Figure 7.1 illustrates cumulative power distribution in a direct sequence signal's main lobe and compares it with the cumulative power distribution for flat noise such as atmospheric or receiver noise. For receiver bandwidth equal to main-lobe bandwidth the receiver would accept all the noise and the desired signal. Narrowing receiver bandwidth to 60% of the main-lobe bandwidth, on the other hand, would reduce the desired signal to 90% of its original power, a loss of 0.46 dB. The noise contribution, however, would be reduced to 60% of its original power, down 2.2 dB. This simple expedient (reducing receiver bandwidth) thus produces a signal-to-noise ratio gain of approximately 1.7 dB.

For frequency hopping receivers, bandwidth manipulation does not give an advantage because the frequency hopper's power distribution is flat, as is the

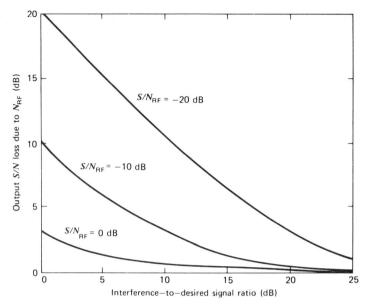

Figure 7.2 Loss in receiver output S/N ratio versus J/S for various RF input S/N ratios.

noise distribution. To be sure, if the frequency hopping signal is truncated, as a filter might, the receiver's error rate would increase, because all the power in any chip transmission lying outside its filter passband would be completely lost.

Assuming that we have chosen a bandwidth for our receiver, what effect has our receiver's noise figure, or cochannel interference? These two noise sources have nearly the same effect on receiver performance. When noise figure or cochannel interference alone increases, then, if output quality is to be preserved, the desired signal must increase a like amount.

When the jamming margin is the important factor, neither the noise figure nor cochannel interference is really dominant, even though both contribute to system performance. Figure 7.2 shows the loss in output signal-to-noise ratio suffered by a spread spectrum receiver as input interference (jamming, etc.) is varied. Various signal-to-noise (exclusive of the jamming signal) ratios are shown as parameters. As an example, consider that N_{RF} (N_{RF} = cochannel interference + environmental + front-end noise) is 10 dB above the desired signal. Then, as the jamming signal is increased, the loss due to N_{RF} decreases to the point at which it is negligible. Thus we realize that even when cochannel interference and receiver noise power equal that of the maximum expected (design level) jamming, the degradation in jamming performance is no more than 3 dB. This is a simple but important result.

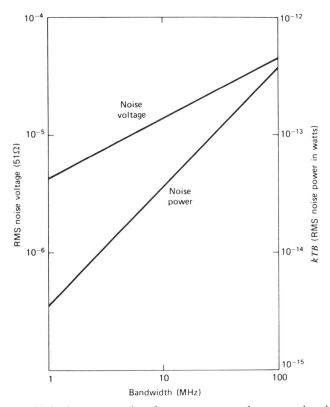

Figure 7.3   Noise input to receiver for common spread spectrum bandwidths.

It should be pointed out that in most cases a spread spectrum receiver should not be expected to give maximum rejection for undesired signals* and simultaneous operation on a minimum RF signal. Maximum sensitivity implies that the desired signal is small enough that combined environmental and front-end noise fall just short of jamming the receiver. Therefore, a spread spectrum receiver should be expected to give optimum interference rejection only when its input signal is at least 3 to 10 dB above the minimal level.

Figure 7.3 is a plot of kTB for bandwidths encountered in spread spectrum systems, to help the designer determine minimum signal operation. The receiver noise factor[†] directly multiplies this kTB.

---

*In consumer systems, the undesired signals that dominate the channel may consist entirely of cochannel users.

[†]Noise factor is the number of times a receiver front end multiplies the noise (kTB) at its input. Its relation to noise figure is: noise figure = 10 log noise factor (dB).

7.2 DYNAMIC RANGE AND AGC

The dynamic signal range a particular receiver must handle depends largely on the type of environment in which the receiver is expected to operate; for example, a satellite-to-ground link is much more stable than an aircraft-to-aircraft or aircraft-to-ground link, for variations of geometry in the aircraft link are extreme and in general unpredictable, whereas satellite location is predictable at all times.

In addition to variation in relative transmitter-receiver location, any earth-bound receiver must contend with variations in signal strength due to the propagation medium. Because the earth's atmosphere is far from homogeneous, it offers various discontinuities that cause diffraction, refraction, and occasional ducting, all of which can produce large variations in signal strength regardless of link geometry, even for fixed-terminal configurations. Even the angle of propagation through the atmosphere can make a significant difference in signal strength and quality; for example, signal strength commonly varies over a range of six or more orders of magnitude in an airborne UHF link. Other types of communication link may vary as much, but few have such rapid variations as the aircraft link. This means that any receiver whose demodulation system is sensitive to input level must, for optimum reception, incorporate some method of compensating for changes in RF signal level seen at its antenna.

The total signal power range for a receiver can be estimated as

$$10 \log \Delta P - 10 \log \Delta M + 20 \log D_{max} + 10 \log L_{max} = \text{signal range},$$

where ΔP = signal maximum/signal minimum,
   ΔM = antenna gain/null depth,
   D_{max} = longest operating distance,
   L_{max} = maximum losses allowed under design conditions (including multipath fading).

These can easily reach up to 120 dB or more, which means that AGC range must be similar. The front end of all stages up to the correlator must have at least 30-dB headroom above the desired signal level to remain linear and provide 30-dB process gain.

One method of holding the signal level constant is to design the receiver system in such a way that clipping or limiting action takes place prior to the demodulator at all levels down to the design threshold. This method is obviously precluded, however, if any amplitude information is to be expected from the signal, for the limiting action reduces all incoming signals to a single value. The limiting approach is restricted to a few types of demodulation systems of the frequency- or phase-modulated variety.

All demodulators, even those designed to demodulate only frequency information, are sensitive to signal level to a certain extent; for example, the phase-lock FM demodulator. Because this device is essentially a servo loop that employs a phase detector as an error-sensing device, any change in the level of input signal varies its effective open-loop gain, thereby affecting the closed-loop bandwidth and damping factor. When the phase-lock loop is used as a threshold-extending demodulator, as is typically the case in spread spectrum communications equipment, these parameters must be held within narrow limits in order to achieve the optimum threshold and processing gain for the overall system.

Direct sequence receivers suffer losses up to 6 dB in jamming margin when operated with limiting front ends[13] and degrade even more when the jamming signal is coherent with the pseudonoise carrier. Indeed, any nonlinearity ahead of a direct sequence system's code-matching correlator can produce spurious interfering correlator inputs and consequent self-jamming. Although limiting amplifiers have been employed in certain direct sequence communications systems, this has been done to avoid the added complexity of an AGC system and at the cost of less than optimum performance.

When signal limiting is not allowable, two general methods are available for controlling the transfer of signal from the antenna to the demodulator in modern communications systems. These are (a) controlled gain amplification and (b) controlled loss attenuation.

Gain control techniques rely on devices in which a DC circuit parameter may be varied to cause a change in the small-signal AC circuit parameters. As an example, a transistor with a nonlinear beta characteristic can be used in a circuit in which the DC operating point may be moved by changing the circuit's DC bias voltage. This change in bias causes a smaller change in output current at one bias point for a given change in input current than results from the same amount of change in input current at another bias point. Various transistor types have beta curves that may provide an increase in gain as collector current is increased or decreased with current, depending on the particular device type.

Other variable-gain devices of note are tetrode transistors, both field effect and bipolar types, which have been used successfully in many communications systems. Attenuating control devices are also used as adjustable shunt or series impedances which cause a variable amount of loss in a voltage-dividing network. Some of the devices that may be used in this way are unijunction transistors and pin diodes, both of which vary inversely in resistance as their DC bias current is varied. Pin diodes have been employed as attenuators at frequencies up to and into the microwave range.

A gain control system, whether of the variable attenuation or variable gain type, is considered to be mandatory for use in most* spread spectrum

---

*Especially direct sequence systems.

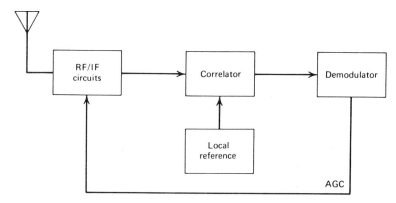

Figure 7.4   Correlation receiver with AGC loop.

systems. The following block diagram (Figure 7.4) shows a typical receiver with AGC that controls the signal level into its correlator and thereby its demodulator input level.

In addition to the problems presented by their requirement for linear transfer function over a large range of values of the desired signal, spread spectrum systems, and antijam systems in particular, are required to control their gain not only as a function of a desired signal but must also have control over the value of all signals that fall within their frequency aperture, including interference. This means that interference present in the absence of a desired signal is allowed to control the system gain. When a desired signal is present, however, interference must not be able to control system gain. If an interfering signal's effect in the presence of a desired signal is not minimized, the desired signal could be depressed simply by increasing the interference level, even within the jamming margin of the equipment. (Interference power may exceed signal power by several hundred times within this margin.) This situation could force the desired signal down in level to the point of demodulator thresholding or cause amplitude modulation of the synchronized signal, which could, in turn, make the jammer even more effective.

The ideal gain control system for a spread spectrum system must not only control the signal level fed to the detector but must also operate under the following conditions:

1. It must control signal level to prevent false synch recognition when the desired signal is not present but a large amount of noise due to interference is being applied.

2. It must control levels only as a function of desired signal input when the desired signal is present and synchronized, ignoring all other signals.

3. It must provide near-critical damping to prevent pulse-stretching effects in a pulsed interference environment.

Conditions 1 and 2 imply the use of two different AGC detectors, one before synchronization and a second after synchronization and a second after synchronization, because the AGC loop bandwidth and gain under the two conditions are necessarily different. Thus the problem of disengaging presync AGC and engaging the postsync AGC is presented. The changeover from unsynchronized to synchronized AGC control must be accomplished with minimum transient overshoot or undershoot, thus requiring that the AGC be properly damped under both noise and signal conditions. Proper damping is also required to prevent stretching of pulsed interference effects, for any overshoot or undershoot in the AGC loop could prevent synchronization over a larger percentage of time than necessary, giving the interferor a larger effective duty cycle. For example, assume a pulse large enough to preclude synchronization during its "on" time and a pulse width less than the fallout time of the sync recognition channel so that once synchronization has occurred the "on" pulse will not cause loss of tracking. Then synchronization can occur at any time when the pulsed signal is in its "off" state, and the probability of success for a single transmission will not be less than $1 -$ interference duty cycle. The effects of interfering pulses may be stretched by improper AGC loop damping, however, and sync probability will be significantly lowered.

Reference to Figure 7.5 points out how a pulsed signal's effect may be stretched by nonoptimum AGC processing. Because the AGC waveform is ideally a reproduction of the envelope of the pulsed signal, any gain controlling signal larger than the amount required suppresses a desired signal

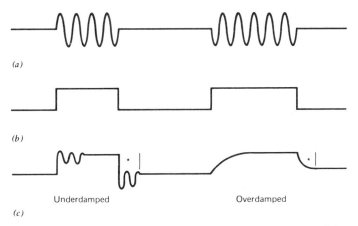

(a)

(b)

Underdamped   Overdamped

(c)

Figure 7.5 Pulse-jamming AGC waveforms: (*a*) pulse-jamming signal; (*b*) ideal AGC waveform; (*c*) poorly damped AGC waveform.

unnecessarily and degrades threshold, whereas control less than optimal permits overdriving of level-sensitive detection circuits and causes false indication of synchronization.*

Processing of the presynchronization AGC envelope by periodic sampling could be used to detect the presence of pulse interference and cause the search process to be speeded up by an amount equal to the interfering signal duty cycle, which could result in a system that searched for synchronization only when the jamming was in the off state, at a higher search rate, thereby completely ignoring the interference. Again, however, near-critical AGC damping would be mandatory.

For a frequency hopping system special consideration must be given to the presynchronization AGC control system. Direct sequence systems permit AGC control by a detector operating within the bandwidth of the signal demodulator. Because all signals applied to the correlator are spread at least to the bandwidth of the local reference signal, a part of the noise power is forced to fall within the bandwidth of the demodulator where it may be processed to generate AGC voltage.

In a frequency hopping system little or no energy may fall within the band of the demodulator because the local oscillator will hop to one of many discrete frequencies, some of which are far enough from the center frequency so that no energy falls into the demodulator passband. Frequency hopping signals at the input to the correlator (see Figure 7.4) lie in one range of frequencies, whereas the local reference signals fall in another range. The difference-frequency output then lies in a range centered at the difference frequency, but twice as wide as the local reference signal, for an unsynchronized frequency hopping input signal. When the input is continuous wave, however, the correlator output is a replica of the local frequency hopping reference translated down to the same band but centered around the difference frequency between the jammer and the local reference. Once synchronization occurs, the input and reference signals correlate and their constant frequency difference produces an output that is entirely within the detection bandwidth of the receiver.

One way to meet the requirements set forth for frequency hopping system automatic gain control, then, is to generate control voltage from a detector that immediately follows the correlator and samples its output over a bandwidth equal to twice the frequency hopping bandwidth. This may be termed a wideband AGC and requires fixed gain IF's following the correlator or a separate AGC loop for control of the stages following the correlator. It would be possible, however, to generate AGC from the frequency hopped energy which does fall in the demodulator information passband at the cost of greatly reduced AGC loop gain. Alternatively, if signal levels permit it, the

---

*Figure 7.5c illustrates the way in which improper damping stretches the effect of pulsed interference. Note that both the under and overdamped AGC waveforms extend beyond the period of the interference.

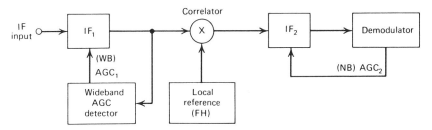

Figure 7.6 Wideband/narrowband gain control network for frequency hopping receiver.

wideband amplifiers preceding the correlating mixer can be left without AGC. In this case narrowband AGC alone, taken from a point following the correlator, is satisfactory. Narrowband AGC is generated as a function of the signal within the information bandwidth.

Figure 7.6 illustrates a two-loop gain control network for frequency hopping systems that could meet the stated requirements. Control signal AGC_1 is generated as a function of the total signal power that falls anywhere within the desired signal band $f_F \pm f_{FH}$. This control signal would set the level of signals applied to the second IF amplifier. AGC_1 loop gain would be set to allow maximum linear output signal level, and IF dynamic range must be such that a desired signal plus the maximum expected interfering signal may be handled simultaneously and linearly.

Control signal AGC_2 is generated as a function of the signal power falling within the desired information bandwidth.

Figure 7.7 shows a single-loop AGC network for frequency hopping. In this configuration no AGC is applied to the wideband receiver sections and an input signal-plus-noise level up to the maximum signal, plus the expected maximum interference, must be tolerated. This implies that IF_1 gain should be low because more than minimum gain would force the frequency hopping local reference to be supplied at a high level, which could be difficult to maintain and could also produce system self-interference.

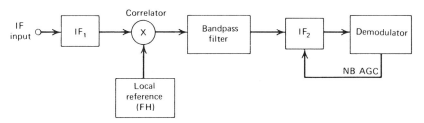

Figure 7.7 Alternative frequency hop gain control.

In the case of a combined frequency hop and direct sequence system, a combination of detection methods might be ideal in which a wideband detector would act to give precorrelator control with a narrowband AGC giving postcorrelator control. This would permit separation of the AGC function and good control of interaction, which is a must for a satisfactory spread spectrum AGC network.

In summary, automatic gain control for spread spectrum systems must provide the following:

1. Linearity over the full range of desired and undesired signal inputs.
2. Effective noise AGC.
3. Close control of demodulator input signal level.
4. Proper signal versus noise AGC takeover characteristics.
5. Near-critical damping of AGC loop response.

Spread spectrum systems using these AGC configurations have been successful in operating in environments in which conventional modulation methods were not satisfactory. Many frequency hopping receivers can operate without AGC. Where it is required, however, the techniques described are useful.

7.3   THE PROPAGATION MEDIUM

Propagation for a spread spectrum signal is much the same as for any other signal, with the exception that the wide bandwidths employed sometimes restrict system use. The wideband modulation that can cause a system to be restricted in some areas can be a boon in others. This section lists some of the advantages and disadvantages of spread spectrum systems when faced with the common perils shared by all RF transmissions.

Line-Of-Sight Loss

Path loss for all kinds of spread spectrum signals is the same as for a narrowband signal operating at the same center frequency. The same physical laws apply. Therefore, if, for instance, an AM signal suffers a given amount of loss in traveling a certain distance, a direct sequence or frequency hopping signal would suffer that same amount of loss.

For convenience, a nomograph of path loss versus distance for frequencies between 100 MHz and 10 GHz is given in Figure 7.8. This nomograph is based on the well-known equation for path loss between isotropic antennas,

$$L = 37.5 + 20 \log \text{distance} + 20 \log \text{frequency},$$

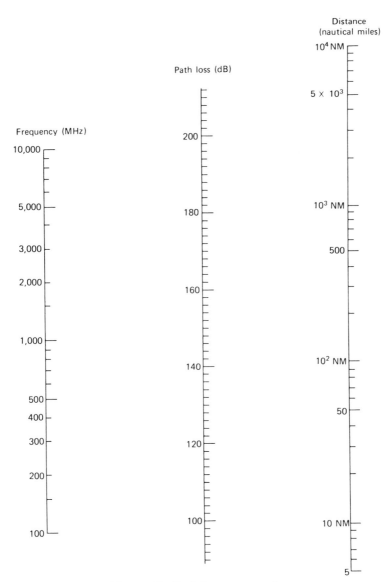

Figure 7.8   Path-loss nomograph.

and does not account for loss except that due to simple reduction in power density with distance.

We note here that for very wideband spread spectrum signals some compensation may be necessary to overcome differences in path loss across the band used. Such a path loss differential would cause a linear decrease in signal strength at a 6 dB per octave rate, which could seriously distort the received signal spectrum and would especially hamper a frequency hopping system in which bit decisions are made on the basis of chip power comparisons. A simple solution is the insertion of predistortion in the transmitter or postdistortion in the receiver before the correlator. It is fortunate that the distortion is such that a simple 6 dB per octave high-pass section can compensate for this frequency-related loss.

Absorptive Losses

At certain frequencies signal propagation is sharply attenuated because of atmospheric constituents.[1302] These frequencies should be avoided unless it is required that for some reason the signal be limited to short range. The frequencies exhibiting highest attenuation are 23 GHz, where oxygen absorption is predominant, and 60 GHz, where water vapor absorption prevails.

Except for a bit of extra care to ensure that the broadband signals common for spread spectrum systems do not overlap one of the highly absorptive regions, there is no difference in spread spectrum and conventional designs as far as the absorptive frequencies are concerned. A broad band signal could, however, suffer differential amplitude loss in the vicinity of the oxygen and water vapor absorptive regions due to the slope of attenuation in the area.

Differential Phase Delay

Differences in the velocity of propagation with frequency can cause severe phase distortion, an effect that is especially pronounced in the ionosphere. Such phase distortion degrades digital signals so that their One-Zero transitions are slowed (see Figure 7.9).

Eckstrom,[1408] Staras,[1417] and others have considered the transionospheric phase distortion effect and have shown that the RF bandwidth of a signal should be less than 5–10% of the carrier frequency to avoid serious differential phase distortion. Sunde[1418] has studied the problem of delay distortion in high-speed data transmission systems (A direct sequence transmitter is cer-

Undistorted
pulse

Phase–distoted
pulse

Figure 7.9   Distortion of square pulse due to differential phase delay.

tainly a high-speed data transmission system, whether or not the high-speed code is recovered as data.) Sunde's results show that there is less-than-2-dB-transmission loss for biphase modulation up to delay differentials (from center frequency to first null) of 1.6 times the transmitted pulse period. This is true for both quadratic and linear distortion. Quadriphase modulation in linear distortion is somewhat worse; it exhibits 2 dB loss for a delay differential of 0.5 T (where T is the period of a data bit).

In the frequency hopping world, differential delay prevents the transmission of a coherent frequency hopping carrier. The difference in delay seen at the various frequencies in a frequency hopped spectrum causes the dehopped carrier at the receiver to be randomly phase modulated at the hopping rate. This phase modulation in turn prevents the use of a phase-lock tracking loop in a frequency hopped system. It also causes the received signal bandwidth to be approximately 0.88 times the hopping rate (see Figure 5.50), which can cause a loss in the receiver of up to 3 dB over a coherent signaling method.

This is not to say that coherent frequency hopping cannot be accomplished, but it requires careful design in at least three areas:

1. Precise phase coherence in the transmitter's frequency synthesizer, with no differential phase shift in the power amplifiers.
2. Precise phase coherence in the receiver's frequency synchronizer, with no differential phase shift in the receiver's RF and precorrelation IF stages.
3. Careful, accurate modeling of the link between transmitter and receiver, with precise delay compensation.

If these can be accomplished, then coherent frequency hopping can be a reality. Until all three can be achieved simultaneously, however, only noncoherent hopping is practical for most applications.

Multipath

In addition to the direct signal path there often exists a reflected path for a signal to follow from a transmitter to a receiver. The receiver then sees at its input two or more signals—some delayed by longer extra path lengths. These signals add according to their phase relationship to produce a signal that contains the modulation of both, and has an amplitude determined by the phase relationship. In television receivers they cause "ghosts" or under extreme conditions complete loss of picture synch. It is possible for a receiver to receive reflected-path signals that are larger than the direct-path signal, especially in aircraft environments in which an antenna may be partly shaded by part of the aircraft structure.

The result of multipath reception is that a moving vehicle is subjected to rapid fading and peaking of its input signal as the direct and reflected signals cancel and reinforce one another. Stationary receivers are also subject to multipath but of a more stable variety, for the only variables are the reflecting surface and propagation medium. (Who has not seen the result of an aircraft flying between his television receiver and the transmitting station?)

Spread spectrum systems are somewhat less subject to multipath signal variations then conventional systems. In a direct sequence receiver, if the reflected signal is delayed (compared with the direct-path signal) by more than one code chip, the reflected signal is treated exactly the same as any other uncorrelated input signal. The higher the code chip rate for a particular system the smaller is multipath problem. With 5-Mcps codes, for instance, the reflected signal path must be 200 ft or less, different from the direct path to have any effect, and as code rate is increased the path length for which the reflected signal interferes is further reduced.

Frequency hopping receivers avoid multipath losses in two ways: (a) by transmitting the same information at more than one frequency, and (b) by comparing the power in alternate mark/space channels for each bit decision. (This considers that frequency hopping systems are used for data transmission. An analog-modulated frequency hopping system would not have these advantages against multipath.)

When a delay-line matched filter or some similar device (chirp filters or digital matched filters, for instance) is used by a receiver, multipath can cause more than one filter output pulse; that is, one output signal is due to detecting the desired input and another is due to late-arriving input generated by multipath delay. This paired output from a matched filter under strong specular multipath conditions is shown in Figure 7.10b. Provision can be made in the matched-filter receiver to overcome any difficulty caused by double-pulse reception by designing the receiver to accept only the first-arriving pulse, which effectively nullifies the effect of late-arriving multipath signals.

Under conditions of diffuse multipath, a matched-filter detector's output is likely to exhibit noncoherent (in time) output such as that shown in Figure 7.10c. In this photograph, the signal was diffused by a hill in the signal path, between the transmitter and intended receiver.

Figure 7.10a shows a normal signal output from the same matched-filter detector as that used in 7.10b and 7.10c, where there is no significant multipath signal present. It should be evident that the receiver's jamming margin may be reduced under such conditions, as the matched-filter output level is reduced by division into one or more signals.

Multipath conditions can also drastically alter the shape of the discriminator function developed in a delay lock tracking loop. Figure 7.11, from C. R. Cahn, shows the effect of multipath on the discriminator function. Tau-dither loops are also affected by multiple signals offset in time. In either case, the tendency of the code clock tracking loop is to follow the energy centroid of

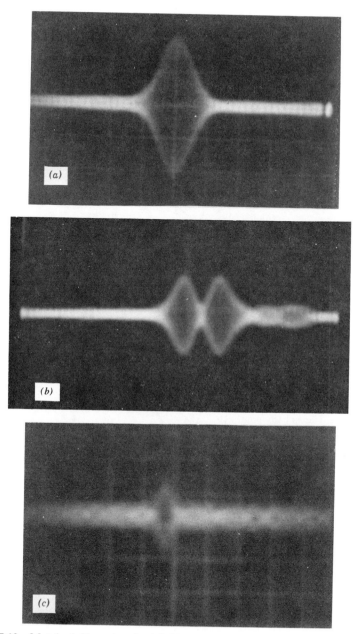

Figure 7.10   Matched filter signals. (*a*) Undistorted matched filter receiver output. (*b*) Receiver output with specular multipath. (*c*) Receiver output with diffraction due to hill in path.

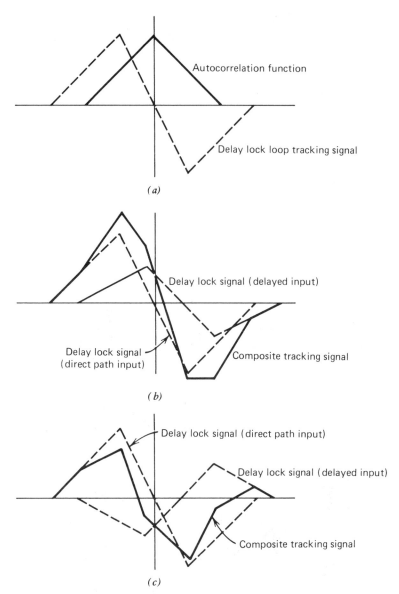

Figure 7.11 Effect of multipath: (*a*) no multipath; (*b*) in-phase multipath; (*c*) out-of-phase multipath.

the arriving signals. The shift in the center (zero crossing of the discriminator curve) is

$$\frac{2\,\Delta t}{1 + \dfrac{S_1}{S_2}} = \text{clock offset,}$$

where Δt = time shift between signals,
S_1, S_2 = signal amplitudes for signals S_1 and S_2.

The clock offset is forward or backward in time, depending on whether the multipath signal is in phase or out of phase with the desired signal.

7.4 OVERALL TRANSMITTER AND RECEIVER DESIGN

Design of transmitters and receivers for use with spread spectrum signals is not basically different from design for any other system in the same frequency range. Bandwidths are obviously wider, however, and other considerations must be made because of the expectation that large signals will be encountered simultaneously with small ones.

Let us trace a signal through a typical system, observing the perils and degrading influences to which it is subjected on its way.

The Transmitter

We assume here that we have a modulated intermediate frequency and carry on from there. A popular intermediate frequency for communications modems (modulator/demodulators) is 70 MHz, which is acceptable for spread spectrum signals as long as their modulated bandwidths are not greater than about 20 MHz. Higher intermediate frequencies are required for the higher code chip rates and/or wideband frequency hopping; 700-MHz IF's are common when wider bandwidth modulated signals are expected.

The transmitter must take a modulated fixed-frequency input signal and move it to the intended operational frequency, amplify it to the desired level, and output it to the antenna. The operation of moving the modulation from an intermediate to a transmitted frequency may be performed in two ways: translation and multiplication. Translation (or conversion) mixes the modulated signal with another signal set to the desired output frequency, plus or minus the intermediate frequency, and selecting (by filtering) the desired sum or difference frequency. The translation method can be used with direct sequence of frequency hopping modulated signals with equal results. Multiplication to the desired output frequency must be employed with a great deal more care because multiplication can change the basic modulation.

TABLE 7.1   Phase Mapping of PSK Signal Following Multiplication

Multiplier	Phase Modulation					
	0	90°	180°	270°	45°	
2	0	180	0	180	90	
3	0	− 90	180	− 270	135	
4	0	0	0	0	180	
5	0	90	180	270	225	(minus sign signifies
6	0	0	0	180	270	phase inversion)
7	9	− 90	180	− 270	− 45	
8	0	0	0	0	0	
9	0	90	180	270	45	

What happens to spread spectrum signals when they are multiplied? Frequency hopping modulation is increased in bandwidth and frequency spacing by exactly the same ratio as the multiplication; that is, a 100-channel frequency hopper with 1-kHz channel spacing before multiplication by 3 would still have 100 channels, but a 300-kHz bandwidth and 3-kHz channel spacing after multiplication. In many applications this multiplication of frequency spacing is quite useful because it allows a frequency hopping synthesizer to be constructed with narrow spacing and overall frequency range. That same synchronizer, multiplied, could supply almost any bandwidth or channel spacing. The same is not true of direct sequence-modulated signals. Multiplication of a direct sequence signal is practical only for certain specific multiplication factors. Others modify or completely remove the direct sequence modulation.

Some care must be exercised when multiplying any angle-modulated signal,* and this care also applies to phase-shift-keyed direct sequence signals. The action of a multiplier is such that not only is the center frequency multiplied but so is the rate of any modulation. Thus a biphase-modulated signal $\cos a \pm 90°$ when doubled becomes $\cos 2a \pm 180°$ (for the term of interest), which is a twice-frequency carrier without modulation because $+180°$ is identical to $-180°$. Multiplication by 5, on the other hand, produces $\cos 5a \pm 450°$, which is identical to $\cos 5a \pm 90°$. Odd and even multiplications of any degree produce similar results; that is, for any even multiplication 180° biphase modulation is lost, but for odd multiplications it is preserved (inversion is usually of no consequence.)

Table 7.1 is a phase mapping for various modulation shifts, with multiplication by factors of 2 through 9 as an operation. Various amounts of phase modulation are shown to be mapped to zero for some multiplication factors, others are phase inverted, and in some cases the resulting modulation is identical to unmultiplied modulation; for example, quadriphase modulation

* See also Appendix 4.

with multiplication by 4 reduces to a four-times carrier with no modulation, whereas multiplication by five produces a five-times carrier with the original modulation.

Table 7.1 shows that all odd multipliers are usable with biphase or quadriphase signals if an output phase inversion is permissible. When phase inversions are not desirable; then $5, 9, \ldots, (4n + 1)$ is the only permissible set of multipliers. For multiplies beyond times 9 the table repeats.

The principle of multiplication of modulation also holds for modulation other than in multiples of 90°, as might be expected. In those cases in which the modulated signal must be multiplied by a factor that would distort or destroy the usual biphase or quadriphase signals this principle has been used to advantage. If, for instance, it is mandatory that a times-four multiplier be used, but $\pm 90°$ PSK modulation is desired, we can choose to phase shift $\pm 22.5°$. After multiplication by 4 the desired $\pm 90°$ biphase modulation is produced.

Power Amplification

At the transmission frequency the spread-spectrum-modulated signal is amplified to the required power level and applied to an antenna. Here we are concerned primarily with the characteristics of the power amplifier, because many antennas are broadband enough to pass spread spectrum signals satisfactorily. It is necessary, of course, to design a power amplifier with sufficient bandwidth to pass the spread spectrum signal. The question is, "What is sufficient bandwidth?"

For a noncoherent frequency hopping signal it is most important that amplitude linearity of the transmitter be as flat as possible; that is, the transmitted power at each frequency channel should be equal to that transmitted for every other frequency. The reason is that a frequency hopping receiver often compares the signal sent in successive channels to make mark/space decisions. Variations in signal power may cause losses in error performance, depending on the receiver's decision criteria. This is especially true when interference is encountered, when the opposite channel from that intended is also occupied (i.e., a "mark" has been sent by transmitting at a given frequency but the frequency channel that would convey a "space" is occupied by interference). Phase linearity is less important in frequency hopping transmitters because the receiver's decisions are usually based only on received power on a channel. If the frequency hopping signal is coherent, however transmitter phase linearity should be well defined.

Direct sequence modulation is somewhat less critical in its demands on the transmitter, although there are definite advantages in keeping the amplification process as linear as possible. In an earlier paragraph in this chapter Sunde's quoted results show that delay differentials (or phase nonlinearity across the transmitted bandwidth) do not introduce significant losses when phase shift is held to a radian or less at the code clock frequency (offset).

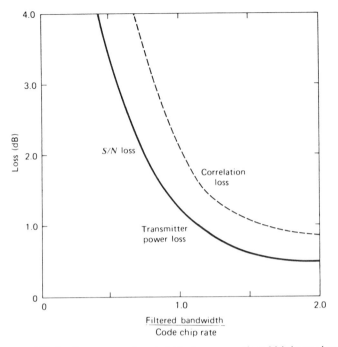

Figure 7.12(a)   Loss versus bandwidth for conventional biphase signals.

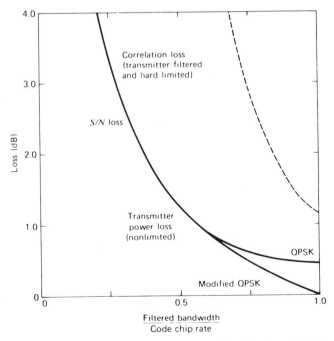

Figure 7.12(b)   Loss versus bandwidth for conventional and modified quadriphase (double-binary) signals.

It is interesting to note that direct sequence can be passed through a band-limiting filter with much less effect than a frequency hopping signal can. The frequency hopper, whose power distribution is flat across its occupied band, loses capability immediately when its passband is truncated; that is, if a frequency hopped signal is passed through a bandpass filter whose bandwidth is β percent of the frequency hopping band, the processing gain of the system is reduced to β percent or less of its original capacity. Figure 7.12a illustrates the power loss for filtering a biphase direct sequence signal. (Quadriphase and double-binary or modified quadriphase* power loss as a function of filter bandwidth is shown in Figure 7.12b. Figure 7.12 applies to systems (except for double-binary) in which filtering is done after the final power amplifier. For conventional direct sequence modulation methods filtering before power amplification is impractical because any nonlinearity in a subsequent power amplifier regenerates the $[(\sin x)/x]^2$ spectrum existing ahead of the filter. Double-binary, MSK (minimum shift keying), or some other form of modulation that has its envelope minimally affected by nonlinear processes is mandatory if the signal cannot be filtered after the power amplifier. The curves shown are adapted from C. R. Cahn.[245]

VSWR

An area of particular sensitivity for spread spectrum signals, when some care must be taken to ensure proper impedance match, is that of VSWR (voltage standing wave ratio). The broadband nature of spread spectrum signals and the acute frequency dependency of VSWR when a termination is improper combine to work against the spread spectrum system. Here we can say only that in both transmitters and receivers for spread spectrum applications extra care should be taken to ensure that terminations are correct. This includes not only impedance matching but connector mating, cable shield termination, and all the other small things that degrade conventional systems. When there is a VSWR effect in the conventional system, the spread spectrum systems is affected more.

Receiver RF Considerations

Design of the RF section in a spread spectrum receiver is somewhat more critical than transmitter design, mainly because of the presence of interfering signals before the processing stages. Ideally, the receiver would accept the desired signal (along with as little as possible of anything else), amplify it, translate it to the proper intermediate frequency, and deliver it undistorted to the correlator. Concurrently it should also deliver any interfering signal that falls in the direct signal band to the correlator.

Common direct sequence systems can suffer a 6-dB or greater loss in jamming-to-signal ratio when passed through a limiter concurrently with a

---

*Offset QPSK, modified quadriphase, and double-binary are the same.

larger jamming signal. This is the reason for such careful attention to linearity in the receiver RF section. Frequency hopping signals also suffer this same degradation in signal-to-jamming ratio, but are unaffected by it from the standpoint of a loss in interference rejection. The frequency hopping receiver does pay for the signal suppression caused by limiting but only in the form of a loss in signal, not in jamming margin. (This can be a significant factor in system design.)

Why does the direct sequence system suffer a loss in interference rejection capability due to limiting but a frequency hopping system does not? First, remember that a direct sequence system is most vulnerable to CW jamming and that the frequency of the CW jamming can vary over more than 10% of the signal bandwidth (or more) with nearly equal effect. [Cahn[13] shows that signal-to-noise loss in a limiter is the worst (6 dB) for nonvarying (or CW) signals. He also shows that when noise is used against a limiting spread spectrum receiver the bandwidth of the noise is of little consequence.]

In a frequency hopping receiver the interfering signal is still effective only in the channel or channels it covers, no matter how great its power advantage; that is, even though the limiter may cause the desired signal to be suppressed with respect to the interfering signal, the interferer has no more effect once his signal is ϵ greater than the desired signal. This is quite different from the direct sequence receiver, in which the interference is spread over the entire signal band and a decrease in signal-to-noise ratio due to limiting is effective overall. When it is necessary for a direct sequence receiver to employ limiting (indeed when there is a limiter anywhere in the link), it is customary to employ multiphase (more than two phase) phase shift keying, which does not suffer the same penalty because of limiting as biphase modulation. Chapter 4 discusses both biphase- and quadriphase-modulation techniques.

Receiver Signal Handling Capacity

In a receiving system in which linearity must be preserved, the linear signal range must be carefully considered. The receiver, from the front end to the correlator, must be designed to handle linearly not only the signal range expected but also the interference. Automatic gain control (AGC) only partly solves the problem in that it can control signal level on a signal, interference, or total power basis. (Because the interference is often much higher than the signal, total power AGC is essentially the same as interference-controlled AGC.)

We are forced to control receiver gain on the basis of correlator output signal level because the level of interference should be ignored as a basis of gain setting. Otherwise a jammer could control signal level or even modulate it by varying his power level. Once the controlled signal level is set each of the receiver stages through which the signal and interference pass together must be able to handle linearly a signal range equal to $(P_{\text{signal}} + P_J)$ dB. For this reason it is often desirable to place the correlator in a spread spectrum receiver as near to the front end as possible, so that fewer high-level stages are necessary before the correlator. (This strategy also reduces the local reference signal level requirement.)

Wideband Front Ends

It is often desirable, when a receiver is to cover a broad band of operating frequencies, to use a wideband front end that covers the entire band (or as much of it as possible). The tuning function is then carried out by mixing the received signal with a variable local oscillator, which causes the desired product to be output at a fixed IF frequency. This is a reasonable and even desirable approach in some cases but of little use in a spread spectrum system when interference rejection is a goal. Consider, for instance, that the interferer has only to transmit two signals offset in frequency by an amount equal to the receiver's IF. Then, no matter what the receiver's desired frequency, the first mixer, or correlator, would have an output that falls in the IF. This, in turn could cause the receiver's interference rejection capability to be useless. Similarly, a wideband receiver front end could allow other products, such as images, to cause interference.

In a receiver specifically intended to reject interference it seems singularly inappropriate to allow any interference to enter the correlator other than that for which the system was designed. Therefore wideband amplification (beyond that required by the signal) is not generally recommended.

The Ideal RF Section

An ideal receiver RF section for a spread spectrum system is one in which the desired signal is amplified to the desired level without distortion, along with any interference lying in the desired signal band, and filtered to the desired signal bandwidth. The combined interference and desired signal is then applied to a heterodyne correlator whose output is at an intermediate frequency. This ideal RF section would pass along to the correlator only the section of the signal band that contains the signal; that is, the front end would be tuned and would have a bandwidth approximately twice the code clock rate for a direct sequence signal, or equal to the signal bandwidth for a frequency hopping* signal. Filtering to this bandwidth before the correlator is important to the direct sequence receiver, for interference at the carrier frequency $\pm$ clock may be most effective.

A tradeoff may be made between a requirement for a high local reference level and a fixed direct sequence correlator prefilter. This is true because a fixed-frequency filter at an IF frequency is attractive, but moving the correla-

---

*The frequency hopping signal bandwidth is equal to the channel bandwidth times the number of channels.

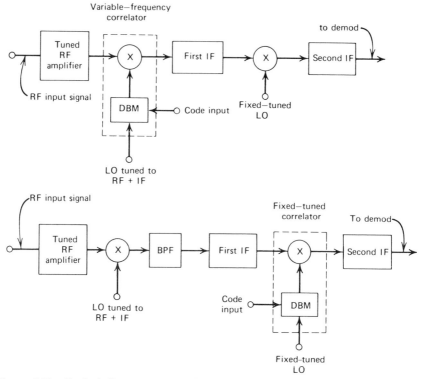

Figure 7.13 Typical direct sequence RF section and correlator configurations for spread spectrum receivers.

tor to an IF requires a larger local reference signal. Alternate configurations are shown in Figure 7.13. Tuned RF amplifiers are shown as the first stage in both configurations. The second, however, which uses a fixed correlator and precorrelation filter, is preferable for direct sequence receivers because the only precorrelation bandpass filtering in the first configuration is in the variable RF stage. Also, carrier suppression in the variable-frequency correlator is more difficult to obtain than in a fixed-frequency correlator.

RF sections for frequency hopping receivers need be no more complex than in Figure 7.13a (in which a frequency hopping correlator is substituted for the direct sequence correlator shown).

Local oscillators and references for direct sequence or frequency hopping receivers must necessarily be larger than the largest expected signal/jamming combination. This is required to preserve a linear transfer function in mixers and correlators; 3-dB-or-greater local oscillator and local reference levels (above the largest expected combined signal/jamming level) should be maintained as a minimum. All other stages before the correlator should also

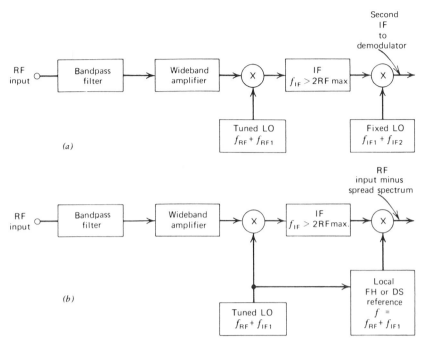

Figure 7.14   Alternative RF configurations with wideband amplification: (*a*) wideband amplifier configuration; (*b*) wideband amplifier configuration adapted to applique use.

maintain a total signal handling capability at least equal to the combined maximum desired and interference signal levels.

A possible alternate to a tuned front end for a spread spectrum receiver is shown in the block diagram in Figure 7.14*a*. In this configuration a bandpass filter covering the expected range of input frequencies precedes the amplifier. Following the wideband amplifier is a frequency converter (mixer) which translates the incoming signal to a fixed IF frequency such that $f_{IF} > 2f_{RF\,max}$. A fixed-tuned IF amplifier and bandpass filter at this frequency then acts to prefilter the signal before application to a correlator. The second multiplier, following the IF, can be the correlator if its locally generated signal is the proper reference. The advantage of this configuration is that RF tuning is completely circumvented. Only a tunable local oscillator input to the first multiplier is necessary. Image rejection (at $2f_{IF} + f_{RF}$) is accomplished by the input bandpass filter's rolloff, and no two frequencies which could give a product equal to the IF frequency can reach the first multiplier. Another advantage is that low-noise wideband amplifiers capable of covering wide frequency ranges are readily available.

On the side of disadvantages, the wideband amplifier must possess sufficient dynamic range to handle all the signals appearing in the bandwidth of
its prefilter. Otherwise, limiting and consequent suppression of the desired
signal can occur, which would give a jammer control of the signal amplitude,
no matter what frequency is transmitted anywhere within the filtered signal
band. Of course, AGC can be added to prevent limiting, but must also be
carefully done.

A modification of Figure 7.14*a* is shown in 7.14*b*. This modification is
specifically oriented to use as an applique to existing receivers to allow their
conversion to spread spectrum use. Here the same prefiltering, amplification,
and conversion to a high IF are used. The same tuned local oscillator,
however, also supplies the center frequency for the local spread spectrum
reference so that the output signal is at the same frequency as the received
signal; only the spread spectrum modulation has been removed, at which
point the receiver to which the tuner/correlator has been overlaid will
perform its usual functions.

In this discussion we have considered only those cases in which the
receiver is a tunable one capable of being set to receive at any one of many
frequencies. This is the most difficult case and therefore of most interest. For
single-frequency receivers filtering and other design problems are often less
severe, although the same principles hold.

Figure 7.15 Correlation loss due to filtering (from Cahn).

TABLE 7.2   Effect of Phase-Shift Distortion on Jamming Margin

Filter BW	Loss (dB)
Phase distortion—1.0 radian at first null	
$2 R_c$	0.96
$1.6 R_c$	1.20
$1.2 R_c$	1.51
R_c	2.27
2.0 radians at first null:	
$2 R_c$	1.2
$1.6 R_c$	1.25
$1.2 R_c$	1.61
R_c	2.29
3.0 radians at first null	
$2 R_c$	1.61
$1.6 R_c$	1.65
$1.2 R_c$	1.83
R_c	2.38

Bandwidth and Phase-Shift Effects

A spread spectrum signal can be transmitted or received through a reduced bandwidth channel with reasonably small losses. The direct sequence signals have somewhat better capability for such operation than the frequency hopping signals, however. For frequency hopping signals, a reduction in bandwidth is a direct loss of hopping frequency channels. Therefore, a ten percent loss in bandwidth is equivalent to having ten percent of the channels jammed, since in either case the demodulator loses signal the same amount of time.

In direct sequence systems, a narrowing of bandwidth reduces the harmonic content of the spectrum, which distorts the signal waveform, but a direct system does not necessarily lose as much capability as it loses in bandwidth. That is, a reduction in bandwidth to one-half the spectral width does not cause a 3-dB loss, typically. The loss that occurs is due to a reshaping of the receiver's correlation function from sharply triangular to rounded. In practice, this rounding causes a loss in jamming margin, but is not evident in any other receiver or transmitter operating characteristic. This loss is shown in Figure 7.15.

Phase shifts in RF stages can also cause a loss in the jamming margin. Table 7.2 illustrates typical loss expected for various types and amounts of distortion. Typical specifications for minimal loss systems, due to phase shift, call for a 0.1 radian or less phase shift between the center frequency and first null of a direct sequence signal.

PROBLEMS

1. Compare the noise power input to a spread spectrum receiver with a 10-MHz bandwidth to a conventional AM receiver with a 50-kHz bandwidth. The spread spectrum receiver has 26-dB process gain. Is the spread spectrum receiver or the conventional AM receiver more sensitive?

2. A spread spectrum receiver is operating in an environment in which the interference present is 20 dB greater than the desired signal. A second interfering signal which is 10 dB greater than the desired signal also appears. What is the effect of the second interferer?

3. The input noise, plus the contribution of a receiver's front end, causes a desired spread spectrum signal to fall 20 dB below the noise level. If the receiver's jamming margin is 23 dB, how much interference can be tolerated?

4. What is the effect of improper AGC loop damping in a spread spectrum receiver?

5. A frequency hopped signal with 100-MHz center frequency and 10-kHz spacing is to be multiplied by 4. What are the new signal characteristics? A 100-MHz direct sequence, biphase-modulated signal is also multiplied by four. What are the characteristics of this new signal?

6. Compare the signal power loss that might be expected when filtering a 5-Mcps code-modulated biphase direct sequence system to a 2.5-MHz RH bandwidth with a quadriphase modulated signal having the same code rate and passing through the same filter.

7. A number of spread spectrum systems are operating in the same frequency band (code division multiplexed), and a large interfering signal comes on the air. Assume that there are 100 simultaneous friendly link users and that the interference is 15 dB above the desired signal at any friendly receiver. What is the effect of the interferer?

8. What noise power might a spread spectrum receiver with a 10-MHz width see? If the system's AJ margin (M_J) is 20 dB, what would the minimum signal be for receiver operation? (Assume a 10-dB noise figure.)

9. What would the effect on minimum signal value be if the above receiver's M_J were increased by increasing its spread spectrum bandwidth to 100 MHz?

10. What would the effect be of multiplying a quadriphase direct sequence signal by 64?

8

NAVIGATING WITH SPREAD SPECTRUM SYSTEMS

Certainly the best-known applications of spread spectrum methods lie in the navigation area. Direct sequence ranging has been applied in space exploration programs since the early 1960s at least, and the methods employed have been well described in the literature.

The coded modulation characteristic of spread spectrum systems uniquely qualifies them for navigation—from the standpoints of both range measurement and direction finding—enough to allow homing or navigation with respect to any other spread spectrum transmitter. Little use has been made of spread spectrum techniques for direction finding,* but spread spectrum ranging, particularly direct sequence ranging, has been well exploited.

This chapter first describes ranging methods for direct sequence systems and for frequency hoppers. The second part of the chapter is a discussion of direction finding as applied in spread spectrum systems.

8.1 RANGING TECHNIQUES

Any RF signal is subject to a fixed rate of propagation (approximately 6 μsec/mi). The signal reaching a receiver at any given instant left the transmitter that sent it some time before. Because signaling waveforms or modulations are also functions of time, the difference in a signaling waveform, as seen at a receiver, from that present at the transmitter can be related directly to distance between them and used to measure that distance.

---

*Systems such as GPS (global positioning system) and PLRS (Position Location Reporting System) are now being configured to employ multiple spread spectrum range measurements (multilateration) for position location.

A common type of echo range measurement method is that used in radar systems. Many such systems simply transmit a pulse of RF energy and then wait for the return of a portion of the energy due to its being reflected from objects in the signal path. The radar marks time from the instant of the pulse transmission until its return. The time required for the signal to return is a function of the two-way range to the reflecting object, and because the signal propagation rate is known, the range is easily derived.

Any signal used is subject to the same distance/time relations. An unmodulated carrier, for instance, is delayed precisely the same amount as a spread spectrum signal traveling the same distance. The spread spectrum signal has an advantage, however, in that its phase is easily resolvable. A direct sequence signal and a simple CW carrier may be likened to a pair of yardsticks, one with feet and inches marked on it and the other blank. With the marked yardstick we can easily resolve distance within about $\frac{1}{8}$ in., but the unmarked yardstick, though just as precise in length, would be of little use for resolving to better than perhaps 1 ft. The marks on a direct sequence signal, of course, are the code sequence modulation, and the basic resolution is one code chip; the higher the chip rate, the better the measurement capability.

Frequency hopping systems do not normally possess high-resolution properties, but this is simply because their hopping rates are not high enough to act as the fine marks on our yardstick. There is no reason, other than that of being less able to construct high-hopping-rate synthesizers, for frequency hopping ranging capability to be less than that of direct sequence.

Let us discuss the operation of spread spectrum ranging systems. In direct sequence ranging, code modulation and demodulation are performed in exactly the same manner as in a system used for communications; that is, the code employed phase shift modulates the carrier. (For ranging, whether biphase or quadriphase modulation is used is usually of little consequence. We assume biphase modulation.) The baseband or information channel is unaffected by the ranging operation and may be employed concurrently. It is possible, in fact, to enhance operation by employing a baseband channel in a ranging system.

Figure 8.1 illustrates two simple direct sequence ranging systems. One we call simple, or transmit-then-receive. The other is called duplex, because it transmits and receives simultaneously. In both, the transmitter sends a pseudonoise code-modulated signal. Considering the duplex method, the signal at F_1 is simply translated to a different center frequency F_2 (preserving the code modulation) and retransmitted. This signal then reaches the first transmitter's site with a time delay corresponding to the two-way signal propagation delay between the two units (see Figure 8.2).

A receiver, located at the f_1 transmitter location, then synchronizes to the return signal, and on measuring the number of chips of code delay between the signals being transmitted and received can determine uniquely the range from itself to the repeating station. This type of system requires that the

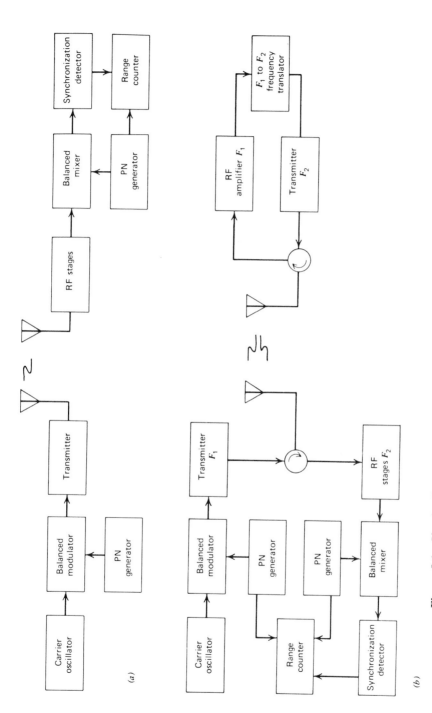

Figure 8.1 Simple direct sequence ranging systems: (*a*) transmit-receive ranging system; (*b*) duplex ranging system.

Code sequence at transmitter, frequency f_1

Code sequence at receiver, frequency f_2

Figure 8.2   Code-sequence comparison at transmitter and receiver.

receiving subsystem employ a code-sequence generator that is independent of the one used to modulate the transmitter, because the code the receiver must use is by definition out of phase with the transmitted code. The repeating terminal, on the other hand, is reduced in its complexity.

In simple time-sharing ranging systems, transmissions are exactly the same as in the duplex systems, except that transmitters and receivers at the two cooperating terminals do not operate simultaneously and, therefore, only one operating frequency is necessary. First, the terminal initiating the range measurement transmits a code sequence for a period long enough for the desired receiver to which a range measurement is to be made to synchronize itself. It is sufficient to say that the code sequences employed are long enough to support unambiguous range measurement over the distances in question, and that an initial synchronization method is provided. The important point is that the receiver matches its internal code reference to the code modulation on the signal it is receiving, and that its internal reference is then offset from the internal reference at the transmitter by the distance (propagation time) between them. The interrogator's transmitter, then, is generating the same code sequence as the responder's receiver but is some number of chips ahead of it.

Once initial acquisition has occurred the transmitter (or interrogator) instructs the receiving system to respond to its transmission by switching to the transmit mode at the reception of a given signal. The interrogating unit then sends a command to the responder to start transmitting, while it goes from the transmission to the receiving mode. Both the interrogator and the responder continue to generate the same code without interruption or resetting throughout the range measurement. The responder (the original receiver) is locked to a coded signal, which was offset from the interrogator's transmitted signal by the amount of the range delay, and transmits by using that same code reference to reply. The interrogator must delay its code phase to match the twice-range-delayed responder's signal. The amount of code-phase change necessary is again equal to twice the propagation delay from the interrogator (who is measuring the range and initiates the measurement) to the responder. In both cases the range measurement is made by counting chips of offset or fractions of chips and is therefore a discrete measurement, inherently accurate within a $\frac{1}{2}$ chip error. In practice measurements are commonly made to within a fraction of a chip period. The highest-resolution spread spectrum systems known can measure range to approximately 1/1000th of a chip period.

Tone Ranging

A technique for range measurement using multiple, coherent tones that modulate a single carrier is also practical. Figure 8.3 shows the phase relationship for these tones. All tones are started at a single point in time (t_0). At a point distant from the point of origin, but within a range delay less than the period of the lowest tone, a receiver that demodulates and compares the phase of the tones can measure range. This is because the three tones have a unique phase relationship that depends on the position of the receiver. Range resolution is limited by the receiver's ability to measure the tone's phases. Thus it is limited in range by signal-to-noise ratio.

The point in discussing tone ranging is that it could be applied to a frequency hopping system, where the tone's separation from the band center would correspond to a particular hop frequency. This capability would depend on having coherence between all of the hopping frequencies used for ranging. Such a method has not been implemented because of the need for coherence, but does offer possible promise for future frequency hopping systems to bring their range resolution more in line with the direct sequence systems.

(a) low—frequency tone
(b) medium—frequency tone
(c) high—frequency tone

t_0 = starting point, all tones at zero phase
Δt = distant point, tone phase relationship a function of distance

Figure 8.3 Tone-ranging technique.

Sources of Range Error

Several sources of range error must be considered when employing a spread spectrum ranging system:

Doppler-induced error
Clock-rate offset
Clock-rate error
Bit-count error

The effects of Doppler-induced error, clock-rate offset (due to an error in choice of chip rate), or clock-rate error (due to drift) are very similar, if not the same; that is, whenever the clock rate for a ranging code is offset from the design rate, the number of chips per mile for which the system is calibrated will also be offset. In Chapter 3 it was shown that the clock rate necessary for one nautical mile range resolution is 161,875 cps, and for one statute mile resolution is 186,333 cps. If a one percent range error is allowable, these chip rates could also be allowed tolerances of ± 1619 and ± 1863 cps, respectively.

Doppler-induced errors accumulate for the code clock just as they do at the carrier frequency, that is, at a 1.102 Hz/MHz/Mach rate. Thus for a 10-Mcps clock rate in a system traveling at Mach 2 (twice the speed of sound) the clock rate would be offset by 22.04 cps. Because a 10-Mcps clock rate would allow a basic resolution capability of approximately 100 ft, $(1 \times 10^7)^{-1} = 1 \times 10^{-7}$ sec (which corresponds approximately to a 100-ft range delay) and each chip represents an error to 100 ft, a possible 2204-ft error exists for every second in which the clocks in two terminals are allowed to accumulate respective differences. If the clocks are locked (one tracking the other), however, the accumulation of range error is greatly reduced.

Consider the two ranging methods we have discussed. In the duplex-ranging method measurements are made at one terminal by counting the number of chips difference between the transmitter's code generator and the local receiver's code generator. This measurement is made while the receiver is locked to the return signal, and the receiver's code rate is shifted by twice the one way Doppler shift (see Figure 8.4). Because the measurement occurs while the two codes are locked together, there is little opportunity for accumulation of offset errors. The only error is due to the difference in received code rate compared with the intended code rate. For our 10-Mcps system with 2204-cps (one-way) rate shift the error would be only $4408/10^6 \times 100 = 0.044\%$, or 0.44 mi in 1000. On the other hand, in the simplex system, in which a ranging unit sends an interrogating signal, then switches to the receive mode for the responder's reply, the system is out of lock while the range count is being accumulated; that is, the range count actually consists of a count of the number of chips that must be searched to bring the interrogator into synchronization with the responder's transmission. Depending on the

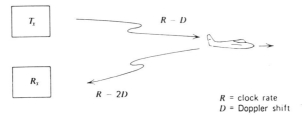

Figure 8.4 Twice-Doppler code-rate-shift accumulation for two-way propagation. (Relative motion closing produces positive frequency shifts.)

number of chips that must be counted, the search process can be quite time consuming—on the order of many seconds. During this time the responder's code clock can drift with respect to interrogator's clock and accumulate range error.

Range count error which accumulates when the two terminal clocks are operating independently accumulates as a chip error; that is, a one-chip cumulative shift during the search time counts as a one-chip error in range. Therefore, if a Doppler clock-rate shift is caused by relative terminal motion during a range measurement in the simplex (transmit-then-receive) ranging systems, range error accumulates at the Doppler-offset rate during the entire time that the interrogator is searching.

One method that could overcome this Doppler and/or clock drift accumulation is to design the responding system in such a way that it shifts its code clock an amount equal to that of the Doppler shift on the interrogating signal it receives but opposite in sense (see Figure 8.5). In other words, if the clock received from the range interrogator is offset by 10 cps in the high direction (signifying that the ranging units are closing on one another), the responder would reply with a code rate offset by 10 cps on the low side. Thus the signal arriving at the interrogator, where the range measurement is made, would be at the nominal code rate and would not contribute a cumulative error. (Of course, there is no way of telling how much of the offset in the interrogator's clock is due to Doppler shift versus the amount due to simple clock drift.)

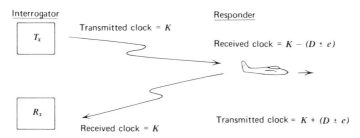

Figure 8.5 Clock-compensation method for range Doppler correction.

One source of error would remain: a possible error in generating the transmitted offset clock which can occur in the responding system. This error is a function of the accuracy of the responder's clock and would not be resolvable by either interrogator or responder. This returns us to the basic need in direct-sequence ranging systems—that of maintaining the most accurate possible clock frequency source.

Frequency Hopping Range Measurement

Range measurements can be made with frequency hopping systems just as with direct sequence systems, though they are not often done because of the general lack of resolution available with frequency hopping systems. Remember that frequency hopping systems use code rates in the kilochip to hundred kilochip (per second) range, whereas direct sequence codes start at a low of about 1 Mcps.* This gives the direct sequence-modulated system at least a ten-to-one resolution advantage in most applications. For this reason frequency hopping is not specifically discussed here as a technique to aid in navigation. It should be realized, however, that the same methods can apply for frequency hopping as for direct sequence systems.

More on the Selection of a Clock Rate[†]

We have stated that the factors affecting the accuracy of a spread spectrum ranging system are chiefly clock rate, clock drift, and Doppler shift of the clock rate (ignoring possible errors in the digital counting and readout circuits, which should be negligible).

At least two methods of attack are possible in choosing clock rate. In one the clock rate is chosen such that the number of chips per mile is an integer, whereas the other chooses a convenient frequency and corrects for errors.

Assume that we wish to design a ranging unit capable of measuring range within 0.1 nautical mile (1 nm = 1852 m). Therefore, we want to count 10 chip/mi of propagation delay; 0.1 mi corresponds to a chip rate with a wavelength of 185.2 m, or a repetition rate of

$$\frac{c}{R} = \frac{2.997925 \times 10^8}{1.852 \times 10^2} = 1.618750 \text{ Mcps},$$

where c = speed of light,
        R = resolution.

---

*The best current frequency hopping systems could not resolve range closer than approximately $\pm 10,000$ ft, based on their chip times.
[†] See also Chapter 3.

A system that uses a code sequence chip rate of 1.618750 Mcps would therefore have a propagation delay of 10 chip/mi, and a receiver 10 mi away would see a signal delayed by exactly 100 chips; that is, the signal appearing at the receiver was transmitted at a time that, on the transmitter's terms, was 100 chips earlier. Therefore, the receiver's sequence generator must be exactly 100 chips behind the transmitter to achieve maximum correlation and between 99 and 101 chips behind the transmitter to be within the bounds of correlation; 1.618753 MHz is certainly not a standard frequency, but an accurate crystal-stabilized source at this frequency is readily obtainable. (As an aside it should be noted that multiples of 0.161875 MHz may be used as clock rates to gain whatever resolution is desired; that is, one-fiftieth of a mile would require a clock frequency of $50 \times 0.161875 = 8.093750$ MHz. Of course, the number used for c, the speed of light, is that for light traveling under vacuum, and in any other medium some error will be seen. This is not significant, however, as long as all clocks operate at the same rate; 0.1% error limit due to common frequency offset would allow the clock rate to change up to 1618 bps.) It can be recalled that in both the simple- and duplex-ranging systems the number of code chips of offset due to range is actually equivalent to twice the range being measured. Therefore, in these and similar systems, clock rate required to give a desired resolution is halved.

Once the nominal clock rate is chosen, both transmit and receive systems are expected to operate at the chosen frequency (within their error limits) except when a receiving unit is in the "search" mode, looking for correlation with an incoming signal. In the interrogate-respond ranging system this search procedure occurs twice: first, at initial transmission and, second, when the responding unit replies.

Direction of search (forward or backward in code phase) is usually not important to a spread spectrum system, but in transmit-receive ranging systems it is necessary that the unit making the range measurement (the interrogator) search backward; that is, the interrogator's receive clock should run at a rate lower than the responder's transmit clock, so that its code sequence would be slipping backward with respect to the incoming signal. This is important in that the responder's code sequence is retarded with respect to the interrogator's sequence generator at the time of the transmit-to-receive switch, and the interrogator's code sequence must also be retarded a number of chips equal to twice the propagation time in order to recorrelate.

The period during which the interrogating unit is searching for recorrelation with the responding transmitter is the most vulnerable for error entry. During this time any relative clock drift or Doppler shift is accumulated as a range error by the chip search counter in the interrogator.

On the basis of clocks operating near their design frequency, range error for an integral frequency clock system may be expressed by

$$\text{error (in chips)} = \left(|K_I - K_R| \pm 1.7\, V_R K_R \times 10^{-3} \right) T_S,$$

where K_I = interrogator clock rate (cps),
       K_R = responder clock rate (cps),
       V_R = relative velocity (nmph),
       T_S = recorrelation search time (sec).

This expresses the accumulated errors due to an interrogator's counting at an assumed clock rate K_I, whereas a responder's reply actually is at a clock rate K_R, offset by the Doppler-frequency shift referred to the responder's clock signal. This is multiplied by search time, for the longer the interrogator's search for correlation, the longer the time that errors are accumulated. Doppler-frequency errors may reduce or add to any errors due to clock offset, depending on the direction of errors and the relative velocity. In some systems it is not practical to use a clock that gives an integral number of clock pulses per mile of delay. These systems may correct for the errors generated by an offset clock frequency. To give an example, with a clock frequency of exactly 2.0 Mcps, the nearest integral multiple of 0.161875 is 1.942500 Mcps. Therefore, use of a 2.0-Mcps clock would produce an overall error of approximately 3% if 12 chips/mile were used as a measure of range, but because this is a constant and precisely known error rate, it can be corrected readily.

Selecting the Ranging Code

Code sequences used in ranging should be chosen on the basis of consideration for their auto-correlation and cross-correlation properties and in most ranging systems, for sufficient length. Interference between different sequences can nullify selective addressing, or minor correlation in a correct sequence because of false recognition of synchronization can cause an erroneous measurement.

A ranging code used by a system that has no secondary resolving capability must be long enough not to repeat over the maximum distance measured; that is, to measure a range of 1000 mi with a code giving 50-chip/mi resolution would require a code length of at least 50,000 chips. Otherwise, the code sequence would repeat, and the interrogator could recognize synchronization at more than one range. At the 8.093750-Mcps rate required to give 50-chip/mi resolution, the repetition period of the $56,535(2^{16} - 1)$-chip code required would be $65,535/8,093,750 = 8.1$ msec, and the bandwidth of the correlation detector should be about 40 Hz. This narrow bandwidth would restrict the search rate severely unless it were possible to accept correlation averaging across less than the complete code length (with the attendant risk of falsely recognizing synchronization at the wrong position of the correct code or even crosscorrelating with a different code).

If a 40-Hz recognition bandwidth is used, searching out a range at the maximum distance could require almost 20 min. Obviously, then, in a simple ranging system high resolution and long distance must be traded off against measurement time. The major problem in direct sequence ranging for aircraft and maneuverable spacecraft has been the requirement for measurement at long range in a reasonably short time. Reduction of ranging time is limited by the maximum search rate a receiving unit is capable of achieving and the length of the code to be searched. Maximum search rate, in turn, is limited by the recognition time of the receiver's correlation detection circuits. (The receiver must be able to recognize acquisition and stop the search process before the point of code synchronization is passed.) To complete the circle the bandwidth of the correlation detectors must be commensurate with the autocorrelation requirements of the codes used.

What techniques are available to reduce ranging time while maintaining accuracy? Much work has been performed in the area of developing special "acquirable" codes which have the required length for long-range measurement but which also have synchronization properties that permit a range to be searched out without traversing the entire code length. Jet Propulsion Laboratories has had great success with code sequences made up of three component codes assembled in such a way that the overall ranging sequence has a length that is the product of the three component code lengths.

Hybrid-Ranging System

Another ranging technique that solves the problem of fast range measurement at long distances is called a hybrid approach, since more than one kind of modulation is used to measure range. In this technique a *PN* code, a few thousand chips long, and a digital "range message" (whose bit rate is the repetition rate of the *PN* sequence) simultaneously modulate the transmitter. The *PN* code does not determine maximum range and therefore its length can be short to reduce synch time. The only bound on reducing code length is that system cross-correlation requirements must be met.

The ambiguity in range caused by using a short *PN* code is resolved by modulating with a "slow" digital pattern whose length is such that its repetition period is longer than the propagation delay of the longest range to be measured. No separate search process is needed to synchronize this pattern as it is synchronized by the initial search process. All that is needed is a measurement of the relative phase of received range messages, which is as simple as counting the number of *PN* code repetitions between range message markers.

The range message itself is the simplest kind of digital signal, made up of a number of square waves whose half-wave period is the code repetition rate, inverted in phase after a time greater than the maximum expected propagation delay. This phase inversion is the marker that resolves the range ambiguity. Modulation is such that the *PN* code biphase modulates the transmitter carrier at its basic rate, permitting high resolution. The range message, however, is sent as a baseband signal at a comparatively low bit rate

(a)

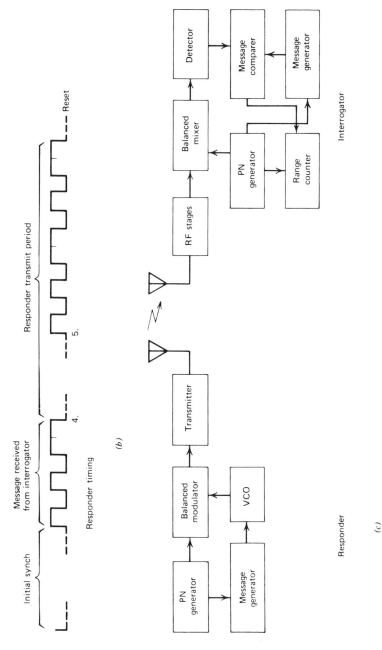

Figure 8.6 Hybrid ranging system; (a) interrogate mode; (b) message timing diagram; (c) respond mode.

(the repetition rate of the code), frequency shift keying the same carrier that is modulated by the *PN* code. The composite transmitted output then is a $[(\sin x)/x]^2$ spectrum which shifts frequency at the range message rate.

Figure 8.6 shows simplified block diagrams of a hybrid ranging system and illustrates the timing of transmissions in interrogating and responding systems. The interrogator and responder units may actually be identical; range may be measured by either one. One range cycle is described as follows (see timing diagrams):

1. The interrogating unit transmits a *PN*-modulated signal for a period long enough to ensure that the intended responder has synchronized.
2. The range message is sent consisting of a series of one-zero bits (a square wave) at the *PN* code repetition rate, and is followed by a phase inversion of the message. This information FSK modulates the *PN*-modulated carrier.
3. At the *PN* code repetition marker following the phase inversion several things occur: (a) the responder starts generating an identical range message to that sent by the interrogator (i.e., a given number of square wave periods, an inversion, repeating); (b) the interrogator switches to receive; and (c) the responder switches to transmit. (Search processing for resynchronization is delayed in the interrogator for a period long enough to exceed the *T-R* switching time. In this way switching time in the two units is removed as a source of range error.)
4. The responder now transmits a *PN-* (only) modulated signal for a synchronization period and the interrogator counts the number of chips searched to regain synchronization, storing the chip count.
5. Responder then FSK modulates the *PN*-modulated signal with the range message. (Remember that it has been generating the range message since the *T-R* switch.)
6. Interrogator receives the range message and counts the number of message bits between the first phase inversion in the received message and the next occurring in its own range message (which has also been generated since the *T-R* switch).
7. Each message bit counted is weighted, added to the *PN* chip count already entered, and read out as a range.

Selection of parameters used in hybrid systems is illustrated by the following example.

Suppose (again) that ranging is to be performed at distances up to 1000 mi and with resolution of 0.1 mi but only 1 sec is allowed for the measurement. We wish to complete a range measurement in 1 sec or less, which means that we must

transmit a code sequence to allow the intended responder to synchronize, add FSK information to the transmitted signal, telling the responder to transmit,

switch to receive and count the number of bits searched over,

receive and accumulate the range message information.

Because the two synchronization search processes consume most of the overall measurement time, let us allow 0.45 sec for each synchronization. A 2047-chip *PN* code would have to be searched at a rate of approximately 4600 chip/sec to ensure recognition in this time.

A 5600-cps search rate would require synchronization recognition in less than 1/2300 sec (because the correlation function is two chips wide). Therefore, the recognition bandwidth should be 805 Hz or greater. Because

$$\frac{\text{code length}}{\text{chip rate}} = \text{repetition period} = \frac{0.35}{\text{recognition BW}}$$

the chip rate should be

$$\frac{1}{\tau} = \frac{\text{BW}_R \times L}{0.35}$$

where τ = chip period,

BW_R = recognition bandwidth,

L = code length,

$0.35 = \text{BW} \times T_R$ for the synchronization recognition circuit.

The code chip rate, using a 2047-chip code, searching 4600 chips/sec, and using a synchronization recognition circuit 806 Hz wide, would be at least

$$\frac{805 \times 2047}{0.35} = 4.7 \text{ Mcps.}$$

The nearest multiple of 0.161875 is 30 × 0.161875, or 4.856250 Mcps. (Resolution capability would be one-sixtieth of a mile.)

Next we construct the range message, which must have a period greater than the greatest propagation delay expected (more than 6.3 msec). The repetition period of the code sequence is 0.42 msec; therefore, the code repeats almost 15 times in this range.

Making use of the property of maximal code-sequence generators that the all-one's vector is repeated once and only once in each code sequence, we use this repetition to generate a word made up of 15 or more bits and a bit period equal to the *PN* sequence repetition rate. This word is the "range message" and consists of a square wave (one-zero sequence) 15 bits long, a phase reversal, 15 bits, and a phase reversal.

In measurements that use the range message, the phase reversal marks the beginning of a sequence and each bit is weighted $2047/30 = 68.23$ mi. Therefore, for each bit of offset in the range message the interrogator adds 68.23 mi to the range. The overall range measurement time, then, using the parameters described, would be approximately 950 msec. The range resolution would be one-sixtieth of a mile, and the maximum range would be 1091 mi. (Adding bits in the range message would increase the maximum range capability by 68.23 mi/bit, and measurement time 0.42 msec/bit. Length of the range message does not affect resolution.)

Accuracy of the hybrid approach is excellent in that resynchronization search time is short and clock drift is minimized. Any error in counting range message bit difference, however, will produce errors in range at the weighted bit rate.

8.2   DIRECTION FINDING

For navigation purposes ranging capability is, of course, an extremely valuable tool. But what of the companion problem—knowing the direction to the system supplying the range response? Let us examine some of the ways in which a spread spectrum system can derive a bearing on, as well as range to, a desired responder, so that a user can home in on or navigate with respect to another spread spectrum system.

Special Antennas

Antennas designed for direction are available for all types of use. Here we consider only one example to describe typical use with a spread spectrum receiver. Figure 8.7 shows an antenna pattern typical of a direction finder designed to operate with a separate receiver. (In fact, it is a composite of two antennas with their cardioid-shaped patterns superimposed.) This type of

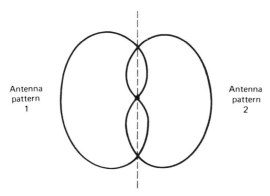

Figure 8.7   Dual cardioid pattern set for typical direction-finding antenna.

direction finder operates by switching alternately between a pair of opposed antennas, detecting (usually by envelope detection) the signal, and comparing the levels from the two antennas. The entire antenna assembly is then rotated until the transmitting station being received lies on the centerline between the two antenna patterns. At this point the signal received by both antennas is equal because both lie at the same angle with respect to the received signal.

This type of direction-finding approach works as well with spread spectrum systems as with any other because the antenna switching operation effectively amplitude modulates any incoming signal, whether spread spectrum or CW. Because envelope detection might not function quite so well on a spread spectrum signal as on others, the AGC signal internal to the spread spectrum receiver is normally taken as an antenna control signal. Two difficulties exist with this two-pattern switched-antenna direction finder:

1. Poor sensitivity, for the antenna directs its lowest response point toward the desired receiver.
2. Two separate null points exist so that the antenna could point toward the transmitter or in the exact opposite direction. This ambiguity is usually corrected by spoiling the composite pattern; the sharply concave section then exists on only one side.

When code-sequence rates are high enough, a different two-antenna approach is feasible; code phase is used as the criterion for measurement, rather than comparative amplitude. For code-phase direction finding the antennas used are conventional communications types. No special provision is necessary. A block diagram of a two-antenna, code-phase difference measuring direction finder is shown in Figure 8.8. It is effectively two receivers both locked to the same received signal; a measurement of phase

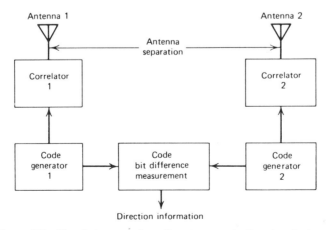

Figure 8.8 Simple two-receiver direct sequence direction finder.

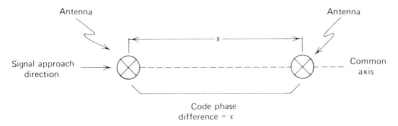

Figure 8.9   Signal approach along a common two-antenna axis ($\theta = 0$).

difference in their local code reference generators determines direction. Alternatives to this configuration are possible and are discussed shortly, but first let us describe the operation of this one.

A signal appearing at antenna 1 causes the code sequence at antenna 1 to be synchronized with the signal's code modulation. The same signal at antenna 2, which is some known distance s away, also causes the second code sequence to be synchronized. The two code sequences are then synchronized with the incoming signal but not with each other, because the signal phase is different at each antenna location. The amount of difference is dependent on the antenna separation and the angle of signal approach; that is, if the antennas are separated by a distance corresponding to two code chips, and the phase difference measured is two code chips, the signal can approach only along a common axis passing through both antennas. (Figure 8.9 further clarifies this point.)

When the signal approaches the two antennas from an angle θ, however, the angle measurement is somewhat more complicated.* Because the distance between antennas is accurately known and the distance to each antenna from the signal source is measured, the three sides of an extremely elongated triangle are defined and so are its included angles. Therefore, the angular direction to the transmitter is known to be one of two nearly supplementary angles θ and (assuming that the distance to the transmitter is many times greater than the antenna to antenna distance). The direction finder must then determine which of the two angles is correct. (See Figure 8.10 for illustration.) This angle determination can be made by observing which of the distances measured is shortest, for the shortest distance must be opposite the desired angle.

A second angle estimation method simply makes use of the observation that the difference in distance measurement at the two antennas, compared with the distance between the antennas, varies approximately as the cosine of the angle to the signal source (for small antenna separation compared with the distances measured).

---

*Except, of course, when $\theta = 90°$, for which condition the range measured at both antennas would be identical.

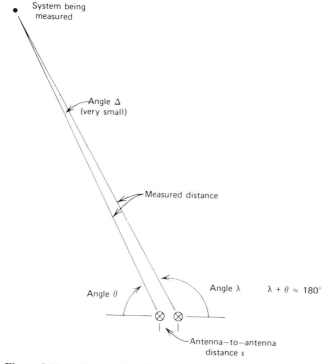

Figure 8.10a Illustration of measurement geometry, $s \ll D$.

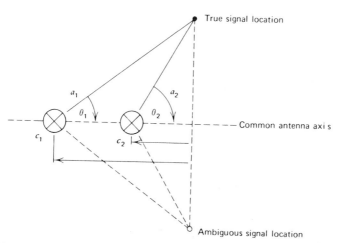

Figure 8.10b Illustration of distance difference measurement geometry, showing possible ambiguity in transmitter location.

Referring to Figure 8.10*b*, in which the angle being measured is the angle enclosed by lines *a* and *c*,

$$\text{distance } a_1 = \frac{c_1}{\cos \theta_1},$$

$$\text{distance } a_2 = \frac{c_2}{\cos \theta_2},$$

$$a_1 - a_2 = \frac{c_1}{\cos \theta_1} - \frac{c_2}{\cos \theta_2},$$

but

$$\theta_1 \sim \theta_2 = \theta$$

(see the geometry of Figure 8.9) and

$$a_1 - a_2 \sim \frac{c_1 - c_2}{\cos \theta},$$

$$\cos \theta = \frac{c_1 - c_2}{a_1 - a_2}.$$

The distance $c_1 - c_2$ is just the antenna separation, and $a_1 - a_2$ is the difference measured by the two antennas. This angle estimate is simple to make and requires little hardware, compared with the previously described method, which requires an estimate based on knowing the three sides of a triangle.

In either of these two-antenna methods an ambiguity exists in that the receiving system cannot tell if a transmitter is in the top or bottom hemisphere, as separated by the common antenna axis. When such ambiguity can cause difficulty, three antennas are needed to resolve it. The third antenna is required to show only whether the signal time of arrival has been advanced or retarded with respect to the signal arriving at the other two antennas.

The accuracy of direction measurement when a two- or three-antenna array is used is determined by two parameters:

1. The distance between antennas used for time of arrival comparisons (assuming the distance is accurately known, of course.)
2. The degree to which distance can be resolved as received by the processors attached to the two (or more) antennas.

Many direct-sequence systems can resolve range to within one-tenth clock chip, and some can do as well as one-thousandth chip. Consider a system which has one-tenth chip capability, say a system whose code rate is 10 Mcps. For ten-chip antenna separation, the distance between antennas would be approximately 1000 ft—not compatible with aircraft mounting. Yet, the maximum distance difference measurable would be ± 10 chips, for 200 resolvable increments (100 increments per 90° quadrant). Direction measurements would be accurate to approximately 0.9°. Of course, a higher code rate or larger antenna separation would improve the measurement capability. What about the receiving system which cannot afford the 100-ft separation we mentioned above? Many aircraft installations cannot afford a 10-ft antenna separation, to say nothing of 1000 ft, and 10-ft separation allows for only one chip, even when code rate is 100 Mcps. It is true that a 100-Mcps code, coupled with 1/100th chip resolution, and 10-ft antenna separation could give 0.9° resolvability, but such systems are in general too advanced to consider for airborne use, at least until 100 Mcps processing systems are proven.

A possible alternative is to make a pair of measurements, separated in time, so that on a moving aircraft or other vehicle the effect is that of having a pair of antennas separated by the distance desired. In this way a 1-Mcps code could be used, with an effective 10,000-ft separation to give the same resolution as higher rate codes with fixed antennas. Also, a single pair of antennas would be able to resolve the hemispherical direction ambiguity, previously discussed, because measurements would be made at two locations.

The same principles are applicable to frequency hopping systems but with the general penalties imposed by the slower frequency hopping code rates. Spread spectrum signals can, in general, perform as navigational signals, comparing very favorably with other signal formats. Although acting as distance or direction-finding signals, none of the interference rejection or other properties is impaired. Therefore, spread spectrum signals offer perhaps a new dimension to navigational signals. Viewed in another way, spread spectrum communicating signals offer a new dimension for communications systems—ranging and direction-finding capability.

It is worthy of note that a spread spectrum system can provide range measurements over any range at which the receiver will synchronize. Furthermore, the range resolution is never worse than a small fraction of a code chip period. This exceeds the performance of many other ranging methods, since they tend to degrade with distance because of their greater dependence on a good signal-to-noise ratio to measure signal rise time or phase.

PROBLEMS

1. What basic range resolution may be expected from a direct sequence system with a 10-Mcps code rate?

2. What sequence generator length is the minimum that should be employed for unambiguous ranging at 350 mi if the code employed is at 20 Mcps?

3. If exactly 25-chip/mi range resolution is desired, what code chip rate is required?

4. Two range measurements are 191.03 and 192.05 mi, made by systems whose antennas are separated by 0.1 mi. What is the direction to the system being ranged on with respect to the common line between receiving antennas?

5. What directional resolution can be expected from a receiving system for 50 Mcps coded signals if its effective antenna separation is 0.5 m?

6. A 100-khop/sec frequency hopping system uses its hopping interval to measure range. Give an estimate of its range resolution capability.

7. In a direct sequence system, what is the effect on a range measurement of a carrier-frequency offset; and a clock rate offset?

8. What effect has distance inaccuracy on angle measurement when distance measurement is used for direction finding?

9

APPLICATIONS OF SPREAD SPECTRUM METHODS

This chapter is intended to show a few of the ways in which spread spectrum techniques have been and are being used in the hope that those shown will spur new ideas and further applications. It is also hoped that the reader will be encouraged by the knowledge that spread spectrum methods have provided practical, real-world solutions to some tough, real world problems.

As far as general communications are concerned, applications of spread spectrum techniques are relatively obvious. Certainly the usual voice, data, and other common information transfer uses are practical, though in some cases restrictions like frequency allocation might prevent the use of a wide-band signal. Some areas in which spread spectrum methods (or at least system components) have already been put to good use are given in Table 9.1, and in the remainder of this chapter we describe a few of the specific ways in which these methods are employed.

Space Systems

In space systems, especially communications satellites, which may be stationary and therefore continuously accessible to interference, spread spectrum methods have proved effective. In general, however, satellites do not employ processing on board because the added complexity is significant and the number of possible users may be sharply reduced.

In most applications of satellites for communications there is little advantage to placing a spread spectrum processor on board. A simple repeating satellite is more economical and does almost as well. In those systems in which nonprocessing satellite repeaters are installed all spread spectrum modulation and demodulation must be performed on the ground, as in Figure 9.1.

TABLE 9.1

Areas	Application	Type of System	Primary User
Space systems	Communications, ranging, multiple accessing, jamming protection	Direct sequence, frequency hopping	Military
Avionics systems	Communications, position location, discrete addressing, jamming protection, low detectability, radar, collision avoidance	Direct sequence, frequency hopping, hybrid direct sequence/frequency hopping chirp	Military
Test systems and equipment	Bit-error detection, noninterfering in-service testing, signal correlation, privacy, (pseudo-) random selection, number generation	Direct sequence	Commercial
Signal protection (nonmilitary)	Interference rejection, voice and data scrambling	Direct sequence	Commercial
Position location	Collision avoidance, direction finding, ranging	Direct sequence	Many

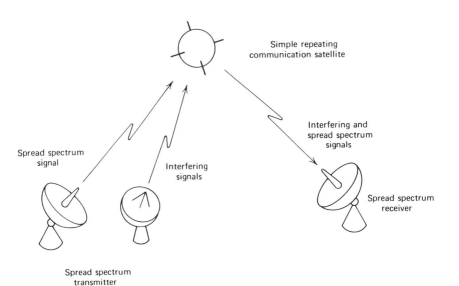

Figure 9.1   Satellite relay configuration for a simple repeater.

Basic advantages and disadvantages to on-board processing are:

The complexity of a spread spectrum demodulator is avoided on the spacecraft without on-board processing.

The satellite is forced to retransmit an uplink interference signal that reduces the spacecraft transmitter power available for sending the desired signal without on-board processing.

Every would-be receiver must have a spread spectrum demodulator. Large numbers of downlink users are therefore precluded, in general, because of the cost of the equipment with on-board processing.

Typical spread spectrum systems for operation through a repeating type satellite are:

1. URC-55
   (a) 9.6-mcps direct sequence (QPSK) modulation.
   (b) 1, 2, or 4 analog voice or TTY signals; one digital data channel, 75 bps to 4.9 kbps; one FSK orderwire.
   (c) Long code synch acquisition with real time start.
   (d) Time transfer capability.
2. URC-61
   (a) Compatible with URC-55.
   (b) 9.6-Mcps direct sequence (QPSK) modulation.
   (c) One analog voice or TTY signal; one digital data channel, 75 bps to 4.9 kbps.
   (d) One FSK orderwire.
   (e) Long code acquisition plus frequency hopping acquisition capability.
   (f) Range measurement.
   (g) Time transfer capability.
3. USC-28
   (a) 40-Mcps-direct sequence (QPSK) modulation.
   (b) Multiple multiplexed users; total data rate up to 5.0 Mbps; net control orderwire.
   (c) Software implementation of communication and fault location functions.
   (d) Multiple acquisition modes.
   (e) Doppler and range offset compensation.
   (f) Time transmit capability.

Figures 9.2, 9.3, and 9.4 show three of these systems. The most complex is the USC-28, which spreads baseband-multiplexed user signals (total data rate $\leq$

TRANSMITTER UNIT

RECEIVER UNIT

REMOTE CONTROL PANEL

Figure 9.2 URC-55 modem (courtesy Magnavox Co.).

MOD-DEMOD GROUP

167-192
UNCLASSIFIED

Figure 9.3 URC-61 modem (courtesy Magnavox Co.).

5.0 Mbps) to a bandwidth of 35 MHz using quadriphase modulation. This system also automatically adjusts its baseband rate as a function of channel performance, which allows optimization of system capacity under varying conditions. Its processing gain varies between approximately

$$10 \log \frac{3.5 \times 10^7}{75} = 56.7 \text{ dB}$$

and

$$10 \log \frac{3.5 \times 10^7}{5.0 \times 10^6} = 8.5 \text{ dB}$$

Figure 9.4   USC-28 modem (courtesy Magnavox Co.).

as the necessity arises. Many of the demodulation and tracking functions of the USC-28 are implemented in software.

A commercial use of spread spectrum modulation has been accomplished in a direct sequence system developed by Equatorial Communications, which operates in the 3.9–4.2-GHz satellite band. This system is used for transmission of business information. It allows use of satellite receiving antennas as small as 2 ft in diameter.

For comparison purposes Figure 9.5 shows the configuration of a spread spectrum system that employees on-board processing. What are the advantages of on-board processing? When on-board processing is used, the satellite receives the signal on the uplink, demodulates it, and uses the information to modulate the downlink. In the demodulation process any uplink interference is suppressed by an amount approximately equal to receiver process gain. The downlink transmitter is not required to retransmit signals other than the desired signal. Therefore, when on-board spread spectrum processing is employed, effective transmitter power requirements can be reduced because it is not necessary to retransmit an interfering signal.

Of course, the downlink can also be spread spectrum modulated, but this is not so useful as on-board processing with conventional (AM, FM, etc.) modulation on the downlink. Little advantage is gained in having a spread spectrum downlink because the distances involved are such that a spread spectrum signal could not protect the receiver from interference that is close

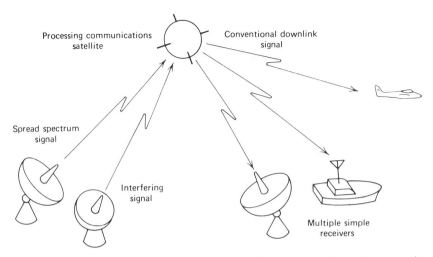

Figure 9.5 Satellite relay configuration using on-board processing with conventional narrowband downlink.

by when the satellite downlink transmitter is often 20,000 mi away. A great deal is gained, however, by transmitting a spread spectrum unlink signal to take advantage of the interference protection afforded while sending the information to the receivers on an easily accessed downlink.

One space-oriented system that makes use of spread spectrum modulation to advantage is the TDRSS (Tracking and Data Relay Satellite System) system. This system is one of the few that make use of spread spectrum modulation for some reason other than antijamming. The reason for its use in TDRSS is to reduce the power density of the signals as seen at the earth's surface. Signal levels from space are limited to a density of 145 dBm in a 4-kHz bandwidth, in a square meter aperture, and TDRSS accomplishes this by transmitting a 3.08-Mcps direct sequence-modulated signal.

Two separate spread spectrum signals are transmitted simultaneously in TDRSS. One employs a data modulated 1023-chip Gold code biphase modulating carrier. The second signal is also biphase modulated, but with a linear maximal code whose length is 256×1023—261,888, truncated from a normal $2^{18}-1$ code length (262,143).

Figure 9.6 shows a block diagram of the TDRSS modulation and demodulation subsystems. One of the most interesting characteristics of the TDRSS spread spectrum subsystem is its synch acquisition technique. The two transmitted signals are sent simultaneously at the same frequency, with the long-code ($2^{18}-1$)-modulated carrier orthogonal to the Gold-code-modulated carrier, and attenuated some 10 dB below it. The receiver then treats the two signals as independent biphase signals. (Alternately, they could be considered to be an unbalanced quadriphase signal. See Figure 9.7.)

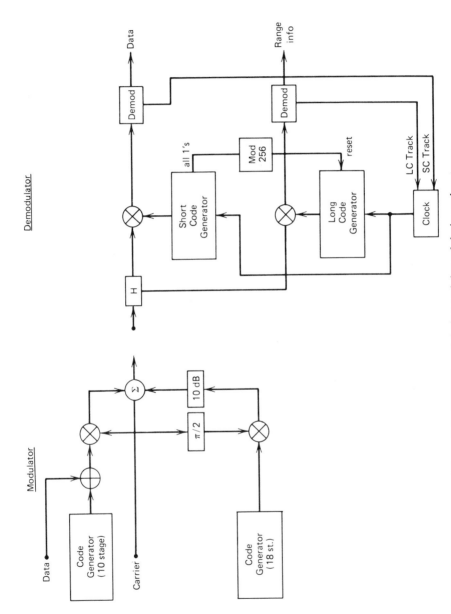

Figure 9.6 TDRSS modulation and demodulation subsystems.

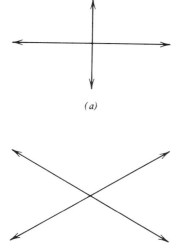

(a)

(b)

Figure 9.7 TDRSS-type signals as (a) independent biphase signals, and (b) composite unbalanced quadriphase.

Code-synch acquisition is greatly accelerated in TDRSS by synchronizing
the start and end of the long code with the repetition period of the 1023-chip
Gold code. Figure 9.8 illustrates this relationship. In the transmitter, the
short and long codes are started at the same time, and the long code is
truncated and started again after 256 repetitions of the short code. In the
receiver, short-code synch is acquired by a search through the total uncertainty (1023 chips maximum). Then, the receiver resets its long-code generator each time the short code repeats and integrates over the short-code
period. If it recognizes long-code synch between short-code repetitions, it
continues the long code without resetting. If long-code synch has not been
recognized by the time the short code repeats, it resets and tries again. In
this way, the long-code synch occurs in not more than 256 repetitions of the
short code, which is approximately $1023/(3 \times 10^6) \times 256 = 88$ msec maximum.

Another unusual feature of the TDRSS system is that the data transmitted
are added to the short code and not to the long code (see Figure 9.6). The
long code is used primarily for range ambiguity resolution.

The TDRSS system will provide relay of signals from orbiting spacecraft,
where 100% coverage is provided by only two TDRSS spacecraft for all users
whose orbits are at least 1200 km above the earth's surface.

Another satellite-type spread spectrum system is the Global Positioning
System, or GPS. This system will make use of a constellation of 18 satellites,
each in a 12-hr orbit, with up to six satellites in view at any time over a large
portion of the earth's surface. Each of the satellites transmits three signals.

1. A long code (one week repetition period) biphase-modulated 1227-MHz
   signal.

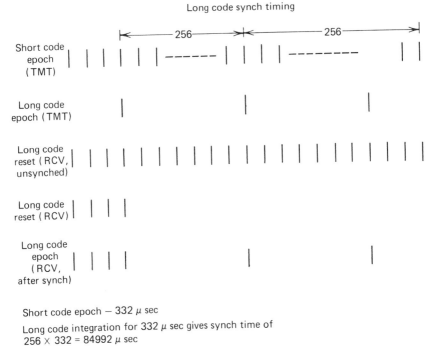

Short code epoch — 332 μ sec

Long code integration for 332 μ sec gives synch time of
256 × 332 = 84992 μ sec

Figure 9.8 TDRSS long-code synchronization timing.

2. A long code (same code) biphase-modulated signal at 1575 MHz.

3. A short code (1023-chip Gold code) biphase-modulated signal on an orthogonal carrier at 1575 MHz.

The 1023-chip Gold code (called a *C/A* code) transmitted by each satellite is different from all others. All satellites use the same long code (called a P code), however. Thus initial acquisition is accomplished using the 1023-chip short code, which is modulated with 50-bps data that provide information to allow the receiver to synchronize to the long code. Figure 9.9 shows the basic modulator configuration used in GPs. Note that the GPS modulator adds the same 50-bps modulation to both long and short codes, so either long code or short code receivers can demodulate the same information.

The GPS signal structure is similar in some ways to the TDRSS signal structure; that is, both employ a pair of quadrature biphase-modulated carriers, with different codes modulating them. The similarity does not end there, as TDRSS and GPS both employ 1023-chip Gold codes for initial acquisition. The long codes used in the two systems are quite different,

Figure 9.9   GPS modulator configuration (L_1 signal)

however. A comparison of the two can be summarized as:

GPS	TDRSS
Short code (clear access or C/A):*	Short code:
1.023 Mcps	3.08 Mcps
1023 chips long	1023 chips long
Gold codes	Gold codes
period, 1.0 msec	period, 0.33 msec
Long code (protected or P):†	Long code:
10.23 Mcps	3.08 Mcps
one week long	$256 \times 1023 = 261,888$ chips long
6.19×10^{12} chips	linear maximal code
mildly nonlinear code	period, 85 msec
Modulation:	Modulation:
1575 MHz	I carrier, short code
I carrier, C/A code	Q carrier, long code
Q carrier, P code (suppressed 3 dB)	(suppressed 10 dB)
1227 MHz	Several frequencies
P code only	

*Also called the Coarse Acquisition Code.
†Also called the Precision Code.

Figure 9.10 Four-channel GPS receiver (courtesy Magnavox).

The idea behind GPS is to transmit spread spectrum signals that allow a range measurement, from a known satellite location (the transmitter on board the satellite tells the receiver the satellite location). Then, with a knowledge of the transmitter location and the distance to the satellite, the receiver can locate itself on a sphere whose radius is the distance measured. After receiving signals and making range measurement on other satellites, the receiver can calculate its position based on the intersection of several spheres.

It is interesting to note that position location can be accomplished using either the short or long code modulated GPS signals. Figure 9.10 illustrates one typical GPS receiver configuration. Receivers are of two basic designs, one of which synchronizes to up to four satellite signals at the same time (satellite signals are differentiated by having different Gold codes). The alternate, lower-cost GPS receiver configuration uses only one receiver, which must sequentially acquire and make range measurements on at least three satellite signals. The second receiver approach is slower and less accurate than the first, but offers more commercial applicability because of less complexity and cost.

In other space applications spread spectrum signals have been employed for ranging at interplanetary distances.

Avionics Systems

Avionics systems can employ spread spectrum methods in any of the ways available to satellite and other systems. Ranging, direction finding, discrete addressing, and communicating are all within the realm of practicality. The major unique consideration for avionics applications is the large and unpredictable range of Doppler-frequency offsets. A pair of supersonic aircraft

FREQUENCY CHANNEL INDICATOR ID-783/ARC-50(V)

TRANSLATOR CV-8I5/ARC-50(V)

TRANSLATOR CONTROL C-2947/ARC-50(V)

*DISTANCE INDICATOR ID-782/ARC-50(V)

*RECEIVER-TRANSMITTER RT-525/ARC

*RECEIVER-TRANSMITTER CONTROL C-2948/ARC-50(V)

INTERCOM SET AN/ARC-IO

INTERROGATE RESET RESPOND

*UNITS FOR WIDE-BAND OPERATION

Figure 9.11   ARC-50 system (courtesy Magnavox Co.).

present Doppler shifts corresponding to a worst case of as much as relative Mach 10 (plus or minus, because the two may be flying directly toward or away from one another) or any amount up to that approximate maximum. The worst part is that, in general, the amount and sign of the shift is not predictable and any equipment used in such an application must be capable of covering the entire range of frequency uncertainty. Although satellite systems can generate higher Doppler shifts, both position and velocity of satellite systems are normally well defined and frequency uncertainties encountered may be compensated out as a demodulator design consideration.

Avionics systems have received heavy spread spectrum emphasis. Several systems have been developed for avionics use, or are currently under development. Some of these are:

ARC-50. The oldest of the avionics systems. First produced in the early 1960s. This is a direct sequence modem for voice communications. Its code rate is approximately 5 Mcps and its modulation format is biphase. The ARC-50 is also capable of ranging, and reads range to $\frac{1}{10}$/mile at ranges of up to 300 mi. Figure 9.11 illustrates the ARC-50 system, whose spread spectrum modem originally contained some 650 discrete transistors. The system is also capable of output power control over a wide range. The spread spectrum modem for a later version of the ARC-50 (the ARC-148) is shown in Figure 9.12 for comparison.

Have Quick. A frequency hopping UHF AM radio adapted from the ARC-164 radio. Have Quick will be the first widely used spread spectrum radio. The

Figure 9.12 ARC-148 spread spectrum modem (part of ARC-148 system).

Have Quick configuration will be a standard for at least the next 10 years. (See Figure 9.13.)

JTIDS. Joint Tactical Information Distribution System. Although really only a data transmission system, JTIDS has been touted for every function in modern aircraft except radar. The system is the heir to many previous programs (previous names were PLRACTA, SEEK BUS, CNI, UCNI, MYSTIC LINK, ITACS, and many others). Operating in the 960-1215 MHz band, JTIDS must provide compatibility with IFF and TACAN systems, which share the band.

The signal format of JTIDS employs frequency hopped direct sequence modulation; the direct sequence modulation is MSK at 5 Mcps. This direct sequence signal is then frequency hopped over a 52 frequency set, with separation of 3 MHz between frequency choices.

Three classes of JTIDS systems are under development. Their capabilities are summarized as follows:

Characteristic	Class 2	Class 1A	Class 1
Data rate, kbps			
transmit	80	160	280
receive	140	140	280
transmit and receive	160	160	300
Open message start rates, Hz	50/100	50/100	100/200
Number function channels	64	64	128
Number voice channels	2	4	10
Number conferencing voice channels	1×3	1×3	2×3
TACAN	Yes	Yes	N/A
Number metachannels	6	8	16
Quaternary function channel	2	2	4
Number of users	512	512	512
Antennas	2	1	2
RF power, watts	200	200/400* 1600[†]	200/400* 3200[†]
Range, n.mi.	300	300	300

*Military power mode.
[†]Burn-through power mode.

Note that the voice channel capability of JTIDS is very limited, even though an overall JTIDS network is designed to handle in excess of 96,000 data users. Figure 9.14 illustrates one of several JTIDS formats.

Figure 9.15 shows the three classes of JTIDS equipment as configured by the current equipment developers.

Figure 9.13   ARC-164 modification to Have Quick configuration.

Another system currently under development, though not strictly for avionics use, is SINCGARS (Single-Channel Ground-Airborne Radio System). SINCGARS is a frequency hopping system that hops over the 30-88 MHz band, with 25-kHz frequency spacing. Though only two candidate systems are under contract, several frequency hopping alternates for the same band have already been developed. A summary of these 30-88-MHz-band frequency-hopping systems is:

Unit	Developer
SINCGARS	Cincinnati Electronics
SINCGARS	ITT*
JAGUAR	Racal
SCIMITAR	Marconi
PRC-117	Harris
PRC-118	Tadiran
PRC-77	Rockwell

All of these units have 25-kHz frequency spacing, and are generally intended for voice or data. Their hopping rate is in the range of a few hundred hops per second. Some have the ability to hop over any of the 2320 channels available, and some hop over a set of 256 of the 2320 available.

Figure 9.16 shows the ITT SINCGARS radio.

---

*Current SINCGARS contractor.

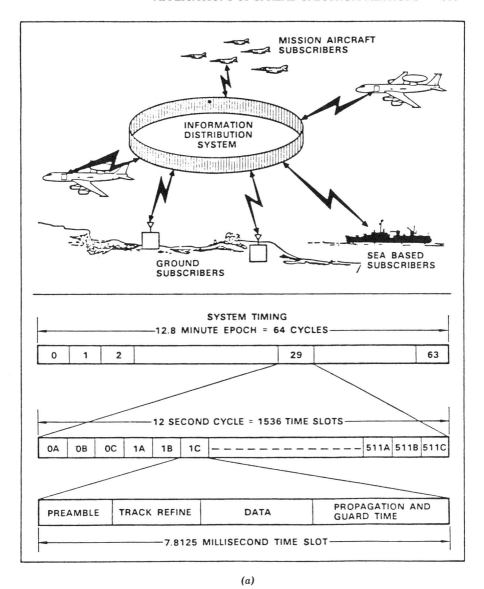

(a)

Figure 9.14   JTIDS format (a) Information distribution system.

Test Systems and Equipment

Spread spectrum techniques (or at least techniques common to spread spectrum systems) have recently become popular in test systems. References 216, 217, 235, and 237–240 describe a few of the spread spectrum techniques or subsystems. Transmission test sets[217, 237, 240] (actually almost complete

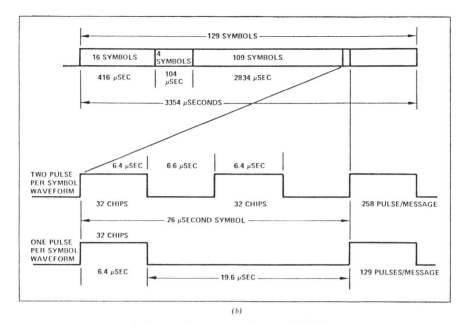

Figure 9.14   (*b*) Waveform modulation.

direct sequence modulated transmitters and receivers) characterize data transmission systems. In such applications a pseudonoise code sequence modulates the system under test. The coded signal then passes through the transmission path to the test set, which locks on to the received code sequence and compares it with its local reference for differences. Any chip difference between the received and locally generated signals is recorded as an error.

Another type of transmission test set allows the transmission system to be under test at the same time it is used for normal operation. This application makes use of the low power density of direct sequence modulated signals by transmitting the direct sequence signal simultaneously with the desired information-bearing signal. Mutual interference is low because the wideband test signal can be transmitted well below the desired signal, and the desired narrowband signal looks like just another noncoherent jamming signal to the spread spectrum receiver. Noise gating[239] and similar methods are related techniques. Figure 9.17 shows block diagrams of typical test setups employing direct sequence methods.

The pseudorandom codes used in spread spectrum systems have also been applied alone for "random" number generation, sources of "noise" modulation, data scrambling, and in most other areas in which noise or random sequences can be put to use. Commercial code sequence generators[216] have been introduced for these uses.

Figure 9.15   Three JTIDS configurations: (*a*) class 1; (*b*) class 1A; (*c*) class 2.

Figure 9.16   ITT SINCGARS radio.

Message Protection

Most of the newer spread spectrum systems provide for fully secure, crypto-type codes to be used for generating their signal spreading functions. Thus the systems are capable of resisting code breakage and spoofing to as high a level as required.

Position Location

The advantages of position-location methods are obvious, whether the system is a spread spectrum system or not. Collision avoidance, homing, ranging, surveying, navigation, and many others depend on position location. Spread spectrum methods are not always best for every application, but they are often capable of providing precision where other techniques fail. One spread spectrum system, for instance, is capable of resolution of range to 3 in. This capability is rivaled only by laser rangefinders, which cannot operate in inclement weather.

Spread spectrum methods certainly must be considered wherever position measurement is to be carried out.

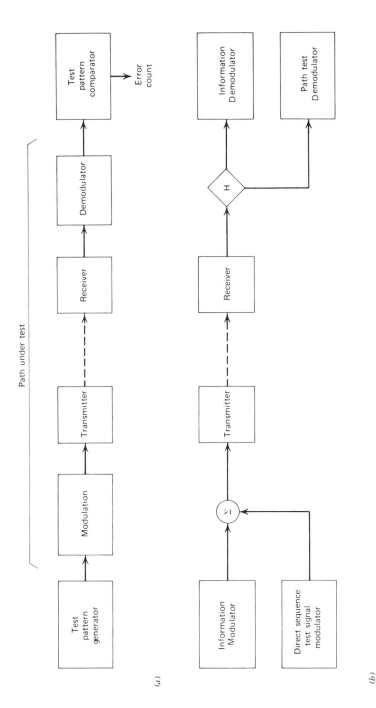

Figure 9.17 Typical transmission path-test setups by direct sequence techniques: (*a*) code pattern transmission test; (*b*) suppressed signal transmission test.

Figure 9.18   PLRS user unit (Courtesy Hughes Aircraft Co.).

Real Systems

We have already mentioned GPS, which is an important example of a spread spectrum position location system. Another spread system position location system is PLRS (Position Location Reporting System). This system employs a format very similar to that of JTIDS, but operates in a different frequency band and with a different number of hopping channels. The number of hopping channels is eight, with direct sequence modulation at 5 Mcps using minimum shift keying. A PLRS user unit is illustrated in Figure 9.18.

10

TEST AND EVALUATION
OF SPREAD SPECTRUM SYSTEMS

Spread spectrum systems may be tested in much the same way as other radio-frequency systems. They should be tested for sensitivity, bit error rate, and signal-to-noise ratio, just as other systems are tested. Parameters such as jamming margin, probability of synchronization, and loss of synchronization in the presence of interference, however, must receive special attention. Here we will concentrate only on those characteristics of spread spectrum systems that require this special attention in testing.

10.1 SENSITIVITY

Sensitivity of a spread spectrum receiver is determined by measuring the minimum signal required into the receiver to produce a minimum output signal-to-noise ratio or maximum bit error rate. Figure 10.1 illustrates the test setup used. The procedure is:

1. With the signal generator set for standard modulation by the information source, and with a given code providing spectrum spreading

2. a. Measure receiver output signal level.

   b. Remove information modulation and measure output again.

   c. The difference between 2a and 2b is the output signal-plus-noise to noise ratio.

a. Analog-Information Sensitivity

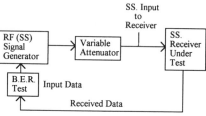

b. Data-Information Sensitivity

Figure 10.1   Receiver sensitivity test setup.

      d. Increase or decrease the RF signal level from the spread spectrum signal generation until the output $(S + N)/N$ ratio is 10 dB. This RF signal level is the receiver sensitivity.

3. a. Measure receiver bit error rate.

      b. Increase or decrease the RF signal level from the spread spectrum signal until the bit error rate equals the maximum allowed. This RF signal level is the receiver sensitivity.

One significant problem for both spread spectrum and narrowband receivers is that leakage can prevent the measurement of sensitivity when the receiver is highly sensitive. In the case of a narrowband receiver, such a measurement must employ a screen room to shield the receiver from measuring leakage around the signal attenuator.

Spread spectrum receivers may be tested without a screen room, even though their sensitivity is similar to a narrowband receiver. All that is necessary is to add a delay (see Figure 10.2) whose length is significant with respect to a code chip time, between the variable attenuator used to control signal level, and the receiver input. (Attenuation of the delay element should first be carefully measured.) Then, by observing the relationship between codes in the receiver and transmitter, one can observe whether there is a delay (signifying that the signal is passing through the correct measurement path) or no delay (which means that the signal is leaking around the delay).

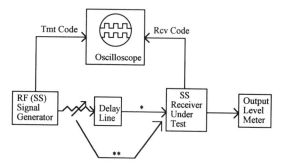

Figure 10.2 Test setup for measuring SS receiver sensitivity using a delay element when no screen room is available. (Codes are delayed when path is correct; undelayed when leakage path is in effect.) *Direct signal path through attenuator and delay line. **Leakage path around attenuator and delay line.

The two codes can be observed on two channels of an oscilloscope, and when the receiver is synchronized, delay or lack of delay can be seen.

10.2 SELECTIVITY

To a direct sequence receiver, selectivity has little meaning. In a narrowband receiver, selectivity is a measure of the response of the receiver to a signal at a frequency other than that to which it is tuned. The difference is in the way that the two types of receivers select a signal that they will accept or reject.

In a narrowband receiver, bandpass filters provide the determination of acceptance or rejection, while direct sequence receivers rely on codes and demodulator lock-in range to deny reception to unwanted signals. It is true that direct sequence systems can and do employ channelization to separate signals, but a DS receiver does not respond to a signal in an adjacent channel, even though there may be interchannel interference. Selectivity therefore does not have the same meaning to a DS receiver as it does to a narrowband receiver, excepting perhaps a narrowband phase lock receiver.

Frequency hopping systems, on the other hand, may have selectivity performance identical to that of a fixed-frequency narrowband receiver. There is, after all, no difference between them except that the frequency hopping receiver changes frequencies at relatively short intervals. A problem with frequency hopping systems is that the receiver cannot adequately reject signals in adjacent channels. Therefore, jamming and other users' signals that

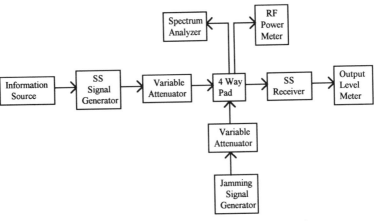

Figure 10.3   Jamming margin test setup; analog information systems.

happen to fall in an adjacent channel can interfere with a desired signal to almost the same degree that they would if they fell directly into the desired signal's channel.

10.3   JAMMING MARGIN

Two test setups for measuring the jamming margin of spread spectrum systems are shown in Figures 10.3 and 10.4, for systems sending analog and digital information, respectively. Both direct sequence and frequency hopping systems (as well as chirp and hybrid systems) can use these same test configurations.

Tests that must be made with these test setups fall into these categories:

1. Tests of the probability of synch acquisition under jamming conditions.
2. Tests to determine the jamming to signal level that causes a receiver to lose synchronization, when it has already acquired synchronization.
3. Measurements of output signal-to-noise ratio and/or bit error rate versus jamming to desired-signal ratio.

It should be noted that a spread spectrum receiver's sensitivity is that point at which the receiver is being jammed by its own noise (KTB). Therefore, any interference or jamming level measurements should be made with the desired input signal level above the minimum signal level required (receiver

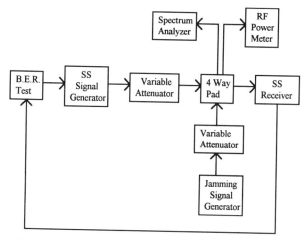

Figure 10.4 Jamming margin test setup; digital information systems.

sensitivity) by at least 10 dB. This is to safeguard against errors in measuring jamming margin due to receiver self-noise input.

We also note that to the receiver, any undesired input, whether deliberate and malignant jamming or friendly interference, is jamming. That is, those systems that are designed to support multiple users must have jamming margin sufficient to operate when the system is fully loaded with users. Also, it must be recognized that in the current spread spectrum world, many uncooperative users may be present in a given frequency band. This means that any system that cannot operate when being jammed by other systems of any type (friendly or unfriendly—deliberate or inadvertent) has no choice other than go out of business.

Further, spread spectrum receivers should be tested with several types of jamming signals, including those that are calculated to be most harmful. It is suggested that the jamming signal types listed in Table 10.1 be employed as a

TABLE 10.1 Recommended Minimum Jamming Types

Type of Receiver	Type of Jamming
Direct Sequence	CW within one percent of center frequency
	Swept CW across at least the main lobe bandwidth
Frequency hopping	Wideband noise
	Multitone (figure 10.5)
Chirp	Swept CW at chirp rate
Hybrid	Wideband noise
	Multitone

Figure 10.5 Block diagram of multitone signal source for test of frequency hopping systems. Sweep rate determines tone spacing. Sweep width determines number of tones.

minimum, with specific interference such as signals identical to the desired signal in all aspects except that the interference employs a different code, to supplement these types of interference.

10.4 SYNCH ACQUISITION

To determine the synch acquisition capability of a spread spectrum receiver, it is necessary to provide a signal generator that goes through its synchronizing process repeatedly. Some spread spectrum systems employ synchronization preambles at the beginning of each transmission. Others transmit a synchronization code continuously. The particular technique is very system dependent. Each attempt at synchronization is registered as a trial, and the number of successful trials is used to determine the probability, where a given time is allowed to achieve synchronization. Then, while making multiple trials at each of the levels (usually in 1-dB steps) from 0-dB jamming to the maximum level at which the receiver is designed to operate, count the number of successful and unsuccessful trials.

The number of successful trials divided by the total number of trials, at each jamming-to-signal ratio is the probability of acquisition at that ratio. It is emphasized, however, that it may be necessary to make a very large number of trials to establish satisfactory confidence in a given measurement. For example, if probability of acquisition to within one percent accuracy is desired, then at least 1000 and preferably many more measurements should be made at each jamming level.

10.5 LOSS OF SYNCHRONIZATION

Capability of a receiver to maintain synchronization in the presence of interference or jamming is measured by allowing the receiver to synchronize and then applying an interfering signal, increasing the interference until the receiver loses synchronization. As this level does not vary significantly for a given receiver and a given type of interference, a smaller total number of measurements than those necessary to measure acquisition probability is necessary. Typically 100 measurements of loss of synchronization for each form of interference is sufficient.

10.6 SIGNAL-TO-NOISE RATIO VS INTERFERENCE LEVEL

Antijamming, or jamming rejection capability is a measure of the ability of a receiver to operate in the presence of interference and to provide a usable output signal. A usable output signal, of course, is one that has an acceptable bit error rate or signal-to-noise ratio. (Remember that bit error rates and signal-to-noise ratios are closely related to one another.) In a typical voice system, the minimum signal-to-noise ratio is in the range of 10–12 decibels, while maximum bit error rate for digital systems is typically one error per 1000 bits to one error per 100,000 bits.

In measuring the receiver's performance, the test setup(s) shown in Figures 10.3 and 10.4 are used. While receiving a signal, an interfering signal is added, and the receiver's output signal-to-noise ratio/bit error rate is measured. For each interference-to-signal ratio, output signal-to-noise or bit error rate should be measured. This provides a measure of the receiver's processing capability and linearity, and a plot like that of Figure 1.4 can be generated. Typically, within the linear range of the receiver, one can expect output signal-to-noise ratio to decrease by 1 dB for each dB that input interference-to-signal ratio increases.

10.7 PROCESS GAIN

In an earlier chapter (see Chapter 1, Section 1.3) it was shown that process gain for a system is the sum total of its signal-to-noise ratio improvement. That is, if a system has an input signal-to-noise ratio of 5 dB and an output signal-to-noise ratio of 30 dB, then its process gain is 30–5 or 25 dB. Therefore, a method of determining the process gain that is simple and straightforward is to measure a receiver's output signal-to-noise ratio, then find the difference between output signal-to-noise ratio and the input signal-to-noise ratio. This method is an approximation to the extent that imperfections in the receiver can cause an error, which is normally in the range of 0–2 dB.

One can also calculate the receiver's process gain as the ratio of the RF bandwidth to the information rate. The difference between this ratio and the measured difference of output and input signal-to-noise ratios is the processing loss due to receiver imperfections.

Figure 10.6   Processing system.

10.8   ON TESTING SPREAD SPECTRUM SYSTEMS TO DETERMINE THEIR PROCESS GAIN

Processing gain, a measure of the robustness of spread spectrum systems (among others), consists of the ability of a receiver to improve the quality of its input signal. This is accomplished by trading bandwidth for improved signal-to-noise ratio or a lower bit error rate. That such a trade off is possible was first quantified by C. E. Shannon in his classic study "Mathematical Theory of Communications," in which he showed that the capacity (c) of a communications system to send error-free information is

$$C = W \log_2(1 + S/N)$$

where $W \equiv$ bandwidth employed for signal transmission,
         $S/N \equiv$ signal-to-noise ratio in the communication channel.

Observation of this simple equation shows that system capacity may be maintained by either employing more bandwidth when signal-to-noise ratio decreases, or by increasing signal-to-noise ratio when the channel bandwidth is reduced.

Both spread spectrum systems and error correction coding systems make use of this property. Let us consider a processing system and the signal relationships within it. Figure 10.6 illustrates such a black box system. (Note that we consider only the receiver, which requires a specially prepared input signal. That is, the receiver requires a signal that employs a wide bandwidth, and a coding method that is known to the receiver and can also be decoded by the receiver.)

If properly implemented, a processing receiver improves the signal-to-noise ratio at its input by an amount equal to its processing gain, less any losses due to imperfections in its design.*

---

*Imperfections in receiver and demodulator design, such as frequency inaccuracy, Doppler uncertainty, and inability to design perfect filters whose 3-dB bandwidth and noise bandwidth are the same, cause typical systems to have losses in the one to three dB range.

Processing gain for a spread spectrum receiver may be estimated as

$$\text{Processing Gain (Gp)} = 10 \log \frac{\text{BW}_{\text{RF}}}{\text{R data}} \text{*}$$

where BW_{RF} = The bandwidth used to send the encoded signal,
      R data = The data rate in bits per second.

When processing gain of a receiver is greater than 1 (Gp > 1) the receiver may be able to deliver some degree of noise and interference rejection. FCC requirements under part 15 rules are for a minimum process gain of 10 dB or more. This means that a receiver meeting this rule should have an output signal-to-noise ratio of 10 dB when its input signal-to-noise ratio is 0 dB (assuming no processing losses). Similarly, if a 15-dB signal-to-noise ratio is required, then that same receiver would provide the required output at an input signal-to-noise ratio of 5 dB. (Note that in this particular example, the input signal-to-noise ratio is never negative.)

When a receiver can operate with a negative input signal-to-noise ratio while still providing a positive output signal-to-noise ratio,[†] it is said to have "jamming margin." The amount of jamming margin is equal to the amount by which the interference or noise exceeds the desired input signal, while the minimum output signal-to-noise ratio is maintained at its minimum level. Figure 10.7 shows the input and output signal-to-noise ratios for a system with 10-dB processing gain and processing loss between zero and two dB. This is a plot described by

$$\text{S/N out} = \text{S/N in} + \text{Gp} - \text{P.L.},$$

where S/N in ≡ input signal-to-noise ratio (dB),
      S/N out ≡ output signal-to-noise ratio (dB),
         Gp ≡ process gain in dB,
        P.L. ≡ processing losses in dB.

Note: This equation is valid only within the linear range of the receiver. Also, if operating near the receiver threshold, the receiver's noise figure may contribute to the overall noise and interference present. Processing gain measurement is a straightforward process, based on the relationship

$$\text{Gp} = \text{S/N out} - \text{S/N in} + \text{P.L. (in dB)}.$$

---

[*]Processing gain may also be considered, for analog systems, to be a function of the RF-bandwidth-to-information-bandwidth ratio.
[†]Receiver output signal-to-noise ratios must always be positive (i.e., signal is always greater than noise). Otherwise, the output signal would not be usable, and the communications system is useless.

Figure 10.7   Input vs. output S/N ratio for system with 10-dB process gain and various processing loss.

That is, the output signal-to-noise ratio of a system is different from its input signal-to-noise ratio by the amount of its processing gain, with processing loss subtracting from the net processing gain.

In the case of data systems, it may be more convenient to measure output bit error rate (as opposed to signal-to-noise ratio) and convert to signal-to-noise ratio using BER versus S/N ratio curves.*

Processing loss should be considered to be 1–2 dB, with very good systems being close to 1 dB, while those that must operate in high dynamics or with other more difficult conditions are typically closer to 2 dB. (It is not unusual for a system to measure 3 dB or more loss due to imperfections in its demodulation process.)

---

*Eb/No curves are not applicable without a correction factor equal to the reciprocal of the number of bits per second per Hertz of bandwidth. That is,

$$\frac{S}{N} = \frac{Eb \times R}{No \times W},$$

$$\frac{Eb}{No} = \frac{S}{N} \times \frac{W}{R},$$

where   S = signal power

N = noise power

Eb = energy per bit

No = noise power per Hertz (noise density)

R = data rate

W = data bandwidth

and R/W is usually termed the signaling "efficiency" of a system.

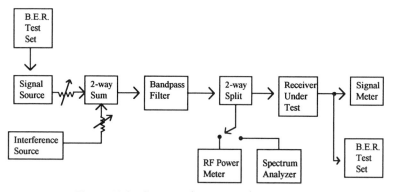

Figure 10.8   Suggested process gain test setup.

Figure 10.8 illustrates a test setup for measuring process gain. The test should be conducted as follows:

1. Measure the signal level from the signal source, and set to the proper level, with the interference source turned off.
2. Measure the signal level from the interference source, and set to the same level as the signal source was previously set, with the signal source turned off. (The source to noise/interference at the receiver's input is now 0 dB when both are turned on.)
3. With only the signal turned on, measure the output signal level (or BER). (The signal generator should be modulated with both the proper information signal as well as the correct spreading signal.)
4. a. To measure signal-to-noise ratio, remove the information modulation from the signal source. Then measure output signal level. The ratio of the output, with the information modulation present, to the output without information modulation is the signal-to-noise ratio.
   b. To measure bit error rate, simply modulate the signal source with a BER test set and measure the output bit error rate.
5. Repeat steps 3 and 4 with the interference source turned on:
   a. Measure S/N out versus signal to interference level.

$$Gp = S/N \text{ out} - S/N \text{ in} + P.L. \qquad P.L. \equiv \text{Processing Loss}$$

   b. Measure BER versus signal to interference level.

$$Gp = (BER \to S/N \text{ out}) - S/N \text{ in} + P.L.$$

Processing gain may vary as a function of type of interference and frequency of interference. The test method, however, remains the same.

Note: It is important to remember that a receiver may not operate when interference is larger than the signal it is receiving. This does not mean the receiver has no processing gain, only that it has no jamming margin (which in effect is simply excess process gain). A receiver with 10-dB processing gain might operate at input signal-to-noise ratio as low as -10 dB, if a 0-dB output signal-to-noise ratio is acceptable. However, if a 10-dB output signal-to-noise ratio is required, that same receiver requires at least a 0-dB input signal-to-noise ratio (actually its requirement would be higher by the amount of its implementation or processing loss).

10.9 THE FCC METHOD

A method for measuring process gain is suggested by the Federal Communications Commission in its regulations for spread spectrum systems usage. The method suggested is one in which the difference in output signal-to-noise ratio from a receiver is measured with and without spectrum spreading. This means that it is necessary to stop the code generator in both the transmitter (or signal generator) and the receiver being tested.

When its code is not operating, a spread spectrum transmitter may modulate its carrier with the unspread data stream (this is not necessarily true of all spread spectrum systems) in the same way it modulates its carrier when both data and code are present. Similarly, a spread spectrum receiver can demodulate a modulated carrier without code, if its own internal code generator is stopped.

Therefore, if output signal-to-noise ratio is measured with and without code operation, the difference in output signal-to-noise ratio is the spread spectrum process gain.

The only drawback to this process gain measurement method is that many systems have no provision for either separating the code from the data transmission method, or any way of separately stopping their code.

10.10 CROSS-CORRELATION

Cross-correlation tests on spread spectrum systems are intended to determine the degree of orthogonality (or lack thereof) between the codes used for spectrum spreading. These tests are similar to those carried out to define interference rejection, except that instead of being used to measure output signal-to-noise ratio or bit error rate as a function of interference level, what is measured is the probability that a receiver will synchronize to a signal having a different spreading code. That is, cross-correlation tests would be carried out by making multiple trials for synchronization, where the spread spectrum signal generator is set to one or more codes other than the one used by the receiver, and the generator's output power is set to various levels, while the receiver's synchronization status is monitored.

Figure 10.9 One-megachip-per-second BPSK with equal-power CW signal. 1 MHz per division horizontal. 10 dB per division vertical.

10.11 TRANSMITTER MEASUREMENTS

Measurements made on spread spectrum transmitters chiefly consist of those having to do with measurement of output power and output signal spectral density. For these, a power meter and spectrum analyzer are the most useful instruments.

Restrictions on the output bandwidth, defined by regulatory agencies such as the Federal Communications Commission, are the most common reason for interest in transmitter output spectra. In military systems, interest in the existence of stationary signals in the transmitter's output spectrum is also important. In the interest of reducing interference to other systems (especially narrowband, conventional systems) even distribution of transmitter output power, without unusual concentration of energy at any point in the output bandwidth, is also important.

When performing tests with spread spectrum systems, it is often necessary to know the equivalent spread spectrum signal power as compared to a narrowband signal. Given that a power meter is available, the measurement can readily be made. However, there are instances in which it is useful to know the equivalent levels without resorting to a measurement. Figures 10.9 through 10.15 show equivalent direct sequence and CW signals for various code rates.

In all of these figures, the CW power and direct sequence power are set to an equal level. The code employed is a 1023-chip linear maximal code, and the center frequency employed is 200 MHz. Also, the spectrum analyzer vertical scale setting is 10 dB per major division.

Spread spectrum direct sequence signal power measurement must be considered from the standpoint of the actual bandwidth used. Very few direct sequence signals are employed without some filtering. For example, if

Figure 10.10   Two-megachip-per-second BPSK with equal-power CW signal. 1 MHz per division horizontal. 10 dB per division vertical.

only the main lobe of a direct sequence signal is used, then the power in the main lobe (compared to the total spread spectrum power) is approximately 1 dB lower. This means that if a wideband power meter is used, a correction should be made.

Observation of the level of the CW carrier (set to near center frequency) with respect to the highest peak in the direct sequence signal gives the results shown in Figure 10.16. (These results are approximate, since the DS peak value is dependent on the specific code used.)

Figure 10.11   Four-megachip-per-second BPSK with equal-power CW signal. 1 MHz per division horizontal. 10 dB per division vertical.

Figure 10.12   Five-megachip-per-second BPSK with equal power CW signal. 3 MHz per division horizontal. 10 dB per division vertical.

Figure 10.13   Eight-megachip-per-second BPSK with equal-power CW signal. 3 MHz per division horizontal. 10 dB per division vertical.

10.12   GENERAL

Other measurements on spread spectrum systems are similar or identical to those made on conventional systems. Code tests are the same as those commonly used in design and application of computer systems.

Therefore we will conclude this chapter having discussed only those tests specifically pertinent to spread spectrum systems. Sufficient references are available for testing of other aspects of both conventional and spread spectrum systems.

Figure 10.14   Ten-megachip-per-second BPSK with equal-power CW signal. 3 MHz per division horizontal. 10 dB per division vertical.

Figure 10.15   12-megachip-per-second BPSK with equal-power CW signal. 3 MHz per division horizontal. 10 dB per division vertical.

Figure 10.16   Approximate CW peak to DS peak level versus chip rate.

With frequency hopping systems, only one relatively narrowband signal (excepting spurious signals) is present at any one time, so little difficulty is experienced in measuring signal power. In fact, typical frequency hopping systems register little difference in signal power whether hopping or sitting at one frequency (with code generator stopped).

Large system tests of spread spectrum systems are often necessary, because of the use of such systems for multiple users. Often, however, it is very difficult to fully exercise a multiple access spread spectrum system, because of the sheer number of spread spectrum transmitters and receivers the system is designed to handle at the same time. (One particular military spread spectrum system is designed to operate with up to 96,302 simultaneous users, for example. Unfortunately, there are not enough such user units in the entire world to fully prove the operation of this system.)

Since, for all practical purposes, it is impossible to fully exercise a system designed to handle hundreds to thousands of users without actually constructing hundreds or thousands of transmitters and receivers, the usual approach is to simulate the system's operation with a computer, or else to construct a partial system with a few transmitters and receivers, and simulate a large number of other users by transmitting a relatively high power interfering signal to represent those other users. Both approaches usually fall far short of fully exercising a multiuser system in a real environment, even though they may be the best that we can do, and the results are sometimes only a little better than no testing at all.

Two cautions are in order:

1. Though it is easy to say, and may seem trivial, computer simulations are no better than the assumptions upon which they are based (garbage in, garbage out). It is very unusual to design a complex system through computer simulation, and not find out a great deal that is unexpected about the system the very first time it is turned on and operated under real conditions.

2. Testing a multiuser spread spectrum system by constructing a few real transmitters and receivers, and attempting to simulate more users by transmitting a single signal at higher power as interference does nothing to show the ability (or lack thereof) of a system and its user components to reject multiple codes that might crosscorrelate and prevent the desired multi-user operation. Such an approach also does little or nothing to simulate the range of varying multisignal conditions that would exist if multiple signals were truly present.

PART II

COMMERCIAL* APPLICATIONS OF SPREAD SPECTRUM SYSTEMS

Part I of this book is concerned with the basics of spread spectrum systems, and is oriented to military applications, or at least a military viewpoint with respect to employment of spread spectrum techniques. In this section we will expand the viewpoint to include those systems that must be much lower in cost, smaller in size, and operate with untrained, inexperienced users. When all factors are considered, such mass production devices also must operate in much more stringent circumstances than their military counterparts.

Before 1985, very few spread spectrum systems had ever been used outside the military bands. However, military spread spectrum systems had been used at frequencies that ranged all the way from ELF through EHF, so there was little doubt that spread spectrum methods could work at almost any frequency with little degradation.

In 1985, the Federal Communications Commission made a historic decision to allow the use of spread spectrum signals in the ISM (Industrial, Scientific, and Medical) bands. The power level transmitted, without license, was set at a 1-W maximum. Of the available spread spectrum techniques, chirp is not allowed, but direct-sequence, frequency-hopping, and hybrid FH/DS are allowed.

Detailed requirements for spread spectrum systems meeting FCC standards will be discussed in the following pages.

---

*Consumer applications are included here as a subset of commercial applications. Strictly speaking, however, consumer, industrial, and commercial products are often considered to be for different markets.

11

TRADEOFFS IN COMMERCIAL* APPLICATIONS OF SPREAD SPECTRUM SYSTEMS

The ISM bands, in which unlicensed spread spectrum systems may be employed, are 902–928, 2400–2483.5, and 5725–5870 MHz. Bandwidths available in these bands are 26, 83.5, and 125 MHz, respectively. As these bands are used by others, however, and spread spectrum users are considered to be secondary users, the spread spectrum systems must not interfere with primary users and must accept without complaint any interference from the primary users.

What are the primary threats to commercial spread spectrum systems (and those primary users that the spread spectrum systems must not interfere with)?

Table 11.1 lists other possible users. It cannot be emphasized too strongly that if one is to operate a system in one of the ISM bands listed, the system must operate when any or all of the other users are present. This means that those who have sold systems without process gain[†] (and those who have purchased them) have wasted their time and money.

Excerpt from FCC regulations concerning spread spectrum and low power operation in the ISM bands

The following paragraphs relate to the use of spread spectrum in the industrial, scientific, and medical frequency bands, and are taken directly

---

*From this point, consumer users will be considered a subset of commercial users.
[†]Some systems have less than zero process gain and they will not operate when their input $S/N < 6$ dB.

TABLE 11.1   ISM Interference Threats

Band	Interference Threat (either way)
902–928 MHz	AVL (Automatic Vehicle Locations) systems
	Up to 2000 W fixed
	Up to 15 W mobile
	Old microwave ovens
	Industrial heaters, Up to 60,000 W
	U.S. Navy Radars, shipboard, Up to 1 MW
	Other spread spectrum ISM users $\leq$ 1 W
	Low power ISM, 0.5–1.0 mW
	Diathermy machines
	Amateur radio operators
2400–2483.5 MHz	Microwave ovens, up to 750 W
	Other spread spectrum ISM users $\leq$ 1 W
	Low power ISM, 0.5–1.0 mW - users
5725–5850 MHz	Other spread spectrum ISM users $\leq$ 1 W
	Low power ISM, 0.5–1.0 mW - users
	Various Licensed users, point-to-point $\leq$ 10 W

from the appropriate FCC regulations. Also are included are the regulations that govern low power operation in the same ISM bands. A summary of the rules for spread spectrum use, with some interpretation, is also given in the following pages.

Amateur radio operators are also allowed to transmit using spread spectrum signals, at much higher operating power but with a restricted set of codes. Information is given here on these requirements. It is extremely important that those who desire to operate using spread spectrum techniques understand exactly their limits and the rules by which they and others must abide.

§15.247 Operation within the bands 902–928 MHz, 2400–2483.5 MHz, and 5725–5850 MHz.

(a) Operation under the provisions of this section is limited to frequency hopping and direct sequence spread spectrum intention radiators that comply with the following provisions:

(1) Frequency hopping systems shall have hopping channel carrier frequencies separated by a minimum of 25 kHz or the 20 dB bandwidth of the hopping channel, whichever is greater. The system shall hop to channel frequencies that are selected at the system hopping rate from a pseudorandomly ordered list of hopping frequencies. Each frequency must be used equally on the average by each transmitter. The system receivers shall have input

bandwidths that match the hopping channel bandwidths of their corresponding transmitters and shall shift frequencies in synchronization with the transmitted signals.

(i) Frequency hopping systems operating in the 902–928 MHz band shall use at least 50 hopping frequencies. The maximum allowed 20 dB bandwidth of the hopping channel is 500 kHz. The average time of occupancy on any frequency shall not be greater than 0.4 seconds within a 20 second period.

(ii) Frequency hopping systems operating in the 2400–2483.5 MHz and 5725–5850 MHz bands shall use at least 75 hopping frequencies. The maximum 20 dB bandwidth of the hopping channel is 1 MHz. The average time of occupancy on any frequency shall not be greater than 0.4 seconds within a 30 second period.

(2) For direct sequence systems, the minimum 6 dB bandwidth shall be at least 500 kHz.

(b) The maximum peak output power of the transmitter shall not exceed 1 Watt. If transmitting antennas of directional gain greater than 6 dBi are used, the power shall be reduced by the amount in dB that the directional gain of the antenna exceeds 6 dBi.

(c) In any 100 kHz bandwidth outside these frequency bands, the radio frequency power that is produced by the modulation products of the spreading sequence, the information sequence and the carrier frequency shall be either at least 20 dB below that in any 100 kHz bandwidth within the band that contains the highest level of the desired power or shall not exceed the general levels specified in §15.209(a), whichever results in the lesser attenuation. All other emissions outside these bands shall not exceed the general radiated emission limits specified in §15.209(a).

(d) For direct sequence systems, the transmitted power density averaged over any 1 second interval shall not be greater than 8 dBm in any 3 kHz bandwidth within these bands.

(e) The processing gain of a direct sequence system shall be at least 10 dB. The processing gain shall be determined from the ratio in dB of the signal to noise ratio with the system spreading code turned off to the signal to noise ratio with the system spreading code turned on, as measured at the demodulated output of the receiver.

(f) Hybrid systems that employ a combination of both direct sequence and frequency hopping modulation techniques shall achieve a processing gain of at least 17 dB from the combined techniques. The frequency hopping operation of the hybrid system, with the direct sequence operation turned off, shall have an average time of occupancy on any frequency not to exceed 0.4 seconds within a time period in seconds equal to the number of hopping frequencies employed multiplied by 0.4. The direct sequence operation of the hybrid system, with the frequency hopping operation turned off, shall comply with the power density requirements of paragraph (d) of this section.

NOTE: Spread spectrum systems are sharing these bands on a noninterference basis with systems supporting critical Government requirements that have been allocated the usage of these bands, secondary only to ISM equipment operated under the provisions of part 18 of this chapter. Many of these Government systems are airborne radio location systems that emit a high EIRP which can cause interference to other users. Also, investigations of the effect of spread spectrum interference to U.S. Government operations in the 902–928 MHz band may require a future decrease in the power limits allowed for spread operation.

[54 FR 17714, Apr. 1989, as amended at 55 FR 28762, Jul 13, 1990]

§15.249 Operation within the bands 902–928 MHz, 2400–2483.5 MHz, 5725–5875 MHz, and 24.0–24.25 GHz

(a) The field strength of emissions from intentional radiators operated within these frequency bands shall comply with the following:

Fundamental Frequency	Field strength of fundamental (millivolts/meter)	Field strength of harmonics (microvolts/meter)
902–928 MHz	50	500
2400–2483.5 MHz	50	500
5725–5875 MHz	50	500
24.0–24.25 GHz	250	2500

(b) Field strength limits are specified at a distance of 3 meters.

(c) Emissions radiated outside of the specified frequency bands, except for harmonics, shall be attenuated by at least 50 dB below the level of the fundamental or to the general radiated emission limits in §15.209, which ever is the lesser attenuation.

(d) As shown in §15.35(b), for frequencies above 1000 MHz, the above field strength limits are based on average limits. However, the peak field strength of any emission shall not exceed the maximum permitted average limits specified above by more than 20 dB under any condition of modulation.

(e) Parties considering the manufacture, importation, marketing or operation of equipment under this section should also note the requirement in §15.37(d).

[54 FR 17714, Apr. 25, 1989, as amended at 55 FR 25095, June 29, 1990]

Summary of Part 15 Spread Spectrum Rules

Operating bands:
902–928 MHZ
2400–2483.5 MHZ
5725–5850

Output power: 1 W into antenna*.

Antenna gain: 6-dBi maximum.
Alternatively, if more antenna gain is used, transmitter power must be reduced by the same amount as the antenna gain exceeds 6 dBi.

Effective EIRP limit: +36 dBm*

Spectrum spreading methods:
Direct sequence. Minimum 6 dB BW 500 KHz in 900-MHz band and 1.0 MHz in 2400- and 5700-MHz bands. *Frequency hop.* Minimum channel separation 25 KHz. Maximum channel 20-dB bandwidth 500-KHz in 900-MHz band, with at least 50 frequency channels. Maximum channel 20-dB bandwidth in 2400- and 5700-MHz bands is 1.0 MHz, with at least 75 frequency channels. *Hybrid FH/DS.* FH must meet average dwell time requirement. DS must meet power density requirement.

FH dwell time:
Average dwell time less than 0.4 sec in 20 sec in 900-MHz band. Average dwell time less than 0.4 sec in 30 sec, in 2400- and 5700-MHz bands.

DS power density:
Power density in any 3-KHz band sector must be less than 8 dBm when averaged for 1 sec.

Processing gain:
DS processing gain for DS systems must be at least 10 dB. For hybrid systems, processing gain must be at least 17 dB.

Minimum hop rate:
Based on maximum dwell time, minimum hop rate is

$$\frac{50 \text{ frequencies}}{20 \text{ sec}} = \frac{75 \text{ frequencies}}{30 \text{ sec}} = 2.5 \text{ hop/sec}$$

with dwell time of $\frac{1}{2.5} = 0.4$ seconds per channel.

Spread Spectrum Operation by Radio Amateurs

Section 97 of the FCC rules allow radio amateurs to operate on all amateur radio frequencies above 450 MHz. The general direct-sequence and fre-

*It is possible that regulations will be implemented that limit transmitter power to a lower level–probably 300 mW, with 6-dBi maximum antenna gain, at least in hand held units.

quency hopping rules apply, except that only specific codes are allowed. These are

$$[7, 1],$$
$$[13, 4, 3, 1],$$
$$[19, 5, 2, 1,].$$

Radio amateurs are allowed to employ up to 100-W peak power, at the transmitter output. They must also identify themselves using non spread spectrum (i.e., conventional narrowband) techniques that are detectable by a conventional narrowband receiver.

11.1 COMPARISON OF MILITARY AND COMMERCIAL ADVANTAGES OF SPREAD SPECTRUM

All of the advantages listed in Chapter 1 of this book are effective for commercial spread spectrum systems, but in different degrees, since the emphasis in commercial systems is usually different than for military-oriented systems. Differences in emphasis for commercial- and military-oriented systems are as listed in the following:

Military Systems	Commercial Systems
Selective Addressing	
Used to direct a signal to a specific receiver. Codes are usually non-linear and very long.	Similar application, but codes are usually less complex and shorter.
Code Division Multiplexing	
Allows multiple simultaneous system applications such as position location and receiving units. Sometimes combined with time division multiplexing to avoid the necessity for highly accurate power control. Some limited power control usually provided, however.	Used to provide networks of interdependent transmitting and receiving systems. Used with time division or power control for voice and local area network applications.
Low Power Density	
Primary application for signal hiding or low probability of intercept. Used in conjunction with power control to minimize signal level and reduce probability of detection.	Important characteristic of spread spectrum signals is reducing interference to other users.
	(Especially interference to licensed systems that have primary use of the spectrum available.)

Military Systems	Commercial Systems
Message Protection	
Important in military applications to protect information from unauthorized parties. Also used to prevent unauthorized use of communication links. Codes used are very long and complex.	Protection for private conversations. Codes less complex than in military systems. Could also be employed in prevention of unauthorized use.
High Resolution Ranging	
Applied in position location systems. Examples are Global Positioning System (also used for commercial applications), PLRS, and other systems.	Position location services being offered for stolen car location, lost child finders, and many other applications.
Multipath Rejection	
Applicable to military system in which multipath delays are long compared to a code chip time period. Aircraft systems and others in which path length is measured in miles to hundreds or thousands of miles are best applications. Military systems also allow wider band spectrum spreading.	Useful in commercial systems, but must be supplemented by methods such as antenna diversity to mitigate multipath effects inside buildings, in downtown-city areas, and other places in which reflecting surfaces are plentiful.
Interference Rejection	
A vital property of military systems to prevent jamming from interrupting command and control links.	Needed in commercial systems to enable the use of multiple uncoordinated systems.

11.2   DIRECT SEQUENCE VS FREQUENCY HOPPING

Three kinds of spread spectrum systems are allowed by current FCC regulations:

1. Direct Sequence-
     minimum bandwidth: 500 KHz,
     minimum code length: 128 chips,
     minimum process gain: 10 dB.
2. Frequency Hopping-
     minimum number of channels*
       902–928 MHz: 50,
       2400–2483.5 MHz: 75,
       5725–5850 MHz: 75,

*Frequency hopping channels are not allowed to overlap one another as in military systems.

maximum instantaneous channel bandwidth: 500 KHz in 902–928 MHz band, 1.0 MHz in the two higher frequency bands.

maximum channel occupancy: 0.4 seconds in any 20 second period in 902–928 MHz band, 30 seconds in the two higher frequency bands.

minimum process gain: 10 dB.

3. Hybrid FH/DS
   restrictions similar to DS & FH
   minimum process gain: 17 dB.

We will emphasize comparison of direct sequence and frequency hopping, with much less consideration of hybrid systems, since hybrid systems will be shown to have less ability to work in the presence of interference than either of the other two types of systems.

Primary tradeoffs in direct sequence and frequency hopping systems are:

Direct Sequence	Frequency Hopping
Good range resolution	Poor range resolution
More noiselike signal	Narrow instantaneous bandwidth
Can operate below ambient noise	Must have positive S/N ratio
Can operate without error correction codes	Require error correction codes
Limited near-far performance	Best near-far performance
Requires linear signal path	Operates well with nonlinearity in signal path
Synchronization more difficult	Easier to synchronize

Range Resolution

Range resolution in spread spectrum systems is a function of code chip rate. Therefore, direct sequence systems inherently possess much better range resolution capability than frequency hopping systems, as their chip rates are typically several orders of magnitude higher. Chapter 8 discusses spread spectrum ranging and position location methods in detail.

Interference to Other Systems

The power density of a direct sequence signal is, on the average, simply the signal power divided by the bandwidth. That is

$$P \text{ average (Watts/Hz)} = \frac{\text{Power in Watts}}{\text{Bandwidth in Hertz}}.$$

Of course, in practice the signals have power distribution that is not flat. Instead, depending on the type of code modulation, the power distribution is similar to $(\sin x/x)^2$, but with 3-dB bandwidth 0.88 times the code rate for BPSK and QPSK modulation, and 0.6 times the code rate for MSK modulation. Thus, power density in the 3-dB bandwidth is

$$\frac{\text{Power in Watts}}{0.88 \text{ R code}} \quad (\text{BPSK or QPSK})$$

or

$$\frac{\text{Power in Watts}}{0.6 \text{ R code}} \quad (\text{MSK}).$$

With a typical 10-Mcps code rate and 1.0-W power, the BPSK power density is

$$\frac{1.0}{8.8 \times 10^6} = 1.13 \times 10^{-7} \quad \text{W/Hz of bandwidth,}$$

while MSK power density is

$$\frac{1.0}{6 \times 10^6} = 1.66 \times 10^{-7} \quad \text{W/Hz of bandwidth.}$$

In both cases, the power density is between 0.1 and 0.2 μW/Hz. Therefore, a narrowband receiver with typical 30-KHz bandwidth would be subjected to only approximately 0.3 mW of the transmitter's overall 1.0-W output power.

   Frequency hopping systems, as opposed to direct dequence systems, simply hop from frequency to frequency, concentrating all of their power in a bandwidth as narrow as 2/R hop* and as wide as 1.0 MHz. (Remember that the widest instantaneous bandwidth allowed by the regulation is 0.5 to 1.0 MHz, at least in the ISM bands.) This means that frequency hop signal power density varies from

$$\frac{1.0 \text{ Watt}}{4.8 \text{ Hz*}} = 208 \text{ mW/Hz}$$

to

$$\frac{1.0 \text{ Watt}}{1 \times 10^6} = 1.0 \times 10^{-6} \text{ W/Hz.}$$

*Minimum hop rate is 2.5 hop/sec.

This means that even the widest-band frequency hopping systems interfere with 30-KHz-bandwidth narrowband systems to an extent $\frac{0.03}{0.003} = 10$ times more than 10 MHz direct sequence systems, or equally at distances 3.16 times as far. For the narrowest band frequency hopping signals, conventional 30-KHz receivers would be interfered with for distances 10 times further than they would be for even the narrowest direct sequence signal because the entire 1-W signal would fall in the receiver bandwidth.

Of course, if we compare the widest allowed frequency hopping signal (500-KHz instantaneous bandwidths) to the narrowest allowed direct sequence signal (also 500-KHz instantaneous bandwidth) interference to narrowband users would be the same. There is very little advantage, however, in using direct sequence bandwidth as narrow as 500 KHz, and many advantages in using more bandwidth.

Interference from a frequency hopping system to a narrowband system is not constant, but occurs at a burst rate that averages P · Rhop/M, where P is the number of system users, M is the number of available frequencies, and Rhop is the hopping rate. The duration of a single burst of interference is simply 1/Rhop. Assume, for example, that a network of ten potential frequency hopping users, hopping at 500 hops per second, with 500 KHz-bandwidth per channel, and 40 channels is postulated. A narrowband listener tuned anywhere in the 20-MHz band (500 KHz × 40 = 20 MHz) would hear:

| | Average Interference Rate | | | |
Receiver BW	1 user	5 users	10 users	Duration/burst
≤ 500 MHz	12.5/sec	62.5/sec	125/sec	2 msec
1.0 MHz	25/sec	125/sec	250/sec	2 msec

Incidentally, "interference" is not restricted to those signals that fall within the receiver's IF bandwidth. An interfering signal that falls in a receiver's RF bandwidth (RF bandwidth is typically much greater than IF bandwidth) often desensitizes the receiver to smaller desired signals. The interference potential of frequency hopping systems is always greater than that of direct sequence systems with the same capacity, because they interfere at much greater distances.

Operation Below Ambient Noise Level

A marvelous example of a direct sequence system that routinely works at signal levels below KTB is the GPS system. GPS signal levels at the earth's surface are specified at −165 dBW, or −135 dBm. (This level is approximately 0.035 micro Volts, meaning that GPS receivers must be extremely sensitive.) Thermal noise (KTB) in the GPS receiver's bandwidth (9.0 MHz), however, is

$$1.38 \times 10^{-23} \times 290° \times 9 \times 10^6 \text{ Hz} = 3.6 \times 10^{-14} \text{ W or } -104.46 \text{ dBm.}$$

The GPS receiver operates routinely at a -30.5-dB RF input signal-to-noise ratio! How can it work?

Answer. GPS is a direct sequence system with 10.23-Mcps code rate, 9.0-MHz RF bandwidth, and 50-bps data rate. Its process gain is therefore approximately

$$10 \log \frac{9 \times 10^6}{50} = 52.5 \text{ dB}.$$

Using its 52.5-dB process gain, a GPS receiver takes in a negative input signal-to-noise ratio of -30.5 dB, and outputs a positive signal-to-noise ratio near

$$52.5 - 30.5 = 22 \text{ dB}$$

which is more than adequate for almost any data transmission system. GPS also uses this process gain to advantage in ranging and thereby position location.

Frequency hopping systems must, unfortunately, always have positive signal-to-noise ratios at their inputs. Any hop during which the signal-to-noise ratio is negative can be expected to produce garbage at a frequency hopping receiver's output. In fact, the signal-to-noise ratio in any given channel of a frequency hopping receiver will determine the output bit error rate, as a function of the data modulation and any error correction coding being employed.

Figure 11.1 is a plot of bit error rate (BER) versus signal-to-noise ratio for three common data-modulation methods. It is noted that the most common modulation form used with frequency hopping systems is frequency shift keying. For coherent FSK, please further note that a 9.7-dB signal-to-noise ratio is required to produce a BER of 1×10^{-3}, and 11 dB is needed for noncoherent FSK at the same error rate. (Most frequency hopping systems employ noncoherent demodulation because carrier coherence is lost each time the system hops from one frequency to another.)

Error Correction Coding

Frequency hopping systems must use some form of error correction coding. A pair of simple examples will serve to show why this is so.

Example one: Given a system with a requirement for 100 users, BER less than 1×10^{-3}, 32 Kbps ADPCM* voice, how much bandwidth is required?

---

*Adaptive Differential Pulse Code Modulation required in many telephone applications for voice quality and processing flexibility. 64-Kbps ADPCM often preferred.

Bandwidth per channel (CPFSK with $\beta = 0.5$):
    48 KHz,

Number of channels interfered with:	99,
Number of channels required per interferor:	500,
Total number of channels required:	$99 \times 500 = 49{,}500$,
Bandwidth required:	$49{,}500 \times 48$ KHz $=$
	2.376 GHz.

Just *try* to get a frequency allocation!

Example two: Given a 10-user frequency hopping system with 1000 frequencies spaced 25 KHz apart. Maximum data rate using CPFSK[†] and deviation ratio of 0.5 would be 16.66 Kbps (assuming the main lobe is 25 KHz wide and filtering is used to reduce the sidelobes).

Number of Users	BER (no error correction)[‡]
1	low (depends on signal level only)
2	5×10^{-4}
3	1.5×10^{-3}
5	2.5×10^{-3}
10	5×10^{-3}

While military users might find such a system acceptable (they would also require many more possible users), a consumer would not tolerate such poor performance and would return the system to get his money back.

Since these examples are both clearly unworkable, but the requirements are quite common, what should be done? The answer is to add some form of error correction coding.

Typical error correction coding formats readily available[*] today are

Convolutional codes	Usually rate $1/2$, $K = 7$ coding gain variable.
Golay codes	(24, 12), 3 in 24 corrected 4 in 24 detected.
Reed-Solomon codes	(21, 6), 8-ary 3 in 21 corrected.

Data sheets for convolutional and Golay code I.C.'s are shown in Appendix 8.

[†]Continuous Phase Frequency Shift Key. This would be identical to MSK.

[‡]Assumes that when users collide they will still guess right half the time. If a collision results in a wipeout, then these error rates would double.

[*]Available in I.C. form or easily implemented in a few I.C. chips.

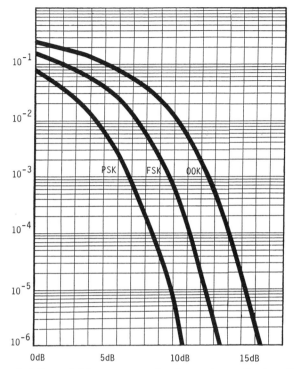

Figure 11.1 Probability of bit error vs signal-to-noise ratio for three different modulation types.

Though the coding technique is very simple, multitransmission methods (replication coding) are not recommended for most commercial applications because of their inefficiency. Also, the hopping rate required with replication coding is often much higher than the rate required for other error correction methods.

In other words, transmitting K times for each bit is a powerful but not too useful error correction method. One error in three can be ignored, or two in five, etc., allowing error rate to approach 0.5. That is, as

$$K \to \infty, \frac{K/2 - 0.5}{K/2 + 0.5} \to 0.5$$

and error rate can approach 50%-requiring only one more correct transmission than incorrect transmission to insure correct decoding. Alas, the required symbol transmission rate is much too high. (It approaches infinity.)

Now, to show the exceptional effectiveness and therefore the importance of error correction coding in frequency hopping, let us revisit one example. Let us assume that a rate 1/2 Golay code, with the ability to correct three bits in 24 is used. The new data transmission rate (called a "symbol rate" once error correction encoding is applied) is twice the original 32 Kbps, or 64 Ksps. Now how much bandwidth would be required?

Bandwidth per channel (CPFSK with $\beta = 0.5$)   96 KHz.
Number of channels interfered with (100 users)   99.
Number of channels required per interferor: 8.
Total number of channels required: $99 \times 8 = 792$.
Bandwidth required: 792×96 KHz $=$
    76.032 MHz.

While 76.032 MHz is still a wide bandwidth compared to a direct sequence system meeting the same requirements (direct sequence systems typically require much less bandwidth than frequency hopping systems for the same performance level) it is much, much less than the 2.376 GHz bandwidth required without error correction coding.

Near-Far Performance

As an illustration of the difference in direct sequence and frequency hopping systems in near-far environments, let is consider the configuration shown in Figure 11.2. In this illustration, transmitter 1 is sending a signal to receiver number 1, ten miles away. Transmitter 2 is only 0.1 miles from receiver 1, however. The distance ratio between the two transmitters and receiver 1 is $10/0.1 = 100$, which means that the difference between the two signals, arriving at receiver 1, is $100^2 = 10,000$ (assuming square law propagation characteristics). Therefore, the interference rejection capability of receiver 1 must be at least 10 log 10,000 = 40 dB, unless the receiver has some way of avoiding the interference from transmitter 2.

If both transmitters are direct sequence transmitters using the same common bandwidth, then the receiver *must* have 40 dB or more jamming

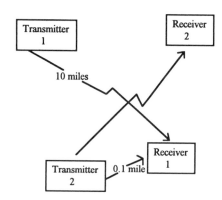

Figure 11.2 Illustration of "near-far" problem (two transmitters).

margin. However, if the two transmitters are frequency hoppers with different codes, then the only time that one interferes with the other is when they happen to hop to the same channel (or close to the same channel). This means that the frequency hopping system can operate under these circumstances with an error rate $1/M$, where the direct sequence could not (unless it has enough jamming margin, i.e., ≥ 40 dB).

The best approach to solving this problem when using direct sequence systems is to employ time division multiplexed direct sequence, since very few spread spectrum systems boast 40 dB or more of jamming rejection capability.

Linear Signal Requirements

In the presence of an interfering signal, when a desired signal is passed through a limiter, the smaller of the signals is suppressed with respect to the larger. For direct sequence systems, this means that the desired direct sequence signal is suppressed with respect to any larger interfering signal, and this is a clear loss in interference rejection capability.

On the other hand, while a frequency hopping signal is also suppressed with respect to interference, the probability of having a particular channel jammed is no higher when the signal is suppressed than when it is not.

The result of this difference is that direct sequence channels should be linear, including the receiver itself, which means that AGC is almost always required in direct sequence receivers that must operate in the presence of interference. Frequency hopping systems can operate without significant degradation in nonlinear channels, and AGC is not always required.

Synchronization

Frequency hopping systems employ code chip rates that are much slower than those used by direct sequence systems. The fastest known frequency hoppers employ codes that operate at rates no higher than 100,000 chips per second, while the slowest codes used in the direct sequence systems operate at least 1,000,000 chips per second. In practice, the difference between real frequency hopping and direct sequence system codes is usually several orders of magnitude.

This difference results in a significant variation in the timing error for synch acquisition and tracking, where the frequency hoppers have the advantage of reduced chip offset, which translates to faster and easier synchronization.

Conclusion

Overall, the conclusion must be that while frequency hopping is generally superior for military systems (and this is the reason that most military spread spectrum systems do employ frequency hopping), direct sequence methods will always outperform frequency hopping in commercial/consumer multiuser applications.

The reason for this is that the criteria for acceptability are very different in the two classes of applications. In military systems the important thing is the ability to get a message through under adverse conditions, whether or not it contains some reasonable number of errors (up to 10% interruption and/or 10% distortion is often acceptable) where a consumer is unlikely to accept a system that does not "sound good." Therefore, when a frequency hopping system has its number of available channels restricted, and the number of channels occupied is 10% or more of those available, it cannot provide the quality of a direct sequence system, even when it employs error correction coding.

In the pages that follow, we will consider specific approaches to design of spread spectrum systems, including both frequency hopping and direct sequence methods.

11.3 FREQUENCY HOPPING IN COMMERCIAL AND CONSUMER APPLICATIONS

Frequency hopping and direct sequence techniques for implementing spread spectrum systems are often compared on the basis of two criteria that override almost any other consideration, when multiuser random access systems are being designed. These are (1) synchronization time and (2) near/far capability. In a simplistic analysis, especially in military applications, frequency hopping is superior on both counts. However, for commercial and consumer applications, the answer is often quite different.

First, synchronization time in nonmilitary systems is not constrained by the same limitations that exist in military systems. Codes can be shorter (to reduce uncertainty), they need not be crypto codes, which alone exacerbate synchronization problems. Also, synchronization techniques are available which are not practical in military systems. All of these together make it possible to synchronize rapidly when military needs are not the limiting factor.

We do not consider synchronization as a limitation to nonmilitary direct sequence systems, as current techniques permit synchronization in the presence of interference in less than a millisecond. This leaves "the near-far problem" as the prime consideration.

The usual solution is to employ frequency hopping, for the reasons that were explained in section 11.2, namely, that a direct sequence system's bandwidth is such that it occupies the entire band that is available continuously, while a frequency hopping system occupies only a small part of the band, and that small part is continuously changing.

If different codes are used by frequency hopping systems that are using the same band, the systems occupy the same frequency channel only when they happen (more or less randomly) to hop to the same frequency at the same time. *This is the fundamental reason why almost all current military communications systems employ frequency hopping.*

Now let us consider a system with 1000 frequencies available. (26 MHz/ 25 KHz) = 1040. This is a typical number for both military and some nonmilitary applications. In a military application, with 20 users (almost always half-duplex push-to-talk users) the probability of a collision (and consequent loss of data during the collision) is $\frac{20}{1000} = 2 \times 10^{-2}$. The raw bit error rate would be approximately the same, and to a military user under jamming/multiuser conditions this would be not only acceptable, but the user would be delighted that he is being interfered with only $\frac{1}{50}$th of the time. (For voice applications, intelligibility would not be degraded.) A consumer's reaction, however, would simply be to return the unit and get his money back.

In a consumer application, such as a telephone, the signal is full duplex. This means that 40 channels out of the 100 available would be in use by the same 20 users, and the probability of a collision would now be twice as high, or $\frac{40}{100} = 4 \times 10^{-2}$. This is not acceptable to a consumer.

Further, when frequency hopping is applied to multiuser nonmilitary systems, other undesirable characteristics surface.

**Frequency Hopping Systems Interfere With Other Users
Over a Longer Distance**

A frequency hopping system is simply a radio system that changes frequencies periodically. In military applications this means that a would-be interferor must be capable of jamming many frequencies to be effective. (The number of potential users is the same as the number of frequencies the interferor must jam.) The potential interference from a single frequency hopping user on a time basis is such that the interference occurs, at a given receiver, once every M hops (averaged over a long period).

For example, if a system hops at 100 hops per second and has 1000 frequencies available, the interference seen by an unintended listener would occur once every $\frac{1000}{100} = 10$ sec for $\frac{1}{100} = 0.01$ sec. With 20 full duplex frequency hopping signals, the user would have interference (on the average) once every $\frac{10}{40} = 0.25$ sec for 10 msec per occurrence. This applies, incidentally, to other frequency hopping systems as well as narrowband listeners.

The above analysis assumes a 25 KHz bandwidth on the part of the receiver being interfered with. If the receiver has a wider bandwidth, then the occurrence of interference goes up directly as the factor by which the bandwidth is greater than 25 KHz. That is, if a receiver has 1 MHz bandwidth, then a 20-user frequency hopping system would interfere with that receiver ten times per second for 10 msec per occurrence, or 100 msec out of every second. (Note: it is difficult to build a receiver with only 1 MHz of preamplifier bandwidth at 900 MHz.) For every MHz of bandwidth, interference would occur another 10% of the time.

Because a frequency hopping receiver cannot operate when its signal level is less than approximately 6–19 dB above the ambient noise level, while a direct sequence receiver can operate with a signal that is as much as 30 dB below the ambient noise (the GPS system does this as a matter of course, routinely) a frequency hopping system must transmit up to 10,000 times more power per Hertz of bandwidth. That is, under the same circumstances, the interference potential to narrowband users from a frequency hopping transmitter is significantly higher than a direct sequence transmitter at the same distance.

Not Enough Channels Are Available

Assume: 25 KHz channels
           (this gives 1040 channels in the 26 MHz bandwidth)
           20 simultaneous users

Base station transmits on 20 frequencies at once (and also receives at 20 frequencies). *Therefore, the Base station requires 40 frequency synthesizers.*

Other independent FH users cannot operate with a bit error rate better than 1.92×10^{-2}. This is unacceptable.

(If wider bandwidth channels are used, fewer frequencies are available, and the BER goes up.)

Independent direct sequence users in the same band could operate only until the FH system exceeds their jamming margin. Then they have nowhere to go, even if they have partial-band operating capability and frequency agility, as the frequency hopper would occupy the entire band.

(For a direct sequence system with a minimum 500-KHz bandwidth, the probability that a FH signal would be in its bandwidth at any given time is 3.85×10^{-1}. This is also unacceptable.)

Error correction coding cannot solve this problem.

Self-Interference

Filters do not exist that are good enough to prevent a collocated transmitter from interfering with a receiver only 25 KHz away at 900 MHz. (They may even be connected to the same antenna.)

Filters also do not exist that are good enough to prevent crossmodulation and intermodulation of multiple collocated transmitters, which means that multiple spurious signals will be generated. (These transmitters may also be connected to the same antenna).

Given that a tunable, frequency hopping filter with a Q of 1000 could be developed to fit in a small box (this is far beyond today's technology) the filter's bandwidth would be 900 KHz (at its 3 dB points). This means there are really only 28.9 independent channels in the 900 MHz band, assuming the filters used also had brick wall response (say 240 dB per octave). Using real filters, there are actually far fewer independent channels available.

Another self-interference problem that is unique to frequency hopping systems, due to their always requiring a positive input signal-to-noise ratio at every frequency they use, is that in a multicell application every cell may have a dead zone at its perimeter in which it cannot operate with acceptable bit error rate. Typical signal-to-noise ratios required by frequency hopping receivers (for every channel in which there is no interference) are in the range of 6–10 dB. The best operational signal-to-noise ratio ever claimed, to this writer's knowledge, for a frequency hopping system, is 3.5 dB at the frequency hopping receiver's input. This claim, of course, involved the use of very powerful error correction coding.

Let us assume a cellular frequency hopping system with adjacent cells, separated by the use of different codes (i.e., a frequency hopping CDMA system). Every cell would use the same set of frequencies, but depend on having enough frequencies to reduce the probability of a collision to an acceptable level. With a 20 MHz-total bandwidth, and a 500-KHz bandwidth per channel, the system would have 40 channels available. This means that all adjacent cells would have a collision rate of $\frac{1}{40}$ with their neighbor, and a cell surrounded by six other cells would have a collision rate of $\frac{6}{40}$. (The collision rate does not necessarily mean the same thing as an interference rate, however, as long as the distance between a potential interferor and a user is sufficient to provide a minimum signal-to-noise ratio.)

A user at distance D_1 from the base to which it is talking would be at distance D_2 from the nearest potential interfering base station (we will not consider other users to be potential interferors—we will instead assume that they transmit in a separate user-to-base frequency band). The user could not operate when it is equidistant from both base stations ($D_1 = D_2$) because the signal-to interference ratio would be zero dB, and a frequency hopping system must always have a positive signal-to-noise ratio in every channel in which it expects to transmit information, however small that signal-to-noise ratio may be claimed to be.

If square law propagation is assumed, then the ratio of signal to interference is

$$(D_1/D_2)^2.$$

Here, D_1 must be less than D_2, and the difference in distances must be sufficient to produce the required D_1-signal-to-D_2-interference ratio. For various required signal-to-noise ratios and 1mi cell radii, the dead zone produced would be as in Table 11.2.

Here we assume that frequency hoppers with small numbers of cells (100 or less is a small number of cells) do not provide adequate user performance

**TABLE 11.2   Dead Zone (high interference zone) Width Versus Required
Signal-to-Interference Ratio in Frequency Hopping Systems. One Mile Cell Radii.**

S/I ratio required	Dead zone width	D_1/D_2 (miles)
3.5 dB	0.1 m	0.899/1.101
6.0	0.17	0.829/1.171
10.0	0.28	0.719/1.281

when other frequency hopping signals are present. The dead zone then is that region in which a user cannot operate inside the cell perimeter in which it finds itself. It cannot operate with an acceptable error rate with the base station that is nearest, or to any other base station. Its only option is to get closer to the base station in the cell in which it is located. With conventional overlapping cell structures, the problem is even worse. The dead zone extends inside each cell from the maximum radius at which a user can operate without interference, through the overlap region, into the next cell at the maximum radius at which a user can operate with the base station in that cell. Handoff from base station to base station is difficult at best, and soft handoff may be completely impractical.

Of course, if a user will accept high interference rates, then frequency hopping systems might approach or even exceed the performance of direct sequence systems in such applications.

11.4   JUST WHAT IS JAMMING MARGIN?

All receivers, including broadcast AM, FM, and TV receivers have jamming margins. Some receivers, however, have more than others and, in general, the more the better. Only those receivers that should work in the presence of unwanted signals actually require jamming margin, although in some cases jamming margain is specified in terms of selectivity or adjacent channel rejection. (Where the unwanted signal is actually at the same frequency as the desired signal, the degree of performance is specified as jamming margin. The question addressed here is how does jamming margin work, and how much is required?)

Jamming margin, in its strictest terms, is the ability of a receiver to provide a usable receiver output when an undesired signal is present. Consider Figure 11.3. In this diagram, the receiver/processor is shown as a simple block into which a desired signal together with noise and/or interference is input. The output is the demodulated, desired signal together with the noise and interference that results from the noise and interference at the receiver/processor input. In both cases (i.e., input and output) the key to the entire event of communicating through the receiver/processor is to provide a

Figure 11.3 Process gain model.

usable signal to noise/interference ratio at the receiver/processor output. Without this usable ratio, a communications system is useless.

For voice applications, a useful signal-to-noise/interference ratio is usually considered to be 10 dB or more. In data transmission systems, 15–18 dB is often specified. (Error rate is inversely proportional to signal to noise/interference ratio.) To provide a usable digitized voice signal, bit error rates as poor as 1×10^{-2} to 1×10^{-3} can often be tolerated.

In a conventional receiver (such as those used for AM reception) there is little or no difference between the receiver's input and output signal to noise/interference ratios. If anything, the output ratio is worse than the input ratio. Some receivers, on the other hand, exhibit "process gain" which allows an improvement in the signal-to-noise/interference ratio between their input and output ports. The effect of this is to be able (in the proper circumstances) to accept very poor input signals, that are corrupted by noise and interference, while deriving from them usable output signals. Table 11.3 summarizes the process gain available from various types of receivers. Receiver output signal to noise/interference ratio is the input signal to noise/interference, improved by the amount of the process gain in the receiver. That is,

$$\frac{S}{N} \text{ out} = \frac{S}{N} \text{ in} \times \text{process gain},$$

or, in dB,

$$\frac{S}{N} \text{ out (dB)} = \frac{S}{N} \text{ in (dB)} + \text{process gain (dB)}.$$

Our specific interest here is in the jamming margin provided by a direct sequence receiver. Such a receiver might employ a bandwidth of 10.8 MHz (3 dB) and a data rate of 192 Kbps. Thus, its process gain would be

$$\frac{10.8 \times 10^6}{192 \times 10^3} = 56.25 \text{ or } \log 56.25 = 17.5 \text{ dB}.$$

(This compares to $10 \log 64 = 18.06$ dB on a chips-per-bit basis.)

TABLE 11.3   Process Gain

Modulation*	Process Gain	Remarks
AM	≤ 0	≥ 10 dB input $\dfrac{S}{N}$ required for ≥ 10 dB output $\dfrac{S}{N}$
FM	$\left(\dfrac{\Delta f}{fm}\right)^2$	≥ 6 dB input (min) $\dfrac{S}{N}$ required to produce process gain
PPM	$\dfrac{\Delta t}{tr}$	See FM
FH	$10 \log M$	$p(\text{bit error}) = \dfrac{J}{2M}$
DS	$\dfrac{BW}{R \text{ data}}$	$\dfrac{S}{N}$ out $= Gp - \left(\dfrac{J}{S} + \text{losses}\right)$

*AM = Amplitude modulation J = No. of frequencies interfered with
 FM = Frequency modulation M = No. of frequencies available
 PPM = Pulse Position modulation Δt = pulse time shift
 FH = Frequency hop tr = pulse rise time
 DS = Direct Sequence

Now, that same receiver under voice and data conditions would provide acceptable bit error rate, with amplitude versus phase demodulation, at the following signal-to-noise/interference ratios:

demod	voice (BER 1×10^{-2})	data (BER 1×10^{-5})
amplitude	10.3 dB	15.4 dB
phase	4.7 dB	9.4 dB

This means that a system having 17.5-dB process gain will provide jamming margins of

demod	voice (BER 1×10^{-2})	data (BER 1×10^{-5})
amplitude	7.2 dB	2.1 dB
phase	12.8 dB	8.1 dB

Compared to an FSK system that employs precorrelation demodulation, even the worst jamming margin shown is at least 8.1 dB better. (The FSK system requires a minimum input S/N of 6 dB just to work. In practice, higher input S/N in is needed.)

Finally, please note that jamming margin always exists whether it is negative or positive. Typical jamming margins are

AM receiver	-10 dB or more
FM/FSK receiver	-6 dB or more
FH receiver	$\leq 10 \log (0.15 \text{ M})$
DS receiver	$\leq 10 \log \dfrac{\text{BW}}{\text{Rdata}} - \dfrac{\text{S}}{\text{N}}$ out

The spread spectrum receiver (either FH or DS) will *always* provide more jamming margin than AM, FM, or FSK receivers (even though the FSK receiver may try to masquerade as a spread spectrum receiver). Also, the actual performance will vary, depending on the specific type of interference (noise, other similar signals, etc.) and the specific S/N ratio required.

The point is, there is always jamming margin. Even though it may be squandered by poor system design or by overspecifying bit error rate, it is *always* present.

11.5 MULTIPATH REJECTION AND SPREAD SPECTRUM SIGNALS

Both direct-sequence and frequency hopping systems have the ability to operate in the presence of multipath signals to some degree. One cannot say, however, that using a spread spectrum signal does away with all multipath effects. In fact, some spread spectrum systems are degraded more by multipath than their narrower-band counterparts, in the same situations.

Direct Sequence

A typical 902–928-MHz direct sequence signal is shown in Figure 11.4. This particular signal has a main lobe null-to-null bandwidth of 26 MHz, sidelobes that are 20 dB down, and approximately $(\sin x/x)^2$ power distribution. The signal is shown without the effects of multipath, as the receiver would see it under undisturbed conditions.

Figures 11.5 and 11.6 show the same signal as it appears to the receiver under relatively mild multipath conditions. (Much greater distortion of the signal by multipath is very common.) In Figure 11.5, there is only one signal dropout, or fade, within the signal band,. The depth of the fade is at least 30

Figure 11.4   Direct sequence signal with a 26-MHz null-to-null bandwidth, centered at 915 MHz. No multipath.

dB, however, and its bandwidth is approximately

Fade depth	Bandwidth
6 dB	8 MHz
10 dB	5 MHz
20 dB	3 MHz
30 dB	1.5 MHz

as read from a comparison of Figures 11.4 and 11.5, where exactly the same conditions exist except that the first figure shows a signal without multipath

Figure 11.5   Direct-sequence signal with 26-MHz null-to-null bandwidth, centered at 915 MHz. Mild multipath fading—only one multipath-related null.

Figure 11.6   Direct-sequence signal with 26-MHz null-to-null bandwidth, centered at 915 MHz. Multiple multipath-related nulls. Note effect on sidelobes as well as main lobe.

distortion, and the second shows the effect of a relatively mild specular multipath signal.

It should be noted that in this particular case, the receiver continued to operate without any apparent effect from the multipath distortion. If the signal had been narrower (say 1–5 MHz) in bandwidth, the entire signal could have faded by 10–30 dB, which could have caused a narrower-bandwidth receiver to lose the signal during the fading interval.

Figure 11.6 shows the same direct sequence signal under more severe multipath conditions. Now there are three multipath fades within the main lobe, the left (lower frequency) sidelobe is almost nonexistent, the right (upper frequency) sidelobe is distorted, and the main lobe has four lobes with individual null-to-null bandwidths from approximately 5–7 MHz. Fade depths are again 30 dB or more. Width of the fades varies, with the center fade being as narrow as 0.5 MHz (at approximately 20–25 dB) and the other two being 1.0–1.5 MHz (in the same range of fading depth). It is also interesting to note that a portion of the signal (the second lobe from the left) is actually larger than the original signal.

Again we note that the receiver continued to operate without apparent disturbance from the multipath effects on its input signal. We also emphasize that the effects shown here are relatively mild when compared to the effects of severe multipath distortion. Under such severe multipath conditions, the receiver would definitely be affected.

How can one best avoid the effects of multipath? In direct sequence spread spectrum systems, the best approach is usually to employ antenna diversity, wherein the receiver employs more than one antenna in the hope that the signals from two antennas, spatially diverse, will have uncorrelated fading. At least, one hopes that both antennas will not experience severe fading at the same time. Separation between the two antennas is not critical, but should be greater than a wavelength at the operating frequency.

Frequency Hopping

From figures (Figure 11.4 through 11.5) showing fading of direct sequence signals, it is obvious that one can expect frequency hopping signals to experience the loss of band sectors as well. When a frequency hopping system experiences the loss of part of its operating band, however, it is the same as having an interferor cause the information to be lost. It has no way to protect that information except to transmit the information redundantly at more than one frequency.

To accomplish the necessary transmission redundancy, frequency hopping systems use error correction coding and where possible, hop at a rate faster than the fading rate. In this way, one can expect to send a symbol representing the information at more than one frequency, and if the hop rate is faster than the fading rate, at least one symbol representing every bit of information is always sent at a nonfaded frequency.

Fading Rate

The rate at which fades occur is a function of the rate at which the signal propagation paths change. In fixed-position systems, the fading rate is determined by how fast the environment changes. This could be determined by factors as slow as atmospheric disturbances or as fast as aircraft flying between them. Figure 11.7 illustrates the multipath configuration that produces multipath reception of television signals with an aircraft as a multipath source.

The rate of fading due to a vehicular multipath source, or moving transmitter or receiver is

$$\frac{\text{Velocity of vehicle (Feet/second)}}{\text{Wavelength (Feet)}} = \text{Fades per second.}$$

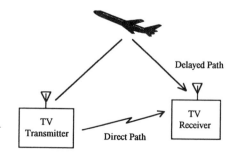

Figure 11.7 Typical multipath configuration near airports.

One can expect that fading rate will vary as follows:

Vehicle	Velocity	Frequency	Fade Rate
Man Walking	5.8 ft/sec	900 MHz	5.3 /sec
		1800 MHz	10.6 /sec
Bicycle	29 ft/sec	900 MHz	26.5 /sec
		1800 MHz	53 /sec
Auto	88 ft/sec	900 MHz	79 /sec
		1800 MHz	158 /sec
Jet Aircraft	880 ft/sec	900 MHz	790 /sec
		1800 MHz	1580 /sec

At the rates of fading expected from everything except aircraft, the hopping rate necessary to hop faster than the fading rate is easy to achieve. Jet aircraft, on the other hand, could force hopping rates that may not be practical.

11.6 FREQUENCY DIVISION IN SPREAD SPECTRUM APPLICATIONS

Spread spectrum multiuser systems, like any other multiuser system, have a limited number of dimensions that can be employed to differentiate one user from another. Spread spectrum systems do not habe code division capability, however, which is not generally available to other types of systems. Also, it is not always employed in spread spectrum applications, frequency division may be employed. This means that spread spectrum systems can employ code division, spatial separation, and also frequency division, all for the purpose of separating one signal from another in such a way that a receiver can be built to selectively reject all signals but the specific one that is desired. In fact, it is advantageous to employ more than one method at the same time to achieve the performance desired in a high-performance multiuser telephone system.

Here we choose to employ a direct sequence approach, in which a direct sequence signal with 9-MHz bandwidth is the basic information-bearing signal. This permits a number of signals to be transmitted simultaneously in the ISM (and other) band(s) using frequency division to enhance their mutual separation. The number of frequency division channels provided is:

Band	Bandwidth	FDM Channels
902–928 MHz	26 MHz	3
1850–1990 MHz	140 MHz	15
2400–2483.5 MHz	83.5 MHz	9
5725–5850 MHz	125 MHz	13

Figure 11.8 Bandpass filter response.

We will concentrate here on the 900-Mhz band, and the use of three frequency division channels, enhanced by code division, time division, and spatial separation. Let us examine the capability provided by each technique, separately and when used together.

Frequency Division. Frequency-Division Multiple Access, or FDMA, is limited by the bandwidth and rolloff of the filters used in the transmitters and receivers employed in a system. Unfortunately, filters in the 900-MHz band are limited (at least for filters with reasonable size and cost) to a 3-dB bandwidth in the range of 10–40 MHz, by the Q of the filter elements necessary. Rolloff (the rate at which unwanted signals are suppressed with respect to their offset from the filter's passband) is a function of the number of filter sections, where each additional filter section provides another 12 dB per octave of frequency separation beyond the 3-dB point (see Figure 11.8). In practice, small two section RF filters are available, which give rolloff (ideally) of 24 dB per octave. IF filters can provide much more.

If such IF filters are used in transmitters and receivers, and three channels are set alongside each other, with center frequencies chosen to meet FCC requirements,* the frequency plan would would appear as in Figure 11.9. (The filters employed in Figure 11.9 have a 3-dB bandwidth of 9 MHz, and a 35-dB bandwidth of 13 MHz. Filter crossover point is at approximately -2 dB.) The interference power from channel 1 to channel 3 is down by 35 dB or more, and the same is true from channel 3 to channel 1.

For adjacent channels (1 to 2, 3 to 2, 2 to 1, and 2 to 3) approximately 32% of a transmitter's power spills over into the adjacent band. This means that the total power from six surrounding cells could be as great as twice that from the cell itself. (Figure 11.10). This assumes that the power adds directly, that there is no loss due to spatial separation, and all transmitters turn on at the same time. This is not the actual case. as will be shown in the subsequent paragraphs.

Spatial Separation. Given that a three-frequency architecture is employed, and we concern ourselves with the worst case, i.e., channel 2 is surrounded by

---

* FCC requirements are such that signal spillover beyond the band must be suppresses by at least 20 dB compared to the signal strength in the band.

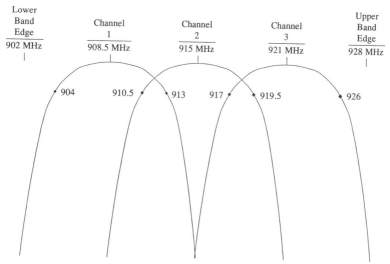

Figure 11.9   Three-frequency spread spectrum plan.

Figure 11.10   Three-frequency cell structure.

three each of channels 1 and 3:

1. A user at the center of the cell using channel 2 would be $2R - D/D$ closer to its base station than to base stations in the adjacent channels, and the power advantage would be approximately $(2R - D/D)^2$ (per interferor). R is the cell radius. D is the user-to-base distance.

2. A single user in an adjacent channel could be colocated with a user in channel 2's cell, so that 32% of its power would interfere. Under this condition, power from other adjacent cells would be minimized because distance to other cells is maximum.

Code Division. In addition to using separate frequencies in the adjacent channels, separate codes may be used, although both frequency and code separation are not mandatory. However, if codes are carefully chosen, their crosscorrelation can be held by design to a practical range of -10 to

-20 dB.* That is, using code correlation and careful design, unwanted codes can be rejected even while desired codes are being accepted, but such rejection ratios do not usually provide more than approximately 20 dB of code separation capability.

Code division alone cannot provide adequate multiple access for most applications.

Time Division. Using time division, the time available is divided among the number of system users. This means that in a 16-user system, each user has access to the system $\frac{1}{16}$ th of the time. Further, if a symbol period is less than a bit time, it is possible to greatly reduce transmit duty cycle, and if a transmitter is not sending a signal, it cannot interfere.

The exceptional advantage of using time division is that no two transmitters within a cell ever collide, since only one transmitter at a time sends a signal. Of course, if other cells are present, and there is no coordination in cell-to-cell timing, then collisions from one cell to another can occur. The probability of a collision from one cell to another is actually reduced as the number of users per cell increases.

Now we postulate a system that handles 32 users per cell, three frequencies, 9-MHz 3-dB bandwidth, 512-Kbps throughput rate, 8-Kbps per user data, and cells 500 ft in radius.

Processing gain for each user would be 30.5 dB. ($10 \log [9 \times 10^6 / 8 \times 10^3]$ $= 30.5$ dB.) Each user would transmit with a duty cycle of $\frac{1}{32}$. Therefore the probability of a collision with a user from an adjacent cell is $\frac{1}{32}$, and the probability of colliding with users from two and three cells is $\frac{1}{1024}$ and $\frac{1}{32,768}$, respectively. Therefore the practicality of being interfered with by multiple adjacent cells is not realistic.

The worst case is that in which two users are at the periphery of their adjacent cells. In this case, the probability of interference is $\frac{1}{32} = 3.1 \times 10^{-2}$, and the amount of signal introduced into the other user's channel is 32% of his desired base station power (assuming equal power for users and base stations). This means that even if the user has only 0-dB jamming rejection, the interfering user must be almost twice as close to the user he interferes with as that user is to his base station—and then he produces (no matter how close he is) an error rate of only 3.1×10^{-2}.

If each user has a 10-dB jamming margin, then an interferor must be almost six times closer before any interference is caused, and still it causes only 3.1×10^{-2} error rate.

Where more frequency separation is available, so that channel overlapping is not necessary, the probability of collisions becomes extremely small.

Frequency Reuse. In a three-channel system with frequency reuse, a user in a reused channel is never closer than approximately two cell radii to another

---

*Compared to the desired code.

user at the same frequency. The user is always within one radius of his own base station, however. This means that the power from a desired base station is always at least four times as large as that from the nearest reuse channel.

When coupled with the low probability of collision and the capability to use code rejection, the probability of interference is very low.

11.7   USE OF SPREAD SPECTRUM SIGNALS IN CONJUNCTION WITH CONVENTIONAL CELLULAR TELEPHONE SIGNALS —A WORKABLE EXAMPLE OF SPREAD SPECTRUM OVERLAY

Cellular telephone companies are investigating code division multiple access (CDMA) technology for application in a new form of cellular telephone. Claims for this technology advance systems employing CDMA as having one to four orders of magnitude more capacity than standard cellular systems. At the same time, proponents of "personal communications" systems have applied for (and obtained) experimental licenses to test spread spectrum telephones in frequency bands currently allocated to other services. Also, spread spectrum telephones for the ISM bands have already been demonstrated.

Availability of the cellular radio bands, and the fact that cellular telephone providers have unique control of these bands, makes the possible use of noncellular spread spectrum telephones in the cellular bands very attractive. Such telephones might provide a new service within the cellular radio bands, without significant impact on the current system. That is, the present FDMA/FM network might be overlaid with a low-power spread spectrum telephone network for use within buildings, independent of the normal cellular telephone network, without either network being aware of the other network's presence.

Here we will investigate the practicality of simultaneous FDMA/FM and spread spectrum use of cellular radio frequencies on paper. Appendix 10 gives experimental results for a real test of spread spectrum signals overlaid on a microwave link.

Cellular Telephone Characteristics

Frequency range—824–849 MHz, 870–894 MHz.

Frequency spacing—30 kHz.

Modulation—FM.

Audio bandwidth—300–3000 Hz.

Transmitted power—600 mW.

Receiver sensitivity—1–5 microvolts.

Output signal/noise ratio—12-dB minimum.

Instantaneous RF bandwidth—25-kHz maximum.

Receiver noise bandwidth—25-kHz.

Spread Spectrum Telephone Characteristics

Frequency range—824–849 or 870–894 MHz.

Instantaneous RF bandwidth—25 MHz.

Voice/data rate—9600 bps (16/32 kbps).

Receiver noise bandwidth—25 kHz.

Process gain—34.15 dB (31.95/28.95 dB).

Jamming Margin— $\geq$ 20 dB (19.95/16.95).

Modulation*—BPSK/direct sequence or frequency hop/FSK.

Receiver sensitivity—1–5 microvolts.

Output signal/noise ratio—12-dB minimum.

Audio bandwidth—300–3000 Hz.

Building attenuation, Inside to outside—20 dB.

Hop rate (FH) - $\leq$ 200 hop/sec.

Users per base station—up to 20.

Base station range—30–300 ft.

Transmitted power—0.6–60 mW.

Code rate (DS)—15 Mcps.

Antenna polarity—horizontal.

Cellular / FH Overlay

With a 200-hop-per-second frequency hopping system, and 20 users per base station, the total throughput rate (at 9600 bps per user) per base station would be $20 \times 9600 = 192$ kbps. Thus a frequency-shift-keyed signal would have a 3-dB bandwidth of approximately 211.2 kHz, assuming continuous-phase FSK is employed. Therefore, the transmitted power density of FH/FSK signals would average 2.84×10^{-9} W per Hz to 2.84×10^{-7} W per Hz.

Neglecting antenna-to-antenna losses (which should be of the order of approximately 55 dB), the maximum signal transmitted to a cellular telephone would be:

power density $\times$ cellular telephone bandwidth = 71 microWatts to 7.1 mWs.

Considering antenna losses, this translates to approximately

$$2.2 \times 10^{-10} \text{ to } 2.2 \times 10^{-8} \text{ W.}$$

In microvolts, this amounts to a signal level of

$$105 \ \mu\text{V to } 1.05 \text{ mV (at 50 ohms)}$$

when the cellular telephone and FH system are sitting side by side. This does not mean, however, that the cellular telephone would see continuous interference from the frequency hopping system, as the frequency hopping system would not use the entire band continuously. At a 200-hop-per-second rate, a single FH user network (up to 20 users) would have up to

$$\frac{25 \text{ MHz}}{211.2 \text{ kHZ}} = 118.4 \,(\text{use } 118)$$

channels available. 8.4 of those would be interfered with at any given time $(211200/25000 = 8.4)$. Therefore, the rate of interference from the frequency-hopping system would be

$$\frac{118}{8.4} = 14 \text{ times per second}$$

for a duration of 5 msec per interference burst.

Such interference might be acceptable, depending on the low-pass frequency rolloff of the cellular telephone receiver, which would normally be at approximately 300 Hz. Further, as the systems are separated, the amount of signal coupled into the cellular telephone decreases as the square of the distance. At 200 ft, one could expect the signal from the frequency hopping system to be attenuated by another 80 dB, which would reduce the signal to approximately

$$0.01 \text{ microvolts to } 0.1 \text{ microvolts,}$$

which is below the threshold of the cellular telephone receiver. In the other direction, interference from a cellular telephone to the frequency hopping system would cause a bit error rate of approximately

$$3.5 \times 10^{-2} \text{ per cellular telephone,}$$

which would force the use of interleaving and error correction coding in the FH system. (This, of course, presupposes that the cellular telephones are close enough to the FH system to interfere with it.) With FH telephones operating at minimum (0.6-mW) power, a cellular telephone (600 mW) within approximately 800 ft would have sufficient power to interfere with the FH telephone to the extent described above.

Also a consideration in the FH system is that each telephone must process and store 1/200th of the 192,000 bits per second, or 960 bits, with an attendant delay of 5 msec. This must be taken into account in the frequency hopping system design (especially in the design of the interleaving subsystem).

Cellular / DS Overlay

When combining cellular and direct sequence systems, two factors are very different when compared with a frequency hopping approach:

1. Direct sequence power density is much lower.
2. Direct sequence signals occupy the entire band when transmitting, but are much more noiselike.

Also, the direct sequence system receiver has the ability to reject cellular transmitter signals, which the frequency hopping system cannot do.

Considering power density, the direct sequence system could employ an entire 25 MHz band segment which would give it a power density (average) of

$$2.4 \times 10^{-11} \text{ W per Hz to } 2.4 \times 10^{-9} \text{ W per Hz.}$$

In a cellular telephone, the amount of signal seen when sitting alongside a direct sequence telephone network would be approximately

$$1.9 \times 10^{-12} \text{ to } 1.9 \times 10^{-10} \text{ W.}$$

In microvolts, this is a signal level of 9.7 to 97 microvolts. At 200 ft, this signal level would be reduced to approximately

$$0.00097 \text{ to } 0.0097 \text{ micrvolts}$$

and would appear (when heard) to be an increase in the background noise level. In the worst case, the cellular telephone receiver's threshold would be increased from 1.0–1.6 microvolts (at the 200 ft distance).

In the other direction, interference from the cellular network to the direct sequence network would be rejected by the jamming margin of the direct sequence telephones (about 20 dB). At the same distance (i.e., with the cellular telephone alongside the DS transmitter) a DS receiver could still operate when the DS transmitter sent at least 6 mW of power.

With the cellular telephone at a 200 ft distance from the DS telephone network, the DS network would be completely unaffected. This would be true for all cases, as long as the combination of difference in distances and power levels does not exceed the 20 dB receiver jamming margin. Examples are

DS transmitter power	(20-dB jamming margin) Distance Ratio*	Operate
0.6 mW	1:1	no
6.0 mW	1:1	yes
0.6 mW	3:1	no
6.0 mW	3:1	no
30 mW	3:1	yes
0.6 mW	10:1	no
6.0 mW	10:1	no
30 mW	10:1	no
60 mW	10:1	no
600 mW	10:1	yes

*Cellular telephone closer to DS receiver than DS transmitter.

This means that, if DS telephones are set up for a maximum 300-ft-radius network, a cellular telephone could be closer than 100 ft to a DS receiver before its performance was degraded. Also, only that telephone would be degraded, while all others in the DS network (at greater distance from the cellular telephone) would be unaffected.

In light of these calculations, it appears that either approach could work, but the direct sequence methods offer the best performance. Certainly, a trial under realistic operating conditions is worthwhile.

Interference from Cellular Towers

Assumed power 500 W (100 W).

Ten transmitters.

Vertical polarity.

Attenuation due to frequency	20 log 850 = 58.6 dB.
Attenuation due to coupling and polarity	57.5 dB.
Attenuation due to structure	20 dB.
Total attenuation (except for distance)	136.1 dB.

Signal at microcell in building (500/100 W transmitted, 20-dB loss for distance).

$$1.23 \times 10^{-13} \text{ W } (2.46 \times 10^{-14} \text{ W})$$

= approximately 7.5 microvolts (3.35 microvolts).

Power at microcell receiver due to desired microcell transmitter

$$60 \text{ mW } (+17.8\text{-dBm}) \text{ transmitted maximum.}$$

With attenuation due to frequency, coupling, and 300 ft distance, 116.1 dB, desired signal at microcell receiver

$$= 2.7 \text{ microvolts.}$$

Signal to cellular interference ratio

$$20 \log \left[\frac{2.7 \times 10^{-6}}{7.5 \times 10^{-6}} \right] = 8.87 \text{ dB}$$

$$20 \log \left[\frac{2.7 \times 10^{-6}}{3.35 \times 10^{-6}} \right] = -1.87 \text{ dB.}$$

Excess margin $= 20 - 8.9 = 11.1$ dB

$$(20 - 1.9 = 18.1 \text{ dB}).$$

In other words, as many as 128 (676) cellular transmitters, each with 500 (100) Watts, could be tolerated. Under varying conditions, the following performance could be expected:

Operation of Spread Spectrum Microcell Receiver (Inside Building)

Cellular Transmitter Power	Number of Transmitters	Microcell Power	Distance from Cellular Transmitter	Distance from Microcell Transmitter
500 W	128	60 mW	1 m	up to 300 ft
500 W	55	60 mW	0.65 m	up to 300 ft
100 W	55	60 mW	0.29 m (0.29/0.29/0.41) m	up to 300 ft
100 W	10	60 mW	0.12 m (0.12/0.12/0.17) m	up to 300 ft

For the above, the following conditions apply:

- Microcell is within building with 20-dB attenuation to outside signals.
- Cellular transmitter power is additive. (Actually, those transmitters nearer the band edge will be much less effective in interfering with microcell receivers).
- Microcells are capable of 20-dB interference rejection.

It is expected that such spread spectrum systems could operate quite well in the presence of multiple high power cellular telephone transmitters, with the conditions stated.

Interference from 7 Watt Mobile Cellular Telephones

Power 7 W (interference),
        60 mW (desired).
Building attenuation 20 dB.
Distance from microcell receiver to interferor 30 ft.
Distance from microcell receiver to microcell transmitter 300 ft.
Polarity attenuation 20 dB.
Desired signal at receiver 2.7 microvolts.
Interference at receiver:

$$
\begin{aligned}
&\text{transmitted power 7 W, 38.45 dBm,}\\
&\text{attenuation: frequency} \quad \text{58.6 dB,}\\
&\qquad\qquad\quad \text{coupling} \quad \text{37.5,}\\
&\qquad\qquad\quad \text{polarity} \quad \text{20.0 dB,}\\
&\qquad\qquad\quad \text{building} \quad \text{20.0/136.1 dB,}\\
&\text{interference level } 1.7 \times 10^{-13} \text{ W} = 2.9 \text{ microvolts.}
\end{aligned}
$$

Signal to interference ratio

$$
20 \log \left[\frac{2.7 \times 10^{-6}}{2.9 \times 10^{-6}} \right] = -0.62 \text{ dB.}
$$

Margin 19.3 dB.

That is, a 7-W mobile transmitter sitting outside a building containing a spread spectrum microcell would not interfere with the microcell, even if the distance was significantly less than 30 ft. (Calculations show the distance could be less than one foot, but this is not a practical condition, since the mobile 7 W transmitter would be in a vehicle.)

Interference to the mobile unit from the microcell would be:

$$\text{microcell power} = 60 \text{ mW},$$
$$\text{attenuation} = 136.1 \text{ dB},$$

Signal from the microcell at the mobile cellular receiver 1.47×10^{-15} W (0.27 microvolts in 25 MHz BW).

Interference in mobile receiver bandwidth (25 KHz) = 0.0085 microvolts.

Therefore, with a 1.0-microvolt cellular signal and a 0.0085-microvolt interfering signal from the microcell, the mobile cellular receiver would have a signal to noise ratio (at its input) of 41.4 dB. It would never even notice the effect of the microcell in an adjacent building.

12

MULTIPLE ACCESS, SNAKE OIL, AND FAIRY TALES

A number of claims surround spread spectrum systems and their capabilities that are at best only partly true, at worst, they are completely false. In this chapter we will examine methods of providing multiple access to a band of frequencies by use of frequency division (FDMA), time division (TDMA), and code division (CDMA). Each of these multiple access methods has advantages and disadvantages. We will compare them in the following pages.

2.1 THE NUMBER OF SIGNALS IN A BANDWIDTH

Spread spectrum techniques are often touted as a solution to the lack of frequency spectrum availability, as every bit of the frequency spectrum from ELF through EHF is allocated to some use. In some isolated cases, spread spectrum techniques may be at least a partial solution, but they are not in any way a panacea. In fact, the number of spread spectrum signals that can be accommodated in a given bandwidth is similar to (if not less than) the number of narrower-bandwidth signals accommodated in the same bandwidth.

There is, in some cases, an ability to operate in the same band as narrower band signals, at the same time, with minimal mutual interference. Let us examine this ability, as well as compare the number of users that are practical in a given bandwidth.

Frequency-Division Multiplex

Frequency-division multiplexing, or FDMA (frequency-division multiple access) systems take advantage of the ability, built up over many years of research and development, to construct accurate frequency sources and highly selective bandpass filters. Historically, FDMA is the basis upon which all radio communications have been built. This includes military and commercial communication systems, not to mention all television and broadcast radio systems.

The question at hand is, how many users can occupy a given bandwidth? Here we will determine how many users per MHz of bandwidth are practical. The bandwidth for a single user is determined by the baseband bandwidth of the information being sent, together with the type of modulation used, and the modulation index (percentage of modulation). Of course, if a "digital modulation" form is used, the modulation index is constant.

Typical analog modulation techniques have bandwidths as follows:

AM	2 × information bandwidth.
DSBAM	2 × information bandwidth.
SSBAM	1 × information bandwidth.
FM	2 fm + fm ($\beta^* \leq 1$).
	β for broadcast FM stations is 5, with 15-KHz modulation bandwidth. Peak to peak deviation is 150 KHz.

And "digital" modulation:

OOK	1.25 Hz per bit per second.
FSK	1.25 Hz per bit per second.
BPSK	1.25 Hz per bit per second.
QPSK	0.53 Hz per bit per second.
MSK	0.53 Hz per bit per second.

The number of channels available using frequency-division methods is simply the bandwidth available, divided by the channel bandwidth. Standard military systems, for example, when using AM voice provide 25-KHz channels, even though the amplitude modulated signal has only 6-KHz bandwidth. This means that for such systems, the number of channels per MHz is

$$\frac{1 \times 10^6}{25 \text{ KHz}} = 40.$$

*β is defined as $\Delta f / fm$, Δf is peak deviation, fm is modulating frequency.

If 6 KHz, the actual modulated signal bandwidth, is used, the number of channels per MHz is

$$\frac{1 \times 10^6}{6 \text{ KHz}} = 166.6.$$

Cellular telephones employ a 30-KHz bandwidth per channel, which allows

$$\frac{1 \times 10^6}{30 \text{ KHz}} = 33.3$$

channels per MHz of bandwidth.

The real limitation on the number of users in a frequency-division system is not necessarily the bandwidth of the modulation, as is evidenced by AM systems in which the signal bandwidth is much narrower than the allocated channel bandwidth. The limitation is in the characteristics of the bandpass filters used in the transmitter and receiver, and the frequency uncertainties they must contend with. We will assume accurate frequency generation* and no significant Doppler shift here.

In the transmitter, the filter used must suppress signal power in adjacent channels. Once the transmitter outputs unwanted power in an adjacent channel, it cannot be suppressed without affecting signals that are supposed to be in that channel. A receiver's filter can suppress signals outside its desired signal bandwidth, but not those within that bandwidth.

Therefore, once a transmitter has done its job of suppressing signals outside its assigned band, the receiver's selectivity must do the rest of the job. Specifically, a receiver's adjacent channel rejection is its most important characteristic with respect to number of users in a band of frequencies.

Military receivers typically have adjacent channel rejection specified at 80 dB or greater (though many do not meet the specification). At the same time, typical transmitter output power is 10 W or +40 dBm. Thus, a receiver with −107-dBm sensitivity must have adjacent channel suppression of up to

$$40 + 107 = 147 \text{ dB}$$

just to keep a receiver's response to adjacent channel signals (at least when the transmitter and receiver are very close together) at a usable level.

If the distance between transmitter and receiver is significant compared to a wavelength, then receiver adjacent channel rejection is very important in determining the number of channels. The distance necessary to reduce the receiver's response to an adjacent channel signal, in the 850-MHz range,

---

*One part per million or better accuracy.

based on free-space loss, is as follows:

TABLE 12.1   **Distance Required for Sufficient Adjacent Channel Rejection.**

Adjacent Channel Rejection	Distance for Signal Rejection
80 dB	0.035 mi. (185 feet)
70 dB	0.11 mi. (586 feet)
60 dB	0.350 mi. (1851 feet)
50 dB	1.109 mi. (5856 feet)
40 dB	3.507 mi. (18520 feet)

The result is that we can see that, if distances are such that the adjacent channel rejection capability of a receiver is sufficient, then the number of channels supportable using FDMA is

Channel bandwidth	Users per MHz
25 KHz	40
30 KHz	33.3
6 KHz	166.6

We also hasten to add that these users can only be supported if the frequency accuracy of transmitter and receiver are adequate. This is the primary reason for the use of 25-KHz channel separation in military systems, where the information-modulated signal is only 6 KHz wide. At 800–900 MHz, it is often necessary to provide transmitter and receiver accuracy of 1×10^{-6} or better. In aircraft systems, it is also necessary to accommodate Doppler-frequency shifts, although this is not necessary for most commercial systems.

Time-Division Multiplex

Time-division multiplexing is used every day in telephone systems. The well known "T-1 carrier" is a good example of a time-division-multiplexed approach, which may or may not be transmitted on a radio link. A typical T-1 signal consists of 24 digitized voice signals, each at 64 Kbps, multiplexed together on a single carrier, with 8-Kbps framing and control overhead added. This makes a total of 1.544 million data bits per second (i.e., 64 Kbps × 24 = 1,536,000 and 1,536,000 + 8,000 = 1.544 Mbps).

To compare TDMA with the FDMA approach in the previous section, let us consider a voice signal with a bandwidth of 300–3000 Hz. If we digitize this voice signal to accommodate digital multiplexing, the sampling rate must be at least twice 3000 Hz, or 6000 samples per second, and oversampling by one third takes us to 8000 samples per second. Then, with eight-bit quantization we have 8 × 8000 = 64 kilobits per second.

Now, putting N users together, each at 64 Kbps, the required data rate is $N \times 64$ Kbps, plus overhead bits.

Today's state-of-the-art systems employ voice digitization rates of 8 Kbps or less. This means that the T-1 carrier that supports 24 64-Kbps users could support 192 users at 8 Kbps.

In a 1-MHz bandwidth, the number of bits that can be transmitted is a function of the modulation technique. With QPSK, for example, the number of bits per second Hertz of bandwidth is 1.9. Therefore, the number of multiplexed users per MHz is

$$\frac{1.9 \times 10^6}{R \text{ user} + \text{OH user}} = \text{users per MHz},$$

where

R user $\equiv$ the number of bits per second per user,

OH user $\equiv$ overhead rate per user (usually in the range of 1–30%)

Given a data rate of 8 Kbps, 20% overhead, and 1-MHz bandwidth,

$$\frac{1.9 \times 10^6}{9600} = 197.9$$

voice users could be supported.

The primary difference between FDMA and TDMA systems is, of course, that TDMA *requires* digital signals, while FDMA systems can use digital signals but do not require them. For example, a voice signal can be digitized to the same 8 Kbps previously discussed, and then QPSK modulated onto a carrier, with bandwidth $8000/1.9 = 4210$ Hz. The subtle difference is that multiple digital signals can readily be transmitted with a single carrier, and a FDMA system that digitizes and then transmits multiple signals on its carrier *is* a TDMA system.* (Multiple analog signals can also be transmitted on a single carrier using FM/FM techniques as well.)

Even though it is possible to digitize information and then transmit one signal on a carrier in a frequency-division format, it is much more effective to multiplex more than one digitized signal on a single carrier. The reason for this is because the time-division approach is much more effective at separating users than anything that can be done with bandpass filters at higher frequencies. That is, it is much more practical to assign bits numbered 8000 through 16,000 (for example) in a digital data stream to a user, than to separate one 8-KHz bandwidth from the 8-KHz band on either side of it. A

---

*A good example of such an approach is the common "T-1 carrier."

very significant property of TDMA is that users can operate together in a TDMA mode with no near-far problem at all.

Transmission of single-user-per-carrier TDMA is also a practical technique, and is the basis for "digital cellular" systems. In such systems, much of the overhead required is not taken up by actual transmission of data, however. Instead, to prevent signals that are transmitted by multiple users at varying distances from a receiver, from arriving at the same time, a guard space is added to each time slot.

In small cells, guard time may be insignificant, but in large cells, guard time may actually take up more time than is alloted for data transmission. Since radio signals propagate at approximately six microseconds per mile, any distance uncertainty in the position of a transmitter with respect to a receiver must appear in the guard time allowance in every time slot. Time slot size and guard time are both very important, as larger time slots mean fewer guard times, but processing delay may be longer.

With the 197.9 users we previously postulated, let us consider an example:

- Suppose each 8-Kbps user has 100 time slots per second. He must transmit his data in 80-bit increments, with eight bits of overhead.
- There are 19,790 time slots per second available, each with period 50.5305 microseconds.
- Data is transmitted in 88-bit chunks, at a rate of 1.741520 megabits per second. Thus 4.6 microseconds per time slot is left for guard time, and this in turn allows approximately a 0.75-mile range uncertainty.
- If longer time slots are used, with fewer per user, longer guard time and therefore larger cells could be accomodated.

This example is intended merely to show the relationships involved in TDMA systems (single user per carrier).

Code-Division Multiple Access

CDMA, or code-division multiple access systems, use codes to separate one signal from another (as we have previously discussed in Chapter 1). Either direct sequence or frequency hopping systems can employ CDMA, so we will compare both methods to FDMA and TDMA approaches.

CDMA systems are dependent on the design of codes that are "orthogonal" to one another, at least within the set of codes employed in a network. Orthogonality in this context means simply that all of the codes used must have low enough mutual cross-correlation that they do not significantly interfere with one another over the dynamic range of the signals presented to any receiver in a CDMA network. This is the purpose of the Gold codes, Kasami codes, and Bent codes discussed in Chapter 3. Unfortunately, there is no known set of codes that is completely orthogonal when used in the conditions* that exist in a randomly distributed, highly mobile multiple access system with a large number of users. The best that one can hope for under these conditions is that the coded signals will add together in the same way as additive white Gaussian noise, i.e., an added transmitter would appear to add no more interference than a transmitter sending AWGN at the same power level. If the codes used are capable of supporting enough users at the same time, then the limit on the number of users is system jamming margin, because a receiver will work as long as the composite power of all unwanted signals does not exceed the power of the desired signal by more than the jamming margin. Therefore, the maximum number of users can be supported if all users' signals are equal at the receiver's input.

Jamming margin in a direct sequence system is

$$\frac{\text{RF bandwidth}}{\text{data rate}} - [S/N \text{ out} + \text{losses}].$$

In a 1-MHz bandwidth, with 8-Kbps data rate, jamming margin is approximately

$$10 \log \frac{1 \times 10^6}{8 \text{ Kbps}} - [4 \text{ dB} + 0.5 \text{ dB}] = 21 \text{ dB},$$

and the maximum number of equal-power user signals arriving at the receiver, each of which meets the receiver's sensitivity, is $\log^{-1}(21/10) + 1 = 126.^{\dagger}$

Improvements might be made by decreasing data rate (which increases processing gain) reducing required output signal-to-noise ratio, or reducing losses.

In frequency hopping, the maximum number of users is also limited by the receiver's jamming margin, but this is evidenced in terms of the number of frequencies interfered with, that can be tolerated by a receiver.

In the ISM bands, at least 50 frequencies must be available in the 900-MHz band, and 75 frequencies in the 2400- and 5700-MHz bands. If up to 20% of these frequencies may be interfered with by other users, then the number of users possible would be

$$50/5 = 10 \text{ users for the 900-MHz band,}$$

---

*More than 100-dB dynamic signal range, rapidly varying multipath, limited processing gain, varying signal blockage, etc.

†This is an extremely optimistic estimate. A 4-dB S/N ratio assumes that the 8-Kbps includes error correction, and this in turn means that the unencoded voice must be at a lower rate (say 1/2 or 4 Kbps). With the same addition of error correction coding, the number of FDMA and TDMA users would go up. 0.5-dB losses is also somewhat optimistic. Without these optimistic estimates, CDMA would look much worse by comparison, but these numbers are typical of those claimed by some vendors of spread spectrum systems.

and

$$75/5 = 15 \text{ users for the 2400- and 5700-MHz bands.}$$

Limiting a frequency hopping system to a 1-MHz bandwidth, and considering an 8-Kbps user data rate with MSK modulation, the number of channels available would be

$$\frac{1 \times 10^6}{12 \text{ Khz}} = 83.3$$

and, again, considering 20% to be the maximum number of occupied channels, the number of users would be

$$\frac{83.3}{5} = 16.7.$$

(Frequency hopping systems typically support fewer users in a given bandwidth than direct sequence systems do.)

However, assuming that a frequency hop channel is not completely blotted out by a collision, and that a binary digital system can guess right half the time, we will give the frequency hopping system the benefit of the doubt and claim twice the number of users, or 33.3.

Summary—Bits Is Bits, or There Is No Magic (Too Bad)

From the preceding paragraphs, in which we see that the number of users per MHz is actually less for spread spectrum CDMA methods than for either TDMA or FDMA. It is, in fact, more effective when using any multiple access technique, to simply reduce the transmission rate per user (which in turn reduces each user's required bandwidth) than it is to go from TDMA or FDMA to CDMA.

Table 12.2 is a summary comparison of multiple access techniques. This comparison shows that, on paper, both FDMA and TDMA can support more users than CDMA. This does not mean that spread spectrum methods do not have advantages in specific applications. Indeed, they do. They do not solve all problems, however.

Some examples of advantageous applications for spread spectrum methods are:

1. Position location. The GPS system would not work without spread spectrum. Neither would the Pactel Teletrac (previously North American Teletrac) or other vehicle location systems that employ direct sequence modulation.

TABLE 12.2   Summary Comparison of Multiple Access Techniques

Multiple Access Technique	Information Rate per Voice User	Number of Users per MHz	Primary Limitation
FDMA	3 KHz analog	166.7 (AM)	Filter shape factor
		333.3 (SSB)	Frequency accuracy
	8 Kbps digital	237.5 (QPSK)	Filter shape factor
			Frequency accuracy
TDMA	8 Kbps digital/ 20 percent overhead	197.9 (QPSK)	Inefficient without multiple users per carrier
			Equalization
CDMA			requirements
Direct Sequence	8 Kbps digital (includes EDAC)	126 (BPSK/MSK)	Jamming margin, code cross correlation
Frequency Hop	8 Kbps digital (includes EDAC)	33.3 (CPFSK)	Number of frequencies available

2. Spectrum sharing. It is possible to overlay a spread spectrum signal on a narrower-band signal with minimum mutual interference between the two. An example of this is point-to-point microwave systems overlaid by direct sequence signals. Omnipoint Corporation has shown that a 10-Mhz bandwidth direct sequence transmitter can be up to 50 times closer to a microwave receiver than a narrowband transmitter can be, while meeting the maximum 1.0-dB degradation of the microwave system's link margin. (See Appendix 10).

3. Use of otherwise unwanted spectrum. Spread spectrum systems can operate in bands (such as the ISM bands) in which there are many thousands to millions of powerful interference generators. In the ISM* bands, there are also licensed users that must not be interfered with, while the spread spectrum user must accept any and all interference from the licensed users. Some examples are:

902 − 928 MHz −

    Vehicle location systems (licensed)

        up to 2000 W fixed

        up to 15 W mobile.

    Older microwave ovens

        up to 750 W.

    Other spread spectrum systems

        up to 1.0 W.

* Industrial, scientific, and medical.

Industrial heaters

50,000–100,000 W typical.

Military radars

up to one MW.

Harmonics of lower-frequency systems

450-MHz-band mobile radios, etc.

2400 – 2483.5 MHz-

Microwave ovens

up to 750 W

(output of typical microwave ovens is in the range of −40 to −60 dBm at a range of 30 m).

Other spread spectrum users

up to 1.0 W.

5725 – 5850 MHZ-

Many licensed users

(primarily point-to-point links).

Other spread spectrum users

up to 1.0 W.

4. Privacy. Current cordless telephone and cellular telephone conversations can easily be overheard. (In some cases it is illegal, but this does not protect the information being carried.) Spread spectrum methods inherently provide privacy to the extent that the code being used protects the information. Some spread spectrum systems also encrypt the information being sent.

5. Availability of unlicensed spectrum. The three ISM bands in which spread spectrum systems are allowed to operate, at up to 1.0 W, without license (on a secondary basis) are extremely valuable for spread spectrum applications, and many devices using spread spectrum are being and have been developed. Sadly, some of these are and have been little more than poorly conceived attempts to cash in on a new market through loopholes in the rules. Some are already out of business. It is hoped that these systems will not give spread spectrum an undeserved bad name. Systems have also been built that operate under special, low-power rules, and may or may not be spread spectrum systems. These also operate in the ISM bands, but their output power is typically less than 1.0 mW.* When other legal users with 1.0 W or more of legal power are present, the low-power systems cannot be expected to operate.

---

* FCC regulations call for maximum field intensity of 50,000 microvolts per meter at a distance of 3 m.

For these reasons, spread spectrum use will continue to grow, and it is expected that spread spectrum systems will be commonplace in the home, in public locations, and in offices.

Receiver Sensitivity in CDMA Systems

CDMA systems that employ power control for intracell solution of the near-far problem suffer from significant loss in their effective receiver sensitivity. The reason is that

1. All signals must be at equal power levels at the receiver, to maximize the number of simultaneous users.
2. Each signal must also be at a level that is above kTB by an amount

$$NF + S/N + 10\log(U - 1),$$

where NF = the receiver noise figure,
S/N = the receiver's required signal-to-noise ratio,
U = the number of user signals.

Table 12.3 shows the effective signal level that must be input to a receiver whose sensitivity is -105, -100, or -95 dBm, where various numbers of signals are being received simultaneously from multiple user transmitters, and each signal is being controlled to arrive at the receiver with exactly the same amount of power as every other signal present. This table shows that the effective sensitivity of the receiver is degraded (if its single-signal sensitivity is -105 dBm) from -105 (1.26 microvolts) to -87 dBm (7.94 microvolts) if 63 signals other than the one it wishes to receive are also present. This is exactly the situation in which a power-control CDMA receiver (base station) is expected to work. That is, even though the receiver sensitivity itself may not be poor, it might as well be. This is because the multiplicity of other signals present prevent small signals from being demodulated. This also assumes that the receiver must have at least

$$10\log 63 = 17.99 \text{ dB}$$

of jamming margin in every one of its processing channels for the system to operate under these conditions.

It also means that every system is limited by the weakest user transmitter. As a poorly performing unit nears the edge of a cell, every other user must reduce its transmitted power to accomodate the weak user, and this in turn could cause every receiver channel to reach its threshold. Either that, or users must be dropped by the cell. (An easy solution, but perhaps unpopular

TABLE 12.3   Signal Level Required Versus Single Receiver Sensitivity and Number of Signals Present

Single Receiver Sensitivity	Number of Signals	Interference-to-Signal Ratio	Signal Level Required by Each Receiver
− 105 dBm	1	—	− 105 dBm
	2	0 dB	− 102
	4	4.8	− 100.2
	8	8.5	− 96.5
	16	11.8	− 93.2
	32	14.9	− 90.1
	64	18.0	− 87.0
	128	21.1	− 83.9
− 100 dBm	1	—	− 100 dBm
	2	0 dB	− 97
	4	4.8	− 95.2
	8	8.5	− 91.5
	16	11.8	− 88.2
	32	14.9	− 85.1
	64	18.0	− 82.0
	128	21.1	− 78.9
− 95 dBm	1	—	− 95 dBm
	2	0 dB	− 92
	4	4.8	− 90.2
	8	8.5	− 86.5
	16	11.8	− 83.2
	32	14.9	− 80.1
	64	18.0	− 77.0
	128	21.1	− 73.9

with the weaker users: drop the weakest user. If the weak user is not capable of working well in the fringes of one cell, it will not work well in the fringes of the adjacent cells either. It could be called a Pariah unit—capable of operating only near the center of a cell, and the first to be dropped whenever it wanders further away.)

12.2   CDMA AND JAMMING MARGIN

In Chapter 2, jamming margin is defined as the amount of interference a receiver can withstand while producing a useful output signal. When expressed with respect to the ratio of interference power to signal power needed to produce the required useful output, one can derive the CDMA (code division multiple access) capability in terms of the number of users possible in a particular application.

Jamming margin for a system is a function of process gain, required signal-to-noise ratio at the system's output, and processing losses. That is,

$$M_J = G_p - [S/N \text{ out} + \text{losses}].$$

Therefore, given a system with a 1.25-MHz spread spectrum bandwidth and a data rate of 9.6 Kbps, with required output signal-to-noise of 6.0 dB, and zero losses, jamming margin would be

$$10 \log \frac{1.25 \times 10^6}{9.6 \times 10^3} - 6.0 = 15.1 \text{ dB},$$

and the maximum number of users would be

$$\log^{-1} \frac{15.1}{10} = 32.35,$$

if and only if the signal power arriving at a receiver from every user is precisely the same as the signal power arriving at that receiver from every other user.

This can only be accomplished by

- Precisely controlling the transmitted power from every user, so that the signal power from each and every user arriving at a given receiver is equal.
- Controlling the distance of all users so that the signal power arriving at a given receiver from every user is equal.
- Some combination of the above.

How does one increase the number of users in such a system? By increasing the available jamming margin. This means that the bandwidth-to-information rate ratio must be increased by either increasing the spectrum bandwidth and/or by increasing the data rate. (One can also claim improvements by adding error correction coding,* but only if the net result of increasing the system's symbol rate does not harm the system more the coding gain improves it.) Military CDMA systems with such power-control capability were introduced in the late 1950's (see ARC-50, Chapter 9). Power control was rejected for use in multiuser satellite spread spectrum systems, in favor of time division, because of the complexity of power control for multiple users, and because a single poorly controlled user (or an interferor over whom no power control is practical) can easily wipe out an entire network.

---

*The best known error correction schemes can correct only 16% or fewer errors. This translates to 16% of the available channels being occupied.

If we compare time division CDMA to power-control CDMA, in multiuser networks (for example, a network with a single base station and many randomly distributed users scattered around it), it is easy to see that

- The time division base station requires only one transmitter and one receiver to service all its associated users. (This includes any signal processing that is to be done.) The power-control base station, on the other hand, requires a separate transmit channel* and a separate receive channel for every user. This means that a power-control system with 32 users would require 32 transmit channels and 32 receive channels, while the time division system requires only one.

If more users are accommodated, by either reducing data rate, or by sectoring a base station's coverage, then the number of transmitters and receivers simply goes up by the same amount as the increase in users. Increasing processing gain, and thereby jamming margin/number of users, by reducing data rate would allow

Data Rate	Jamming Margin	Number of Users
8 Kbps	15.9 dB	39.25
4 Kbps	18.9 dB	78.5
2 Kbps	21.9 dB	157

From this we see the value of reducing the data rate in a spread spectrum system. We note, however, that reducing the data rate in non-spread-spectrum systems can produce similar results. (We will see an example of this in the succeeding pages.)

Important Note:

It is of no use whatsoever to consider the number of users possible in a spread spectrum system to be 50, 100, or any other number unless the codes employed (remember that we are discussing *code division multiple access*) are capable of providing sufficiently low cross-correlation between every user and every other user.

Frequency Hopping and CDMA

For some reason which is not obvious, it is often assumed that CDMA and direct sequence methods are synonymous. Our discussion of CDMA in the previous section certainly applies to direct sequence (notice that the words "direct sequence" were never used) but also may be applied to frequency hopping.

---

*A single power amplifier and a single receiver preamp/down converter can be used, but separate processing, modulation, and demodulation is still needed for every user because each uses a separate code to differentiate itself from other user signals.

If one has two or more frequency hopping systems sharing the same bandwidth, but using different hopping sequences, should this not also be called code division multiple access? It should, and we will continue to do so, no matter that some mistaken persons have gone to the extent to declare that only direct sequence with power control is called "CDMA."

How many users are practical with FH/CDMA? Again, the number of users is a function of the receiver's jamming margin. Let us suppose again that we have a 1.25-MHz RF bandwidth available, a 9.6-Kbps data rate with rate 1/2 coding (19.2-K symbols per second), and one symbol in every six correctable. We will also assume CPFSK modulation with deviation ratio = 1/2. (This is MSK modulation.)

Bandwidth per channel $\quad = 1.5 \times 19.2$ KHz
$$= 28.8 \text{ KHz.}$$
Number of channels available $\quad = \dfrac{1.25 \times 10^6}{2.88 \times 10^4} = 43.4.$

Jamming MARGIN (broad band noise)

$$S/N \text{ per channel: 6 dB}$$

$$= 10 \log \frac{43.4}{4} = 10.35 \text{ dB.}$$

Jamming Margin (other users)

$$= \frac{43.4}{6} = 7.2 \times 2 = 14.4.$$

(This assumes five channels for every jammed channel [5 + 1 = 6] and that a jammed channel produces a bit error rate of 0.5.)

In military systems, it is also possible to offset each user in time with respect to every other user. Further, a military system would have twice as many channels available since their channels are allowed to overlap. Offsetting each user with respect to every other user means that each user sequentially uses the same hopping sequence and therefore they do not randomly interfere with one another. The military system would therefore support up to 83.2 users in the same band.

Commercial users are not allowed to either overlap their frequency hopped channels or to employ offset of their code sequences. The reason is that offsetting of code sequences in time, by multiple users, simply causes up to every channel in the band available to be used simultaneously, and the approach is effectively the same as frequency division multiplexing with every channel being used. Interference to other users of the band is therefore greatly increased.

CDMA and "Soft" Handoff

Power-control CDMA (PCDMA) Systems (in which the transmitted power from every user must reach a single receiver at an equal power level*) have a particular problem, in that if adjacent networks of users (called cells) have base stations that can be heard by any single user, then provision must be made for that user to be controlled by both base stations. The reason is that if the user, being controlled by base station A, goes out of sight of A while in sight of base station B (and its group, or cell, of users) then base station A is forced to tell the user to increase its transmit power, which must cause either:

1. cell B to be jammed and its capacity reduced
2. all units in cell B to be forced to increase their power (which has a ripple effect-in turn causing every cell in sight of cell B to increase its power, which causes every cell in sight of C, D, . . . Z to increase its power, etc.)
3. power control and communications for the unit shaded from base station A to be handed over to base station B.

The result of this requirement is that, if the user in question is transmitting to and receiving from every base station in sight, its control can be handed over to the most convenient base station without the noisy and time-consuming process used in cellular telephone systems today.

Soft handoff is, in fact, not a handoff at all, at least not in the usual sense. In a standard cellular system, handoff is accomplished by shifting a two-way communication link with a user, from one base station to another. (This is why it is called handoff.) In the case of soft handoff, more than one base station is in full communication with the user any time it is in view. Therefore there is no handoff. The alternate base stations simply take turns servicing the user, and when one base station goes out of hearing range, another joins the pack serving that user.

This "soft" handoff process has an advantage and some disadvantages.

The advantage is:

The handoff is not discernible to the listener.

The disadvantages are:

• The user unit must receive signals from every base it can hear. This makes the user set much more complex.
• Each base must dedicate a channel to each user it can hear, even if they are in adjacent cells. This reduces system capacity as the base station

---

*This is necessary to maximize the number of users.

must dedicate channels to users for which it is not a conduit or which it is not controlling, as every other base in sight of the user must do.

• Each base must provide a Telephone circuit for each user it hears, whether it is controlling it or not, because it must provide instant access to the line if it takes over control with a seamless handoff.

This major problem, simultaneous multicell connection and signal tracking, has been turned into a primary selling point (soft handoff) by those who are proponents of PCDMA.

Alternatives to PCDMA for multiple access are distance-control CDMA (DCDMA), frequency hop CDMA (FCDMA), and time-division CDMA (TCDMA). DCDMA is used in satellite systems like GPS (see chapter 9) where handoff is accomplished by the receiver dropping one satellite and synchronizing to another. Soft handoff could be accomplished in this system by employing the receiver's and transmitter's known positions to acquire a new satellite signal. As the receiver usually is synchronized to five or more satellites simultaneously in the normal course of its operation, there is little value or interest in soft handoff, as such.

FCDMA systems are not readily adaptable to soft handoff, because it is difficult for a frequency hopping receiver to synchronize to multiple hopped signals at once, without using multiple complex synthesizers. Military frequency hopping systems could accomplish handoff by adjusting their code generator's timing, assuming all users are employing the same code. In any case, because frequency systems can synchronize and resynchronize quickly, there is not particular need for a soft handoff.

TCDMA, or time division multiplexing using direct sequence spread spectrum signaling, solves the near-far problem by allowing only one user in a cell to transmit at a time. In adjacent cells there is also only one user transmitting at a time. When a TCDMA user is out of view of one base station it simply resynchronizes to another base that is in view. Since a TCDMA system inherently must have a very fast acquisition time to maintain time-slot-to-time-slot synchronization, it can perform a handoff from one base station to another so quickly that a listener cannot perceive the handoff. This approach does not have the disadvantages of PCDMA, since it does not require a user to maintain simultaneous contact with multiple base stations. Soft handoff in the PCDMA sense could be easily accomplished if the TCDMA system chose to do so, by simply commandeering time slots from multiple base stations (this, in effect, is what is done in PCDMA systems). Soft (PCDMA) handoff is of little value to a TCDMA system, however, and is viewed as a waste of user time slots.

13

CURRENT CONSUMER AND COMMERCIAL SYSTEMS

Today, there are many spread spectrum devices and systems that have already been developed, or are currently in the process of development. We will not attempt to describe all of them here, but we will list a few representative units of various kinds. Some spread spectrum devices were rushed to the market soon after the FCC regulations first allowed spread spectrum systems in the ISM* bands, but at least some of these could not meet today's rules, some could not compete because of price and/or performance, and some of the companies that built them are no longer in business.

Most of the spread spectrum effort for commercial applications has been concentrated in the lower-frequency bands, primarily at 902–928 MHz. One very good reason for this is that many of the components that can be applied in the 902–928-MHz band were originally developed for use in the 800–900-MHz range, in cellular telephone systems. Some components used in other spread spectrum systems were developed for use in Global Positioning System receivers, that operate at 1227.6 MHz and/or 1575.42 MHz. These GPS components are generally applicable to spread spectrum systems operating in the tentative 1850–1990 personal communications system band as well, and in some cases can be applied to either or both the 900-MHz band or the 2400-MHz ISM bands.

At this time, the greatest interest lies in two general areas for commercial spread spectrum systems:

1. Personal communications systems (sometimes called either PCS or PCN [personal communications networks]).
2. Local area networks.

---

* Instrument, Scientific, and Medical.

416

Other applications are dwarfed in comparison to these two. Personal communications systems are being developed for the ISM bands, as well as the general range of 1700–2500 MHz, and this has led to a level of concentration by integrated-circuit companies, on communications applications, that has never seen before. While companies in Europe have not concentrated on spread spectrum systems, the frequencies used there are much the same as those used in the U.S. for spread spectrum applications, and efforts by European companies have served to reinforce those of U.S. companies in the RF integrated-circuit area. Some companies, both American and European, are presently in the process of developing entire receiver-transmitters on one or two integrated-circuit chips, and some of these chips are well disposed to spread spectrum usage.

A few commercial spread spectrum units typical of the state of the art are shown in the following pages.

Global Positioning System Receivers

Figures 13.1 through 13.4 show several commercials GPS receivers. In Figure 13.1 a complete handheld receiver is shown. This receiver is from Magellan,

Figure 13.1   Handheld GPS receiver. Capable of achieving 10-meter DGPS accuracy, the Magellan NAV 5000D is differential-ready, water proof, and floats. (Courtesy Magellan Systems Inc.)

Figure 13.2 GPS OEM Receiver boards. The Magellan GPS Brain™ receiver modules, available as a full-size board with power supply or a smaller five-volt board, offer firmware options for timing, data collection and standard navigation. (Courtesy Magellan Systems Inc.)

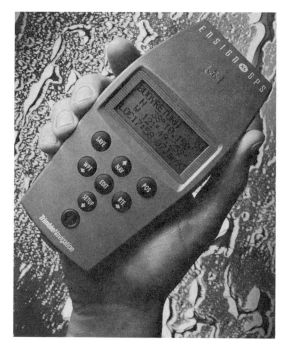

Figure 13.3 Handheld marine GPS receiver. (Courtesy Trimble Navigation).

Figure 13.4   Handheld GPS receiver. (Courtesy Magnavox Co.)

and weighs less than two pounds, including its antenna and batteries. Printed circuit boards, meant to be included in other companies' systems, are shown in Figure 13.2. GPS receivers of these types are also available from a number of other companies, including Rockwell, Ashtech, Magnavox, and Trimble. Figures 13.3 and 13.4 show units from Trimble and Magnavox. GPS receivers are being used in a very large number of applications, many of which were never envisioned by the original system planners and designers, such as leveling of farm fields to minimize water runoff, or monitoring of earthquake fault lines to determine movement and perhaps lead to earth-quake prediction. Surveyors are using GPS receivers to plot position to within millimeters. Of course, the most well-known use of GPS receivers in recent time occurred during Operation Desert Storm (the Gulf War between U.N. and Iraqi forces.) In Operation Desert Storm, commercial GPS receivers were used by both sides for their own advantage.

GPS receivers are also being used inside other systems to provide accurate timing and synchronization. GPS receivers can provide real time to an accuracy of 100 nsec or better, and when a pair of GPS receivers are synchronized to the same set of satellites, can synchronize the two receivers to within 25 nsec. A notable use of GPS for commercial application is that of Qualcomm Corporation, where a GPS receiver is planned for every cellular spread spectrum telephone base station.

Figure 13.5   Spread spectrum telephone handset. (Courtesy Omnipoint Corp.)

Personal Communications

Spread spectrum telephones for the ISM bands as well as the newer personal communications bands are shown in Figures 13.5 through 13.8. The telephone handset shown in Figure 13.5 is from Omnipoint Corporation, and is available in models that operate in the 902–928- and 2400–2483.5-MHz ISM bands, as well as the 1850–1990-MHz personal communications band. It weighs less than 12 ounces, including rechargeable batteries. (Figure 13.6 shows the previous Omnipoint spread spectrum telephone model, which has the distinction of being the first working multiuser handheld spread spectrum telephone ever developed, having been demonstrated first some three years ago, in 1990.) These telephones, in both models, have been used to demonstrate spread spectrum operation inside buildings, intercell handoff in a moving vehicle at freeway speeds, cell-to-cell handoff in New York City, and New York City-to-Chicago wireless telephone calls through the ISDN* network. The same telephones have also been used for the wireless portion of a coast-to-coast telephone call (San Diego to Washington D.C.) over the cable network normally used for home television delivery.

Figures 13.7 and 13.8 show Omnipoint multiuser base stations. (Those shown support eight users and are meant for indoor [3.7] and outdoor [3.8] use.)

---

*Integrated Services Digital Network.

Figure 13.6 Base station and handsets for 1990 model spread spectrum telephone system. (Courtesy Omnipoint Corporation)

The Omnipoint telephone systems employ time-division CDMA, in which the near-far problem is solved through time-division multiplexing (there can be no near-far problem if only one transmitter in a network transmits at a time) and the cross-correlation problem is solved by having all users in a network or cell employ the same code. Between networks, or cells, both code separation (CDMA) and frequency separation (FDMA) are provided.

Figure 13.7 Eight-User spread spectrum telephone base station (office model) with handsets and antennas. (Courtesy Omnipoint Corp.)

Figure 13.8 Eight-User base station for spread spectrum telephone system. (Courtesy Omnipoint Corp.)

Current demonstrable models of the Omnipoint telephones for multiuser applications are available in the /two lower ISM bands, plus the 1800-MHz personal communications band. As of this writing, 5725–5850-MHz ISM-band telephones are in development, and dual-mode (two-band handsets with more than one modulation format) telephones capable of choosing between public and private network operation are being constructed. When fully developed, Omnipoint base stations will support up to 30 handset users apiece.

When compared with the military systems shown in Part One of this book, only the largest, most complex, and most expensive military systems match the performance available in commercial spread spectrum systems today. A prime advantage of the Omnipoint system, for example, is that it is designed to accomodate from two to forty users without adding hardware other than line cards, to interface with added telephone lines. It is far less costly to deploy such a time-division system, which allows all users to employ the same transmitter and receiver, than to deploy systems that require a separate transmit-and-receive processing channel for every user.

Vehicle Location Systems

A very important application for commercial spread spectrum systems is in automatic vehicle location. These systems are considered important enough to be given licenses in the 902–928-MHz band, which means that the unlicensed spread spectrum users have the dual burden of

a. not interfering with AVL systems,
b. accepting whatever interference the AVL systems may generate, without complaint.

At least two companies are licensed to develop and deploy vehicle location systems in major cities across the U.S. These companies, Pactel Teletrac and METS, are in the process of doing so, with Pactel Teletrac being further along in their deployment. These systems are currently being marketed primarily as stolen car locators, but have many other potential applications. During the first month of operation, the Pactel Teletrac system located more than 80 stolen cars within 30 min of their being stolen, and allowed them to be recovered undamaged. (Many others have been recovered since then.)

Pactel Teletrac provides the ability to locate any transponder-equipped vehicle or other entity to within 100 ft. This is done through a trilateration method wherein the transponder's signal is repeated by multiple accurately placed repeating stations. At this time the stated accuracy of the METS systems is not known. Vehicle location units are being made for Pactel Teletrac by a number of companies, including Kenwood, Mitsubishi, Matsushita, Samsung, and Audiovox. Base stations are installed by Pactel Teletrac as a service provider. One of the typical vehicle units, the Kenwood KCS-10, is shown in Figure 3.9. Its size is $1 \frac{9}{16} \times 5 \frac{1}{2} \times 6 \frac{3}{8}$ inches. It should be noted that the Pactel Teletrac and METS systems are not intended to be compatible, as they operate their systems (by license) with their spread spectrum and narrowband signals at opposite ends of the 902–928-MHz band.

Local Area Networks

An example of a wireless spread spectrum local area network component is shown in Figure 13.10. This device, called GINA (Global Integrated Network

Figure 13.9 Pactel Teletrac vehicle location unit made by Kenwood. (Courtesy Pactel Teletrac.)

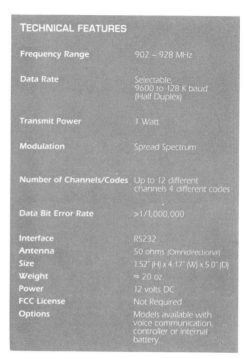

TECHNICAL FEATURES	
Frequency Range	902 – 928 MHz
Data Rate	Selectable, 9600 to 128 K baud (Half Duplex)
Transmit Power	1 Watt
Modulation	Spread Spectrum
Number of Channels/Codes	Up to 12 different channels 4 different codes
Data Bit Error Rate	>1/1,000,000
Interface	RS232
Antenna	50 ohms (Omnidirectional)
Size	1.52" (H) x 4.17" (W) x 5.0" (D)
Weight	≈ 20 oz.
Power	12 volts DC
FCC License	Not Required
Options	Models available with voice communication, controller or internal battery.

Figure 13.10 GINA Local Area Network Wireless Transmitter-Receiver. (Courtesy GRE America, Inc.)

Access) is advertised as having 1-mi range, with 1-W transmitted power. It uses direct sequence modulation, and operates in the 902–928-MHz ISM band. Specifications are also given in Figure 13.10. This unit is made by GRE.

Telesystems, a Canadian company, also manufactures a direct sequence wireless system for local area networks, as does NCR. Many other companies are also said to be in process of developing both direct-sequence and frequency hopping versions of wireless local area networks.

Sure-Comm Voice System

A somewhat unusual frequency hopping system is illustrated in Figure 13.11. This system operates in the 902–928-MHz band, and consists of a small, pocket-size transceiver, with a headset. Six simultaneous users are accomodated, with range of 500 ft (using 50 mW) or 1 mi (with 1 W). This unit is made by Telephonics Corporation. Basic specifications are listed in Figure 13.11.

Specifications

- Complies with FCC Part 15 for Spread Spectrum Radio Operation

- Over 250,000 selectable Frequency Hopping Nets

- Transmitting may be in a fully automatic mode (VOX) or Press-To-Talk Mode

- Size: 6.6" × 3.9" × 2.2"

- Weight-Transceiver 20 ozs
  - Headset 10 ozs

- Includes a 14 hour battery charger

- Five to seven hour operation before recharging

- Selectable low power (500 feet) or high power (up to 1 mile) modes

Figure 13.11 Sure-Comm frequency hopping transceiver. (Courtesy Telephonics Corp.)

14

TURNING THE (SPREAD SPECTRUM) WORLD UPSIDE DOWN

As this book is being prepared, the Federal Communications Commission is preparing rules for auctioning spectrum allocations,* much of which will be employed for personal communications systems (PCS), which include such things as telephones and local area networks. Some of these will employ the spread spectrum techniques that have been discussed here.

Although changes may be made, the frequency allocations that are to be auctioned are the following:

Range	Total Bandwidth
1850–1865 MHz/1930–1945 MHz	30 MHz
1865–1880 MHz/1945–1960 MHz	30 MHz
1880–1890 MHz/1960–1970 MHz	20 MHz
2130–2135 MHz/2180–2185 MHz	10 MHz
2135–2140 MHz/2185–2190 MHz	10 MHz
2140–2145 MHz/2190–2195 MHz	10 MHz
2145–2150 MHz/2195–2200 MHz	10 MHz

These bands are to be allocated to licensed users, with unlicensed PCS allocated from 1890–1930 MHz, and other "emerging technologies" assigned bands at 1970 to 1990 MHz, 2110 to 2130 MHz, and 2160 to 2180 MHz.

---

*Under congressional orders.

426

Further speculation on the specific bands allocated and their use is probably fruitless at this time, as they are likely to change somewhat. It is certain,[†] however, that the near future holds personal communications systems for almost everyone who wants them.

14.1 THE PCS IDEA

Personal communications systems have been envisioned as providing both incoming and outgoing voice, data, and video services to a user, no matter where that user happens to be. These services would be available without the tether of a wired hookup (wireless), on demand.

Ideally, the same portable (preferably pocket-sized) device would be usable at home, in public, or in an office/factory environment. While in public, when the user's device (which we will assume is a handset) would be connected to the public telephone system through a public base station, as is currently the case with cellular telephones, the user would pay for the privilege of using the public base station. At home, or in the office, a privately owned base station would be used. This is illustrated in Figure 14.1.

Home and office/business applications would employ unlicensed frequencies, but public bases and applications would be expected to operate in licensed bands only. These would provide all the current services of the cellular system including handoff, roaming, et cetera.

The home and business systems would operate only within range of their own base station(s) (except for those very large businesses for which roaming and handoff could be provided within their own system) in an unlicensed band. Since businesses and homeowners will be employing their own equipment on their own premises, without using the licensed public systems, *they will not pay for airtime.* This is the reason for their operating in the unlicensed bands, where the signals must be robust and provide for multiple access while furnishing minimum interference to others. This is an ideal use of spread spectrum signaling.

One other point with regard to business and home systems: Cooperation and/or coordination between systems owned by different businesses (or homeowners) is neither practical nor desirable. Multiple systems must work close together but independently, whether they are made by the same manufacturer or not. A business owner does not want to be forced to coordinate his system with the business next door, especially if it means that there is even the remotest possibility that the neighbor can listen in on his conversations, and coordination is a small step from synchronization. Fur-

---

[†]As certain as any quasipolitical process can ever be.

a. Office/Business Base Station - Voice, Video, Data

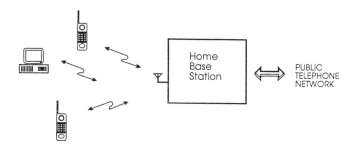

b. Home Base Station - Voice, Data

c. Public Base Station - Voice, Data

Figure 14.1   Various base station service configurations.

thermore, having to recoordinate systems every time a new business comes on line is far too cumbersome, expensive, and time-consuming.

14.2   CHOICES IN PCS DESIGN

The most important choices for personal communications system design include the operating frequency band to be used and the signal format to be employed. In some cases, choosing a band is tantamount to choosing a signal format and vice-versa. In other cases, other occupants in the operating band may force the choice of an operating format. Some of the considerations that must be taken into account are described in the following pages.

Operating Band

Several possibilities exist with respect to operating frequency band in the personal communications system arena. Earlier in this chapter, a listing of new allocations announced by the Federal Communications Commission in September 1993 was given. It must be realized, however, that these are shared bands and will be shared bands until such time as the current occupants are moved to other frequencies. (In some cases this is never expected to happen, as is the case with public safety services, who do not have to move.)

1. Industrial, scientific, and medical (ISM) bands. Up to 1.0 W unlicensed if spread spectrum modulation is used. Up to approximately 1.0 mW if spread spectrum is not used. Must accept interference from other users, some licensed. Must not interfere with licensed users. (Potential interference sources are listed in chapter 11.)

2. Shared bands. The 1800–2200-MHz bands recently allocated are all shared bands, at least at this time. In general, spread spectrum signaling methods outperform all others by far in sharing capability.

3. Cellular bands. Four approaches have been advanced for the purpose of allowing multiple users in the cellular telephone bands. Two of these approaches employ spread spectrum modulation (direct sequence) and the others use narrowband modulation. The conventional cellular approach employs narrowband FM, and the second narrowband method employs digital time division multiplexing. A sharing approach, with wideband spread spectrum signals overlaid on the conventionally (FM) modulated signals, has been proposed, and a spread spectrum PCDMA system for which the band of operation must be cleared of all other users is also in development. In general, the cellular band is not a promising place for PCS applications, if only because of the large number of current narrowband users and the huge investment in conventional cellular infrastructure.

Signal Format

Signaling techniques are often coupled to the band of operation, but the choices available are:

1. Conventional narrowband. These techniques are restricted to the cellular band, to low power ISM operation, or non-interfering 1850–2200-MHz applications. A European system called Cordless Telephone 2 (CT-2) has been tried in the 900-MHz band but with little success.
2. Digital narrowband. This is currently proposed to be employed in cellular applications. Two European systems, Global System for Mobile Telecommunications (GSM) and Digital European Cordless Telephone (DECT) are also remote possibilities for PCS.
3. Spread Spectrum. Frequency hopping, direct sequence, and combinations of these are usable in all of the bands described.

Overall, the spread spectrum techniques in general, and direct sequence in particular, appear to be superior to any other for personal communications system applications.

The Capacity Game

In spread spectrum systems, the amount of excess processing gain, or jamming margin, possessed by a system determines the number of users that it can accommodate at one time. Processing gain, in turn, is a function of the bandwidth available and the data rate per user. Therefore, the most direct path to increasing the number of users, defined as system capacity, is to either increase the bandwidth employed or decrease the data rate per user.* Since the bandwidth available is usually fixed, the only course to increasing capacity is to reduce user data rate or provide more links. This is exactly what has been done.

Average Data Rate

Some considerable effort has gone into reducing *average* data rate employing voice encoding schemes that reduce the coded data rate when a user is not talking (between words, inhaling, listening, etc.). This leads to an average data rate per user of 2–3 kbps. If we compare this to a more common voice data rate, 9.6 kbps, we see that the apparent reduced data rate allows us to claim 3.2 to 4.8 times as many users, *on the average.* Yet, peak capacity (when everyone happens to be talking) is still the same. The question is, can one claim that system capacity is a function of an average data rate when it means

---

*This sounds exactly like an FDMA system.

that under peak (everyone sending 9600 bps) load more than two thirds of every *average* user's data must be dropped?

Admittedly, this is a worst case, but it *will* happen. Every day. Therefore, if average data rates are used to determine system capacity, we should designate that capacity as *average* capacity. Such a system certainly could not support the peak number of users if those users were all 9600 bps data (or music) users, and it cannot support the *average* number of users when all of them (or even a significant number of them) are sending at their peak rate.

A telephone subscriber does not care if *on the average* his telephone sounds good when it sounds bad a significant part of the time.

Sectorization

Another way to increase apparent capacity is to sectorize a cell by employing directional antennas. This is equivalent to increasing the number of communication links available, as long as one sector does not interfere with another.

The same cell sectorization technique is applicable to any other multiple access approach. It is not reserved for PCDMA* systems.

The Impact of "Capacity"

Capacity is a primary benchmark by which personal communications systems are judged today. Much less emphasis should be placed on single-cell capacity as such. The reason is that the capacity of a single cell becomes much less important as requirements for system capacity increase. That is, the urban areas are the ones in which maximum capacity is required, but these are precisely the areas in which propagation range (at the frequencies set aside for personal communications systems) greatly diminishes. Therefore, smaller cells are mandatory, which results in greatly increased number of overall users. An example will serve to show this:

Suppose we wish to cover a 3-square-mile urban area with a wireless PCS approach. If 1/4-mile radii for base stations are used, the area could be covered with 16 base stations (assuming an approximate size of 1 × 3 miles).

$$\frac{3 \text{ square miles}}{0.25^2 \times \pi} = \frac{3 \text{ square miles}}{0.196 \text{ sq. mi}} = 15.3.$$

If each base station has a capacity of 60 users,[†] then the total number of simultaneous users served would be:

$$60 \times 16 = 960.$$

Base stations covering slightly greater radii (say 1/2 mile) could cover the

---

*Power-Controlled Code Division Multiple Access.
[†]Not including sectorization.

3 square miles with only 4 base locations, but each station would have to have the ability to handle 240 users at once.

Now fill the area with tall buildings, and it becomes almost impossible to achieve full coverage without using multiple smaller cells, rather than a few powerful cells. In this kind of situation, time-division CDMA (TCDMA), which can service multiple users with a single transmitting and receiving channel, is far superior to power-control CDMA (PCDMA) that requires a separate transmitter channel and a separate receiver channel for every user. The PCDMA system is much more expensive, and the cost of providing the multiple cells required, because of the propagation conditions, would be prohibitive.

From the example:
Sixteen TCDMA base stations could service the 960 users for a cost that would be approximately

$$\frac{16}{960} \times 100 = 1.2 \text{ percent}$$

of the cost of the 4 cell PCDMA system to do the same job. This is because the PCDMA system requires 960 transmitting and receiving channels, compared to 16 of each for the TCDMA system. Both would be forced to use more and smaller cells by propagation conditions, so that single-cell capacity is almost irrelevant, especially in urban areas.

The point is that capacity expressed in number of users per cell is not the end-all consideration for PCS systems. Far more important is the cost per user, especially when large cells cannot cover an area effectively.

Today's Needs

What is needed most is low rate robust voice encoders that have reduced rates that are real. That is, encoders whose peak rate is the same as their average rate, that operate at 4 kbps or less, require bit-error-rate no less than 1×10^{-3}, have minimal idiosyncrasies,* and are available in integrated circuit form.

At the lower frequencies, components (integrated circuits and other devices that require high investment for development) are readily available. This is primarily due to the market for cellular telephones in the 800-MHz frequency ranges. Similar components must be developed for application in the 1800–2200-MHz, 2400–2500-MHz, and 5725–5850-MHz bands.

A Final Word

I am not convinced that there will ever be a final word on spread spectrum systems. They continue to surprise me, even though I have known and loved

---

*Such as distortion of signals from nonhuman sound sources.

them for more than 35 years. The variety of applications in which spread spectrum systems are used is growing as fast as the techniques are introduced to new discoverers.

The thing that has really made the difference, however, is the change in the attitude of the frequency-regulating bodies around the world—especially the Federal Communications Commission. I am personally grateful to the commissioners and to their technical staff, individually and collectively, for their far sightedness, despite those who could see no good in ever allowing spread spectrum systems to be used outside the military bands.

Now, the cat is out of the bag. Never again, I hope, to be forced back in. I do not expect that spread spectrum techniques will ever replace other kinds of systems for most applications; I never expect to buy a television set that changes codes instead of frequency channels, or a microwave oven with a spread spectrum power stage (even though their spectra may resemble spread spectra even now). I do expect that there will be future spread spectrum books—perhaps even another edition of this one—in which the systems will be smaller, less expensive, and operate better then anything available today. I look forward to that time.

APPENDIX 1

GLOSSARY OF TERMS

This glossary is intended to define the terms used in this book in ways that are meaningful here but do not necessarily carry the same impact as they do elsewhere. Also defined are a number of terms commonly abbreviated without explanation. This is done as much as for any other reason to ensure that what is said here does have some relation to the real world.

AGC. Automatic gain control.

AJ. Antijamming.

AJ Margin. Antijamming margin; the worst case jamming-to-signal ratio at which a receiver will deliver rated performance.

AM. Amplitude modulation.

Antijamming. Capability of withstanding jamming or interference.

Autocorrelation. Multiplication of a signal with a time-delayed replica of itself, with product integration

$$\psi_{\text{auto}} = \int f(t) f(t - \tau) \, dt.$$

Baseband. The basic information channel for a communication system. For voice systems approximately 3-kHz low-pass baseband bandwidths are usually required.

Biphase. A term used to signify two-phase ($\pm 90°$) phase-shift keying.

Carrier. A term used to identify the center frequency of a signal, whether modulated or not.

CDMA. Code division multiple access. Any signal multiplexing technique using codes to separate signals.

Chip. A signal frequency output from a frequency hopping signal generator. Chip time signifies the time occupied by that single frequency. Also the period of a code clock, or the output of a code generator during one clock interval.

Chirp. Term applied to designate pulsed-FM modulated signals.

Clock. The frequency or chip-rate reference used to set the rate of code generation in a spread spectrum system.

Codes. *Composite*; A code made up of two or more distinct codes, usually generated by modulo-2 addition. *Gold*: A particular type of composite code. Oriented to multiple access. *JPL*: A particular type of composite code. Oriented to ranging and rapid synch acquisition. *Linear*: A class of codes in which only linear operations are allowed for generation (e.g., for binary codes only modulo-2 addition or subtraction is allowed). *Maximal*: Codes whose length is $2^N - 1$, where N is the generator length. *Nonlinear*: A class of codes in which nonlinear operations are performed. All codes other than linear. These codes encompass the truly secure codes. *Nonmaximal*: All codes other than maximal. *Quasiorthogonal*: A term used to describe codes with good mutual cross-correlation properties. *Synchopated*: Composite codes generated with nonsynchronous or nonsimultaneous chip timing.

Coherent. A synchronized, phase-matched condition between a receiver's reference and the desired signal.

Correlation. The degree of agreement between a pair of signals.

Correlator. That section of a spread spectrum system in which a received signal and the local reference are compared for agreement. The desired, synchronized signal is despread and undesired signals are spread.

Costas Demodulator. A special compound phase-lock loop intended for demodulation of suppressed carrier, double-sideband signals.

Cross-correlation. The degree of agreement between two unlike signals; that is,

$$\psi_{\text{cross}} = \int f(t) g(t - \tau) \, dt.$$

CW. Abbreviation for continuous wave, signifying an unmodulated carrier signal.

Dehop. The process (dehopping) by which a frequency hopping receiver removes or remaps the spread spectrum frequency hopping modulation from an incoming signal into a single frequency.

Delay Lock. A form of clock-frequency tracking loop which employs two correlators whose references are offset by some fraction of a chip. Used with direct-sequence systems.

Direct Sequence. A form of modulation wherein a code sequence is used to directly modulate a carrier, usually by phase-shift keying.

Direct Spread. Same as direct sequence.

Direct Synthesis. A frequency synthesis method in which all output frequencies are developed by multiplication or division from one or more frequency sources. No phase-lock circuits are employed.

Dither. A form of clock-frequency tracking loop in which the clock train supplied to the receiver's code reference is deliberately jittered a controlled amount—usually a fraction of a chip—generating an error signal for clock control.

DS. Direct sequence.

FCDMA. Frequency hop CDMA.

FDMA. Frequency division multiple access. A signal multiplexing technique using frequency to separate signals.

Feedback Tap. A point of feedback output used in a sequence generator for generating code sequences.

Frequency Hopper. A type of spread spectrum system in which the wideband signal is generated by hopping from one frequency to another over a large number of frequency choices. The frequencies used are chosen (pseudo-) randomly by a code similar to those used in direct sequence systems.

Frequency Shift Keying. A form of modulation similar to frequency hopping, except that normally two frequencies are employed and the choice of frequency is made on the basis of a "One" or "Zero" being sent.

Frequency Synthesis. Generation of multiple frequencies from one or a few sources. See direct and indirect synthesis categories.

FH. Frequency hopping.

FH / DS. Combined frequency hopping and direct sequence modulation.

FM. Frequency modulation.

FSK. Frequency shift keying.

G_p. See Process Gain.

Heterodyne. A process in which a signal is mixed with another or multiplied with it, usually for the purpose of frequency translation.

Hybrid Ranging. A ranging method in which short code sequences are combined with a synchronous tone to give faster acquisition than is available with longer sequences but longer range than available with short sequences alone.

Incremental Phase Modulation. A phase-modulation technique used to shift one code sequence with respect to another, usually for the purpose of a synchronization search.

Indirect Synthesis. A frequency synthesis method in which output frequencies are developed by the use of phase-lock methods.

Integrate and Dump. A form of synchronous detector that integrates over an information bit period and is discharged (dumped) at the end of that period. The integrator is sampled just before discharge to determine its state. When synchronized with a received signal, it is a matched filter.

Interference. Any signal that tends to hamper the normal reception of a desired signal. Equivalent to jamming except considered nonhostile in origin.

IPM. Incremental phase modulation.

Jammer. A source of deliberate interference, usually hostile.

Jamming. The signal produced by a jammer, and intended to interfere with reception of a desired signal.

Jamming Margin (M_j). The amount of interference a system is able to withstand while producing the required output signal-to-noise or bit-error rate.

Limiter. A nonlinear device used to prevent the passage of signals above some chosen limiting level. Below this level signals are passed unimpaired.

Limiting. The action performed by a limiter. Limiting action, because of its nonlinear character, produces signal products similar to those produced by multiplication or modulation.

***m*-ary.** Used to designate signal formats other than binary; multilevel signal formats such as ternary (3) and quaternary (4).

Matched Filter. The ideal filter, designed for reception of a specific signal; passing the signal with minimum loss while passing minimum noise.

Maximal Sequence. A type of code sequence which is the longest that can be generated by a feedback code generator. For binary code generation its length is $2^N - 1$, where N is the generator length in chips.

Mersenne Prime. A particular type of maximal code sequence in which the code length ($2^N - 1$) in chips is a prime number. Important when cross-correlation is a major factor.

M_j. See jamming margin.

Modem. Abbreviation for modulator-demodulator.

Modular Shift Register Generator. A particular type of code sequence generator in which each delay element is followed by a modulo-2 adder.

Modulo-2 Adder. Arithmetic logic element used in binary code generators to form feedback. Sometimes called an exclusive-or gate, or half-adder, because it performs the addition function but does not include a shift output. Logically its output is $1 = \overline{A}B + A\overline{B}$.

MQPSK. Modified quadriphase shift keying. Four-phase PSK modulation modified by shifting chip clocks so that phase transitions larger than 90° do not occur. Same as OQPSK and SQPSK.

MSK. Minimum shift keying. A form of modulation closely allied to FSK and staggered QPSK.

MSRG. Abbreviation for modular shift register generator.

Noise Figure. The amount of noise added (in dB) by a signal-handling stage to that existing at its input. Usually applied to a first RF stage to signify the amount of noise output over the noise input.

OQPSK. Offset quadriphase shift keying. Same as MQPSK and SQPSK.

Orthogonal. Used to signify signals that are mutually transparent or noninterfering. Frequency and amplitude modulation are ideally considered orthogonal.

PCN / S. Personal Communications Network/System.

PCDMA. Power-control CDMA.

PDM. Pulse duration modulation.

Phase-lock Loop. A feedback loop designed to cause an oscillator internal to the loop to track an incoming carrier's frequency. Generally used for angle demodulation.

Preamble. A special code sequence used for acquisition of synchronization in spread spectrum systems. Usually short compared with codes used in communicating.

Process Gain (G_p). The gain or S/N improvement enjoyed by a spread spectrum system due to coherent band spreading and remapping of the desired signal. Expressed by the equation $\mathrm{BW_{RF}}/R_{\mathrm{info}} = G_p = TW$.

Propagation. The process by which a signal proceeds from the transmitter to a receiver. A general term; for example, "propagation path."

Pseudonoise. A term used to signify any of a group of code sequences that exhibit noiselike properties. Also sometimes used as a name for systems that employ pseudonoise code modulation.

PSK. Abbreviation for phase-shift keying.

Pulsed FM. Also called "chirp." A signal format in which a pulsed carrier is transmitted and is swept in frequency during the transmission. This format permits a special type of matched filter (chirp) detector to be used for signal reception.

Quadriphase. Same as QPSK.

QPSK. Four-phase, phase-shift keyed signal format. Phase shifts occur at $0°$, $\pm 90°$, and $180°$, depending on the data being transmitted.

Remapping. The process of correlation in which a spread spectrum desired signal is transformed into a coherent narrowband signal and undesired signals are transformed into wide bandwidths.

SCPDM. Abbreviation for suppressed clock PDM, a signal format similar to common PDM, except that the bandwidth required for its passage is reduced by removing clock information.

Scrambler. One of a group of signal processors that cause an input signal to be incoherent to anyone not having the proper key for descrambling.

Search. A process by which spread spectrum systems acquire sychronization.

Secure. A term used to signify that a signal is protected from both inadvertent and deliberate attempts at decoding the signal. For all practical purposes a secure signal can be decoded only by one possessing the proper key.

Sequence. A code or train of chips.

Settling Time. The time required for a frequency hopping signal to change from one frequency to a new one.

Shift Register. A sequentially connected group of delay elements used to generate code sequences in spread spectrum systems.

Shift Register Generator. The combination of a shift register and modulo-2 addition for code sequence generation.

Simple Shift Register Generator. A form of shift register sequence generator in which the minimum number of modulo-2 addition stages is used to generate the particular code desired.

Spread Spectrum. Any of a group of modulation formats in which an RF bandwidth much wider than necessary is used to transmit an information signal so that a signal-to-noise improvement may be gained in the process.

SQPSK. Staggered quadriphase-shift keying. Same as OQPSK and MQPSK.

SRG. Abbreviation for shift register generator.

SSRG. Abbreviation for simple shift register generator.

Subcarrier. A carrier, modulated with information separate from carrier modulation, which in turn modulates a carrier.

Suppressed Carrier. A particular type of modulation format in which the carrier signal itself is suppressed or held to a level below the modulation components.

Synchronization. Timing agreement or the act of gaining timing agreement between a spread spectrum transmitter and its receiving system.

Tau-jitter. Same as dither or Tau-dither.

TCDMA. Time-division CDMA.

TDMA. Time division multiple access. A signal multiplexing technique using time division to separate signals.

TH. Abbreviation for time hopping.

Time / Frequency Hopping. A combination of time-hopping and frequency-hopping modulations for the purpose of overcoming near-far problems.

Time Hopping. Similar to simple pulse modulation, except that the pulse pattern may be pseudorandom.

TW (TW Product). The equivalent of G_p, or process gain, where T is the period of the information sent, and W is the bandwidth employed to send it; that is, $T = 1/R_{info}$.

APPENDIX 2

BIBLIOGRAPHY ON SPREAD SPECTRUM SYSTEMS AND RELATED TECHNOLOGY

The following bibliography is a listing of articles, memoranda, and technical papers that pertain either directly to spread spectrum systems or to the technology drawn on in implementing them. The papers listed are generally obtainable from the sources given or in many cases from any reasonably good technical library. The range of technical depth should satisfy either those that are just curious or those who are in need of rigorous theoretical and mathematical treatment.

INDEX OF CATEGORIES

1. ANTIJAMMING (INTERFERENCE REJECTION)

11. Aein, J. M. "On the Output Power Division in a Captured Hard-Limiting Repeater," *IEEE Trans. Comm. Tech.*, June 1966.

12. Bohrer, S. A. "Example of Frequency Hopping Performance Analysis: Magnavox Research Laboratories, MX-TM-6755-3006-66, December 1966.

13. Cahn, C. R. "A Note on Signal-to-Noise Ratio in Band-Pass Limiters," *IRE Trans. Info. Th.*, January 1961.

14. Cahn, C. R. "Increasing Frequency Hopping Rate of Digital Synthesizers," MRL STN-14, December 1964.

15. Cahn, C. R. "Data Coding to Reduce Vulnerability to Pulse Jamming," MRL STN-13, November 1964.

16. Cahn, C. R. "Comparison of Frequency Hopping and PN for AJ Data Transmission," MRL STN-15, December 1964.

17. Corneretto, A. "Spread Spectrum Com. System Uses Modified PPM," *Electronic Design*, June 21, 1961.

18. Costas, J. P. "Poisson, Shannon and the Radio Amateur," *IRE Proc.*, December 1959.

19. Friedman, H. J. "Jamming Susceptibility," *IEEE Trans. Aerosp. Elect. Syst.*, July 1968.

110. Gilmore, K. "The Secret Keepers," *Popular Electronics*, August 1962.

111. Goodman, L. M., and Russell, S. B. "The TATS Master—A Net Controller for Tactical Satellite Communications."

112. Jacobs, I. "The Effects of Video Clipping on the Performance of an Active Satellite PSK Communication System," *IEEE Trans. Comm. Tech.*, June 1965.

113. Kadar, I., and Schreiber, H. H. "Performance of Spread Spectrum Systems in a Multipath Environment," WESCON, 1970.

114. Levine, S. L. "Anti-Jam Communications Systems," Astia Document AD445897, April 1964.

115. Mason, J. F. "Air Force Plans Integrated Avionics Network," *Electronic Design*, August 15, 1968.

116. Mattis, T. J. "Tactical Satellite Communications Requirements," *Signal*, September 1969.

117. Miadich, A. J. "On-Board vs. Ground A. J. Processing," TRW Correspondence 7323-2-257, June 1968.

118. Morchin, W. C. "Radar Range in a Jamming Environment," *Microwave Journal*, June 1968.

119. Pfanstiehl, A. "Intelligent Noise," *Analog Science Fiction*, 1961 (approx.).

120. Sevy, J. L. "The Effect of Limiting a Biphase or Quadriphase Signal Plus Interference," *IEEE Trans. Aeros. Elect. Sys.*, May 1969.

121. Sullivan, D. P. "Future Trends in Military Communications Satellite Repeaters," TRW, 1968.

122. Thorensen, R. "On the Theory of Jam-Resistant Communication Systems," MRL Report R-502, April 1962.

123. Viterbi, A. J. "Maximum Problems in Coding for Jammed Channels," MRL STM-27, June 1966.

124. Viterbi, A. J. "Bandspreading Combats Multipath and RFI in Tactical Satellite Net," *Communications Designer's Digest*, December 1969.

125. Shklarsky, D., Das, P., and Milstein, L. B. "Adaptive Narrow-Band Interference Suppression," in *Proc. Natl. Telec. Conf.*, 1979, pp. 15.2.1–15.2.3.

126. Johnston, S. L. "Radar ECCM history," in *Proc. NAECON*, May 1980, pp. 1210–1214.

127. Pettit, R. H. "Error Probability for NCFSK with Linear FM Jamming," *IEEE Trans. Aerosp. Electron. Syst.*, Vol. AES-8, pp. 609–614, September 1972.

128. Lewis, H. D. "Array Radars Solve Communication Jams," *Microwaves*, April 1982.

129. Simon, M. K., and Polydoros, A. "Coherent Detection of Frequency-Hopped Quadrature Modulations in the Presence of Jamming—Parts I and II," *IEEE Trans. Comm.*, November 1982.

130. Milstein, L. B., and Schilling, D. "Performance of a Spread Spectrum Communication System Operating over a Frequency-Selective Fading Channel in the Presence of Tone Interference," *IEEE Trans. Comm.*, January 1982.

131. Bouvier, Jr., M. J. "The Rejection of Large CW Interferers in Spread Spectrum Systems," *IEEE Trans. Comm.*, Vol. COM-26, pp. 254–256, February 1978.

132. Johnson, G. R. "Jamming Low Power Spread Spectrum Radar," *Electron. Warfare*, pp. 103–112, Sept.–Oct. 1977.

133. Matsumoto, M., and Cooper, G. R. "Multiple Narrow-Band Interferers in an FH-DPSK Spread-Spectrum Communications System," *IEEE Trans. Veh. Technol.*, Vol. Vt-30, pp. 37–42, February 1981.

134. Houston, S. W. "Modulation Techniques for Communications—Part I: Tone and Noise Jamming Performance of Spread Spectrum M-ary FSK and 2,4-ary DPSK Waveforms." *Proc. IEEE 1975 Natl. Aerosp. Electron Conf.*, June 10–12, 1975, pp. 51–58.

135. Omura, J. K., Levitt, B., and Stokey, R. "FH/MFSK Performance in a Partial Band Jamming Environment," *IEEE Trans. Commun.* (to be published).

136. Avidor, D. "Anti-jam Analysis of Frequency Hopping M-ary FSK Communication Systems in HF Rayleigh Fading Channels," Doctoral Dissertation, School Eng. Appl. Sci., University of California, Los Angeles, 1981.

137. Milstein, L. B., Davidovici, S., and Schilling, D. L. "The Effect of Multiple-tone Interfering Signals on a Direct Sequence Spread Spectrum Communication System." *IEEE Trans. Commun.*, Vol. COM-30, pp. 436–446, March 1982.

138. Houston, S. W. "Tone and Noise Jamming Performance of a Spread Spectrum M-ary FSK and 2,4-ary DPSK Waveforms," *Proc. Natl. Aerosp. Electron. Conf.*, June 1974, pp. 51–58.

139. Pettit, R. H. "A Susceptibility Analysis of Frequency Hopped M-ary NCPSK—Partial-band Noise on CW Tone Jamming," presented at the Symp. Syst. Theory, May 1979.

140. Kullstam, P. A. "Spread Spectrum Performance Analysis in Arbitrary Interference" *IEEE Trans. Comm.*, August 1977.

141. Ketchum, J. W., and Proakis, J. G. "Adaptive Algorithms for Estimating and Suppressing Narrow-Band Interference in PN Spread Spectrum Systems," *IEEE Trans. Comm.*, August 1977.

142. Li, L. M., and Milstein, L. B. "Rejection of Narrow-Band Interference in PN Spread-Spectrum Systems Using Transversal Filters," *IEEE Trans. Comm.*, August 1977.

143. Matsumoto, M., and Cooper, G. R. "Performance of a Nonlinear FH-DPSK Spread Spectrum Receiver with Multiple Narrow-Band Interfering Signals," *IEEE Trans. Comm.*, August 1977.

144. Lee, J. S., and Miller, L. E. "Error Performance Analyses of Differential Phase-Shift-Keyed/Frequency-Hopping Spread-Spectrum Communication System in the Partial-Band Jamming Environments," *IEEE Trans. Comm.*, August 1977.

145. Simon, M. K. "The Performance of M-ary FH-DPSK in the Presence of Partial-Band Multitone Jamming," *IEEE Trans. Comm.*, August 1977.

146. Baer, H. P. "Interference Effects of Hard Limiting in PN Spread-Spectrum Systems," *IEEE Trans. Comm.*, August 1977.

147. Glenn, A. B. "Low Probability of Intercept (LPI) Performance in Jamming and Nonjamming Environments for SHF and EHF Satellite Communication Systems," *IEEE Milcom. Conf. Proc.*, October 1981.

148. Huth, G. K., and Weber, C. L. "The Performance of Direct Sequence Systems in the Presence of Jammers," *IEEE Milcom. Conf. Proc.*, October 1981.

149. McEliece, R. J. and Stark, W. E. "The Optimal Code Rate vs. A Partial Band Jammer," *IEEE Micom. Conf. Proc.*, October 1981.

150. Blanchard, J. E. "A Slow Frequency Hopping Technique That is Robust to Repeat Jamming," *IEEE Milcom. Conf. Proc.*, October 1981.

151. Stark, W. E. "Coding for Coherent Frequency-Hopped Spread-Spectrum Communication in the Presence of Jamming," *IEEE Milcom. Conf. Proc.*, October 1981.

152. Huang, T. C., and Yen, L. "Error Probability Performance of Noncoherent MFSK/FH Receivers in the Presence of Interference and Gaussian Noise," *IEEE Milcom. Conf. Proc.*, October 1981.

153. Spellman, M. "A Comparison Between Frequency Hopping and Direct Spread PN as Antijam Techniques," *IEEE Milcom. Conf. Proc.*, October 1981.

154. Smith, I. R. "Adaptation of Spread Spectrum Processing Gain and Voice Processing Algorithms in a Jamming Environment," *IEEE Milcom. Conf. Proc.*, October 1981.

155. Simon, M. K., and Wang, C. "Limiter Discriminator Detection of Narrowband FM (CPDSK) in Jamming and Multipath," *IEEE Milcom. Conf. Proc.*, October 1981.

156. Levitt, B. "Comprehensive Analysis of FH/MFSK with Multi-tone Jamming," *IEEE Milcom. Conf. Proc.*, October 1981.

157. Omura, J. "Anti-Jam Performance of FH/MFSK in Rayleigh Fading," *IEEE Milcom. Conf. Proc.*, October 1981.

158. Rafuse, R. P. "Interference Rejection Properties of Pseudonoise and Frequency-Hopping Systems—A Review," *Int. Conf. Comm. Proc.*, June 1980.

159. Scholtz, R. A. "Centered Interface Rejection Using Spread Spectrum Techniques," *Int. Conf. Comm. Proc.*, June 1980.

160. Ward, M. K., and Gibbons, H. M. "The Performance of Frequency Hopped Noncoherent M-ary FSK Modulation in the Presence of Repeat Back Jamming," *Int. Conf. Comm. Proc.*, June 1980.

161. Lang, R. H., and Pickholtz, R. L. "Transient Analysis of an Adaptive Array in the Presence of a Step Jammer," *Natl. Telec. Conf. Proc.*, Nov. 1980.

162. Mahood, S. "Performance of an FHSS Signal in White Gaussian Noise Using a CCD/C2T Implementation," *Natl. Telec. Conf. Proc.*, November 1980.

163. Cooper, G. R. "Narrowband Interference in an FH-PPSK Spread Spectrum Communication System," *Natl. Telec. Conf. Proc.*, November 1980.

164. Milstein, L. B., and Schilling, D. L. "The Effect of a Comb Jammer on a Direct Sequence Spread Spectrum System," *Natl. Telec. Conf. Proc.*, November 1980.

165. Orr, R. S. "Performance of PN/FH Hybrid Spread Spectrum Systems Against Partial-Band Noise Jamming," *Natl. Telec. Conf. Proc.*, November 1980.

166. Lee, S., and Park, J. H. "Spreading Nulling Tradeoffs in an Interference Environment," *Natl. Telec. Conf. Proc.*, November 1980.

167. Bouvier, Jr., M. J. "The Rejection of Large CW Interferers in Spread Spectrum Systems," *IEEE Trans. Com.*, February 1978.

168. Hsu, F. M., and Giordano, A. A. "Digital Whitening Techniques for Improving Spread Spectrum Communications Performance in the Presence of Narrowband Jamming and Interference," *IEEE Trans. Comm.*, February 1978.

169. Simon, M. K., and Woo, K. T. "The Performance of Suppressed Carrier Receivers in a Pulsed RFI Environment," *IEEE Trans. Comm.*, May 1980.

170. Milstein, L. B., and Schilling, D. L. "Performance of a Spread Spectrum Communication System Operating Over a Frequency-Selective Fading Channel in the Presence of Tone Interference," *IEEE Trans. Comm.*, January 1982.

171. Johnston, S. L. "Guided Missile ECM/ECCM," *Microwave Journal*, September 1978.

172. Buskirk, R. L., and Nossen, E. J. "ECM/ECCM Effects on Voice Transmission," *ITC Conf. Proc.*, 1978.

173. Li, L., and Milstein, L. B. "Rejection Pulsed CW Interference in PN Spread Spectrum Systems Using Complex Adaptive Filters," *IEEE Trans. Comm.*, January 1983.

174. Shabsigh, O. A. H. "On the Effects of CW Interference on MSK Signal Reception," *IEEE Trans. Comm.*, August 1982.

175. Cook, C. E. "Anti-Intercept Margins of Relay-Augmented Data Links," *IEEE Trans. Comm.*, July 1981.

2. APPLICATIONS OF SPREAD SPECTRUM METHODS

21. Allain, L. R. "First Steps to Radar-Waveform Design," *Microwaves*, December 1965.

22. Allen, W. B., and Westerfield, E. C. "Digital Compressed Time Correlators and Matched Filters for Active Sonar," *J. Acoust. Soc. Amer.*, January 1964.

23. Anderson, G. C., and Perry, M. A. "A Calibrated Real-Time Correlator/Averager/Probability Analyzer," *HPJ.*, November 1969.

24. Barrow, B. B., Abraham, Jr., L. G., Stein, S., and Bitzer, D. "Tropospheric-Scatter Propagation Tests Using a Rake Receiver," *IEEE Comm. Conv.*, June 7, 1965.

25. Bitzer, D. R., Chesler, D. A., Ivers, R., and Stein, S. "A Rake System for Tropospheric Scatter," *IEEE Trans. Comm. Tech.*, August 1966.

26. Blasbalg, H., Freeman, D., and Keeler, R. "Random Access Communications Using Frequency Shifted PN (Pseudo-Noise) Signals," IBM Corporation.

27. Blasbalg, H., Najjar, H., D'Antonia, R., and Haddad, R. "Air-Ground, Ground Air Communications Using Pseudo-Noise Through a Satellite," *IEEE Trans. Aerosp. Elect. Sys.*, September 1968.

28. Brandon, W. T., Camera, A. G., and Wilson, Q. C. "A New Approach to Tactical Aircraft-Satellite Communications," ESD-TR-68-296, Mitre Corp.

29. Chesler, D. "Performance of a Multiple Access RADA System," *IEEE Trans. Comm. Tech.*, August 1966.

210. Cook, C. E. "Pulse Compression—Key to More Efficient Radar Transmission," *IRE Proc.*, March 1960.

211. Corneretto, A. "Better Radars Designed Using Pulse Compression," *Electronic Design*, August 30, 1963.

212. Cushman, R. H. "Make the Most of Noise: Correlate It," EDM, March 1, April 15, 1971.

213. Dayton, D. S. "Coming to Grips with Multipath Ghosts," *Electronics*, November 27, 1967.

214. Diamond, P. M. "Satellite Systems for Integrated Communications, Navigation and Identification," AF Report SAMSO-TR-70-160, Aerospace Corp.

215. Dixon, R. C. "A Spread Spectrum Ranging Technique for Aerospace Vehicles," SW IEEE CON, April 1968.

216. Donn, E. S. "Manipulating Digital Patterns with a New Binary Sequence Generator, *HPJ*, April 1971.

217. Donn, E. S. "Measuring Digital Error Rate with Pseudorandom Signals," *Telecommunications*, November 1971.

218. Drouilhet, Jr., P. R., and Bernstein, S. L. "TATS—A Bandspread Modulation-Demodulation System for Multiple Access Tactical Satellite Communication," EASCON, 1969 record.

219. Golomb, S. W. et al. *Digital Communications with Space Applications*, Prentice-Hall, Englewood Cliffs, New Jersey, 1964.

220. Huang, R. Y., and Hooten, P. "Communication Satellite Processing Repeaters," *IEEE. Proc.*, February 1971.

221. Hultberg, R. M., Jean, F. H., and Jones, M. C. "Time Division Access for Military Communications Satellites," *IEEE Trans. Aerosp. Elect. Sys.*, December 1965.

222. Judge, W. J. "Multiplexing Using Quasiorthogonal Functions," *AIEEE Winter General Mtg.*, January 1962.

223. Klauder, J. R., Price, A. C., Darlington, S., and Albersheim, W. J. "The Theory and Design of Chirp Radars," *BSTJ*, July 1960.

224. Price, R., and Green, Jr., P. E. "A Communication Technique for Multipath Channels," *IRE Proc.*, March 1958.

225. Price, R. "Wideband Analog Transmission Through Multipath by Means of a Pseudo-Noise Carrier," *URSI Spring Mtg.*, April 30, 1962.

226. Rex, R. L., and Roberts, G. T. "Correlation, Signal Averaging, and Probability Analysis," *HPJ*, November 1969.

227. Roberts, L. G. "Picture Coding Using Pseudo-Random Noise," *IRE Trans. Info. Th.*, February 1962.

228. Shepherd, S. G., and Seman, A. E. "Design with Integrated Circuits, Part 2," *Electronic Design*, July 19, 1965.

229. Springett, J. C. "Pseudo-Random Coding for Bit and Word Synchronization of PSK Data Transmission Systems," *Jet Propulsion Laboratories*.

230. Spoonemore, W. "UHF Propagation Experiment Using a Pseudonoise MODEM," AFAL Report TOD NR. 0510 PROJ. NR. 7662-04, January 1972.

231. Talamini, Jr., A. J., and Farnett, E. C. "New Target for Radar: Sharper Vision with Optics," *Electronics*, December 27, 1965.

232. Ward, R. B. "Digital Communications on a Pseudo-Noise Tracking Link Using Sequence Inversion Modulation," *IEEE Trans. Comm. Tech.*, February 1967.

233. Ward, R. B., and Strubel, F. L. "Remote Clock Synchronization Experiment Progress 1968," LMSC/A946603 Lockheed MASC, December 1968.

234. Wester, W. J., and Carter, H. G. "Spacecraft Sequence Generator Utilizes Integrated Circuitry," *EDN*, December 1964.

235. "Model 3721A Correlator," Hewlett-Packard, August 1969.

236. "Unified-Carrier Link Proposed for Lunar Mission," *Electronic Design*, September 27, 1963.

237. "Range Rider Pseudo-Noise Transmission Test Sets," Data Sheets 1000, 1100, International Data Sciences, September 1970.

238. "PN Carrier Modem Brings Privacy to Commercial Mobile Radio Networks," *Communications Designers' Digest*, January-February 1971.

239. "New Methods for Continuous, In-Service Monitoring of Microwave Systems," International Microwave Corp., Cos Cob, Connecticut.

240. "System 5000/5100 Digital Communication Test Set," Tau-Tron Inc., June 1971.

241. "MX230a Voice Modem," Magnavox Research Laboratories Report R-2081 A.

242. Crush, J. F., and Grossman, B. C. "Applications of Spread Spectrum Technology to a Low Cost Aircraft Traffic Control System," *Proc. Symp. Sp. Spec. Comm.*, March 1973.

243. Cuccia, C. L. "Communication by Satellite: A Status Report," *Microwave Systems News*, April/May 1974.

244. Hoff, L. E. "Design Consideration of Spread Spectrum Communication Systems for the HF Band," *Proc. Symp. Sp. Spec. Comm.*, March 1973.

245. Cahn, C. R. Paper No. 5, AGARD-NATO Lecture Series No. 58 on "Spread Spectrum Communications," May 1973.

246. Cuccia, C. L. "Spread Spectrum Systems Serve Nearly all C^3 Aspects," *MSN*, April 1982.

247. Goggans, P. M., Jelinek, D. A., and Nielsen, W. F. "SECURICOM: An Antijam, Secure-Message, Security Force Radio Communication System, Technical Summary," Sandia Laboratories SAND80-0347, March 1981.

248. Dixon, R. C. "SECURICOM: An Anti-Jam, Secure-Message, Security Force Radio Communication System. Design Feasibility Study," Sandia Laboratories SAND80-7049, March 1981.

249. Spellman, M. "Spread-Spectrum Radios Thwart Hostile Jammers," *Microwaves*, September 1981.

250. Custance, K. "Hop, Skip, and Jump. Who Wins?" *Telecommunications*, August 1981.

251. Ribchester, E. "Frequency Hopping Radios Outwit 'Smart' Jammers," *Microwaves*, November 1979.

252. Melton, W. C. "Time-of-Arrival Measurements Possible with Use of Global Positioning System," *Defense Electronics*, December 1981.

253. Bolen, N. E. "JTIDS Coordinates Tactical Elements," *Defense Electronics*, January 1982.

254. Cooper, G. R., and Nettleton, R. W. "A Spread-Spectrum Technique for High Capacity Mobile Communications," *IEEE Trans. Veh. Techn.*, Vol. VT-27, pp. 264–275, November 1978.

255. Goodman, D. J., Henry, P. S., and Prabhu, V. K. "Frequency-Hopped Multilevel FSK for Mobile Radio," *Bell Syst. Tech. J.*, Vol. 59, pp. 1257–1275, September 1980.

256. Goldberg, B. "Applications of Statistical Communications Theory," presented at the Army Sci. Conf. West Point, NY, June 20–22, 1962 (AD 332048): republished in *IEEE Commun. Mag.*, Vol. 19, pp. 26–33, July 1981.

257. Sass, P. F. "Army Spread Spectrum—Evolution or Revolution?" *IEEE Milcom. Conf. Proc.*, October 1981.

258. Slechta, J. S., and Willman, G. G. "Design Tradeoffs for Frequency Hopping VHF Net Radio," *IEEE Milcom. Conf. Proc.*, October 1981.

259. Leiner, B. M. "Performance of an Advanced ECCM Packet Radio," *IEEE Milcom. Conf. Proc.*, October 1981.

260. Goldman, H. B. "A Comparison of JTIDS and Packet Multiple Access Techniques," *IEEE Milcom. Conf. Proc.*, October 1981.

261. Scales, W. C. "Maritime Uses of Spread Spectrum," *IEEE Milcom. Conf. Proc.*, October 1981.

262. Sant, D. "Spread Spectrum and Satellite Systems," *IEEE Milcom. Conf. Proc.*, October 1981.

263. Howell, W. E. "Civil Aeronautical Applications," *Natl. Telec. Conf. Proc.*, November 1980.

264. Lane, J. "Geophysical Applications," *Natl. Telec. Conf. Proc.*, November 1980.

265. Ligon, J. M. "The GPS Z-Set Goes to Sea," *Natl. Telec. Conf. Proc.*, November 1980.

266. Painter, J. H., Rhyne, V., and Noe, P. "Z-Set Results in the Gulf of Mexico," *Natl. Telec. Conf. Proc.*, November 1980.

267. Yiu, K. P., Lee, F., and Eschenback, R. "Land Navigation of a Low Cost GPS Receiver," *Natl. Telec. Conf. Proc.*, November 1980.

268. Bolen, N. E. "JTIDS—Status and Issues," *Natl. Telec. Conf. Proc.*, November 1980.

269. Freret, P. "Application of Spread Spectrum Radio to Wireless Terminal Communications," *Natl. Telec. Conf. Proc.*, November 1980.

270. Girodano, A., Sunkneberg, H., de Padro, H., Stynes, P., Brown, D., and Lee, S. C. "A Spread-Spectrum Simulcast MF Radio Network," *IEEE Trans. Comm.*, May 1982.

271. Freret, P., Eschenbach, R., Crawford, D., and Braisted, P. "Applications of Spread-Spectrum Radio to Wireless Terminal Communications," *IEEE Natl. Telec. Conf. Proc.*, 1980.

272. Cariolaro, G. L., Cucchi, S., and Molo, F. "Transmultiplexer Via Spread-Spectrum Modulation," *IEEE Trans. Comm.*, July 1982.

273. Goodman, D. J., and Timor, U. "Spread-Spectrum Mobile Radio with Variable-Bit-Rate Speech Transmission," *IEEE Trans. Comm.*, March 1982.

274. Dixon, R. C., and Kaiser, F. P. "Applications of Anti-Jamming Techniques in Voice Communications Systems," NRL Workshop on Technology for Communications ECCM, February 1976.

275. Davenport, Jr., W. B. "NOMAC Data Transmission Systems," presented at the RDB Symp. Inform. Theory Appl. Guided Missile.

276. Green, Jr., P. E. "Correlation Detection Using Stored Signals," M.I.T. Lincoln Lab., Tech. Rep. 33, August 4, 1953 (AD 020524).

277. Eisenstadt, B. M., Fleck, Jr., P. L., Selfridge, O. G., and Wagner, C. A. "Jamming tests on NOMAC systems," M.I.T. Lincoln Lab., Tech. Rep. 41, September 25, 1953 (AD 020419).

278. Green, Jr., P. E. "The Lincoln F9C Radioteletype System," M.I.T. Lincoln Lab., Tech. Memo. 61, May 14, 1954 (not available from M.I.T.).

279. Green, Jr., P. E., Berg, R. S., Bergman, C. W., and Smith, W. B. "Performance of the Lincoln F9C radioteletype system," M.I.T. Lincoln Lab., Tech. Rep. 88, October 28, 1955 (AD 080345).

280. Craig, Jr., J. W. "An Experimental NOMAC Voice Communication System," M.I.T. Lincoln Lab., Rep. 34G-007, August 29, 1960 (AD 319610).

281. Gray, D. J. "A new method of teletype modulation," M.I.T. Lincoln Lab., Tech. Rep. 9, September 22, 1952 (AD 000928).

282. Costas, J. P., and Widmann, L. C. "Data Transmission System," U.S. Patent 3 337 803, August 22, 1967 (filed Jan. 9, 1962). "Reliable tactical communications," General Electric Res. Lab., Schenectady, NY, Final Rep., Contr. DA-36-039sc-42693, March 2, 1954 (AD 30344).

283. Baldwin, Jr., E. R. "SINCGARS: The Army's New Frequency Hopping Jam Resistant Combat Net Radio," *Signal Magazine*, November 1982.

284. Parker, E. B. "Micro Earth Stations Make Business Satcom Affordable," *MSN Magazine*, November 1982.

285. Denaro, R. P. "Navstar: The All Purpose Satellite," *Spectrum*, May 1981.

3. CODING

31. Anderson, D. R. "A New Class of Cyclic Codes," *SIAM J. Appl. Math.*, **16**, No. 1, 1968.

32. Anderson, D. R. "Periodic and Partial Correlation Properties of Sequences," TRW I.C. 7353, 1-01, July 1969.

33. Baumert, L., Easterling, M., Golomb, S. W., and Viterbi, A. "Coding Theory and Its Applications to Communications Systems," JPL Report 32-67, March 1961.

34. Birdsall, T. G., and Ristenbatt, M. P. "Introduction to Linear Shift-Register Generated Sequences," Tech. Report 90, University of Michigan Research Institute, October 1958.

35. Blizard, R. B. "Quantizing for Correlation Decoding," *IEEE Trans. Comm. Tech.*, August 1967.

36. Bluestein, L. J. "Interleaving of Pseudo-Random Sequences for Synchronization," *IEEE Trans. Aerosp. Elect. Sys.*, July 1968.

37. Braasch, R. H. "The Distribution of (n-m) Terms for Maximal Length Linear Pseudo-Random Sequences," *IEEE Trans. Comm. Info. Th.*, July 1968.

38. Curry, R. C. "A Method of Obtaining All Phases of a Pseudo-Random Sequence," Naecon 1967.

39. Dixon, R. C. "100 MHz PN Code Generator," TRW 10C 7325. 3–19, July 1968.

310. Forney, Jr., G. D. "Coding and Its Application in Space Communications," *IEEE Spectrum*, June 1970.

311. Freymodsson, J. B. "A Note on the Run-Length Distribution of Ones and Zeros in Maximal-Linear Shift Register Sequences," unpublished memo dated June 11, 1963.

312. Gold, R. "Study of Correlation Properties of Binary Sequences," Magnavox Research Laboratories Report AFAL TR-66-234, August 1966.

313. Gold, R. "Characteristic Linear Sequences and Their Coset Functions," *J. SIAM Appl. Math.*, **14**, No. 5, September 1966.

314. Gold, R. "Optimal Binary Sequences for Spread Spectrum Multiplexing," *IEEE Trans. Info. Th.*, October 1967.

315. Golomb, S. W. "Sequences With Randomness Properties," Glenn L. Martin Co., June '55.

316. Golomb, S. W. *Shift Register Sequences*, Holden-Day, San Francisco, 1967.

317. Gupta, S. C. and Painter, J. H. "Correlation Analyses of Linearly Processed Pseudo-Random Sequences," *IEEE Trans. Comm. Tech.*, December 1966.

318. Haberle, H. "Modulation, Synchronization and Coding," *Intl. Conf. Digital Satellite Comm.*, London, November 1969.

319. Hansen, J. C. "Modern Algebra for Coding," *Electro-Technology*, April 1965.

320. Horton, W. P. "Shift Counters," Computer Control Corporation.

321. Kahn, D. "Modern Cryptology," *Scientific American*, July 1966.

322. Lindholm, J. H. "An Analysis of the Pseudo-Randomness Properties of Subsequences of Long *m*-Sequences," *IEEE Trans. Info. Th.*, July 1968.

323. Lindsey, W. C. "Coded Noncoherent Communications," *IEEE Trans. Space Elect. Telem.*, March 1965.

324. O'Clock, Jr., G. D., Grasse, G. L., and Gandolfo, D. A. "Switchable Acoustic Surface Wave Sequences Generator," *IEEE Proc.*, October 1972.

325. Painter, J. H. "Designing Pseudo-Random Coded Ranging Systems," *IEEE Trans. Aerosp. Elect. Sys.*, January 1967.

326. Peterson, W. W. "Error Correcting Codes," MIT Press, Wiley, New York.

327. Shannon, C. E. "Communication Theory of Secrecy Systems," Bell Labs.

328. Swick, D. A. "Wideband Ambiguity Function of Pseudo-Random Sequences: An Open Problem," *IEEE Trans. Info. Th.*, July 1968.

329. Tausworthe, R. C. "Random Numbers Generated by Linear Recurrence Modulo-Two," *Math. Computation*, **19**, April 1965.

330. Titsworth, R. C. "Optimal Ranging Codes," *IEEE Trans. Space Elec. Telem.*, March 1963.

331. Toerper, K. E. "Biphase Barker-Coded Data Transmission," *IEEE Trans. Aerosp. Elect. Syst.*, March 1968.

332. Viterbi, A. J. "On Coded Phase-Coherent Communications," JPL Report 32-25, August 1960.

333. Wainberg, S., and Wolf, J. K. "Subsequences of Pseudo-Random Sequences," *IEEE Trans. Comm. Tech.*, October 1970.

334. Watson, E. J. "Primitive Polynomials (Mod 2)," Math. Comp. 1962.

335. White, Jr., R. C. "Experiments with Digital Computer Simulation of Pseudo-Random Noise Generators," *IEEE Trans. Elect. Comp.*, June 1967.

336. Wolf, J. K. "On the Application of Some Digital Sequences to Communication," *IEEE Trans. Comm. Syst.*, December 1963.

337. Zierler, N. "Several Binary-Sequence Generators," Tech. Report 95 MIT, Lincoln Laboratories, September 1955.

338. Zierler, N. "Linear Recurring Sequences," *J. Soc. Indust. Appl. Math.*, 7, 1959.

339. "Surface Acoustic Waves Open New Avenues in Signal Processing," *Microwaves*, July 1971.

340. "Microwave Acoustics; Waves of the Future," *Microwave Systems News*, December/January 1974.

341. De Vito, P. A., Carr, P. H., Kearns, W. J., and Silva, J. H. "Encoding and Decoding with Elastic Surface Waves at 10 Megabits Per Second," *IEEE Proc.*, October 1971.

342. Jacobs, I. M. "Practical Applications of Coding," *IEEE Trans Info. Th.*, May 1974.

343. Lee, J., and Smith, D. R. "Families of Shift-Register Sequences with Impulsive Correlation Properties," *IEEE Trans. Info. Th.*, March 1974.

344. Solomon, G. "Optimal Frequency Hopping Sequences for Multiple Access," *Proc. Symp. Sp. Spec. Comm.*, March 1973.

345. Eckler, A. R. "The Construction of Missile Guidance Codes Resistant to Random Interference," *Bell Syst. Tech. J.*, Vol. 39, pp. 973–994, July 1960.

346. Gold, R. "Maximal Recursive Sequences with 3-Valued Cross-Correlation Functions," *IEEE Trans. Info. Th.*, Vol. IT-14, pp. 154–156, January 1968.

347. Marolf, R. A. "200 Mbit/s Pseudo Random Sequence Generator for Very Wide Band Secure Communication Systems," in *Proc. NEC*, Chicago, IL. 1963, Vol. 19, pp. 183–187.

348. Mathes, R. C. "Secret Telephony," U.S. Patent 3 967 066, June 29, 1976 (filed September 24, 1941).

349. Potter, R. K. "Secret Telephony," U.S. Patent 3 967 067, June 29, 1976 (filed September 24, 1941).

350. Jones, S. V. "After 35 Years, Secrecy Lifted on Encoded Calls," *New York Times*, p. 27, July 3, 1976.

351. Busch, A. J. "Signalling Circuit," U.S. Patent 3 968 454, July 6, 1976 (filed September 27, 1944).

352. Joel, Jr., A. E. "Pulse Producing System for Secrecy Transmissions," U.S. Patent 4 156 108, May 22, 1979 (filed January 21, 1947).

353. Hansen, W. W. "Secret Communication," U.S. Patent 2 418 119 Apr. 1, 1947 (filed April 10, 1943).

354. deRosa, L. A., and Rogoff, M. "Secure Single Sideband Communications System Using Modulated Noise Subcarrier," U.S. Patent 4 176 316, November 27, 1979 (filed March 20, 1953); reissue appl. filed September 4, 1981, Re. Ser. No. 299 469.

355. Chaffee, E., and Purington, E. "Method and Means for Secret Radiosignalling," U.S. Patent 1 690 719, November 6, 1928 (filed March 31, 1922).

356. Shannon, C. E. "A Mathematical Theory of Cryptography," Bell Tel. Lab., memo., September 1, 1945; later published in expurgated form as "Communication theory of secrecy systems," *Bell Syst. Tech. J.*, Vol. 28, pp. 656–715, October 1949.

357. Viterbi, A. J., and Jacobs, I. M. "Advances in Coding and Modulation for Noncoherent Channels Affected by Fading, Partial Band, and Multiple Access Interference," in *Advances in Communication Systems*, Vol. 4, Academic, New York, 1975.

358. O'Clock, Jr., G. D., Grasse, C. L., and Grandolfo, D. A. "Switchable Acoustic Surface Wave Sequence Generator," *IEEE Proc.*, October 1971.

359. Schobi, P. "Memory Structure Speeds Generation of Pseudonoise Sequences," *Computer Design*, March 1981.

360. Pursley, M. B., and Sarwate, D. V. "Performance Evaluation for Phase-Coded Spread Spectrum Multiple Access Communication—Part II, Code Sequence Analysis," *IEEE Trans. Comm.*, Vol. COM-25, pp. 800–803, August 1977.

361. Pursley, M. B., and Roefs, H. F. A. "Numerical Evaluation of Correlation Parameters for Optimal Phases of Binary Shift-Register Sequences," *IEEE Trans. Comm.*, Vol. COM-25, pp. 1597-1604, August 1977.

362. Solomon, G. "Optimal Frequency Hopping Sequences for Multiple Access," in *Proc. 1973 Symp. Sp. Spec. Comm.*, Vol. 1, AD915852, pp. 33–35.

363. Sarwate, D. V., and Pursley, M. B. "Hopping Patterns for Frequency-Hopped Multiple-Access Communication," in *Proc. 1978 IEEE Int. Conf. Comm.*, Vol. 1, pp. 7.4.1–7.4.3.

364. Garber, F. D., and Pursley, M. B. "Optimal Phases of Maximal-length Sequences for Asynchronous Spread-Spectrum Multiplexing," *Electron. Lett.*, Vol. 16, pp. 746–747, September 1980.

365. Pursley, M. B., and Roets, H. F. A. "Numerical Evaluation of Correlation Parameters for Optimal Phases of Binary Shift Register Sequences," *IEEE Trans. Comm.*, Vol. COM-27, pp. 1597–1604, October 1979.

366. Sarwate, D. V., and Pursley, M. B. "Crosscorrelation Properties of Pseudorandom and Related Sequences," *Proc. IEEE*, Vol. 68, pp. 593–619, May 1980.

367. Cooper, G. R. "Covert Communication with a Purely Random Spreading Function," *IEEE Milcom. Conf. Proc.*, October 1981.

368. Gutlebert, F. S. "Spread Spectrum Multiplexed Noise Codes," *IEEE Milcom. Conf. Proc.*, October 1981.

369. Pupolin, S., and Tomasi, C. "Moments of the Weights of Pseudo-Noise Subsequences," *IEEE Milcom. Conf. Proc.*, October 1981.

370. Beeler, Bolt, Beranek, and Newman "Finding Good Signalling Codes with Custom VLSI" *IEEE Milcom. Conf. Proc.*, October 1981.

371. Elhakeem, A. K. "Optimal Codes and Filters for Hard-Limited OQPSK/DS Multiple Access Systems," *IEEE Milcom. Conf. Proc.*, October 1981.

372. El-Khamy, S. E. "Matched Spread Spectrum Techniques," *IEEE Milcom. Conf. Proc.*, October 1981.

373. Dupuis, J. "Hardware and Software Tradeoffs in Spread Spectrum Code Generation," *ITC Conf. Proc.*, October 1981.

374. Schobi, P. "Memory Structure Speeds Generation of Pseudonoise Sequences," *Computer Design*, March 1981.

375. Titsworth, R. C. "Correlation Properties of Cyclic Sequences," Ph.D. dissertation, California Institute of Technology, Pasadena, 1962.

376. Lunayach, R. S. "Performance of Direct Sequence Spread Spectrum System with Long and Short Period Code Sequences," *IEEE Trans. Comm.*, March 1983.

377. Chakrabarti, N. B., and Tomlinson, M. "Design of Sequences with Specified Autocorrelation and Cross Correlation," *IEEE Trans. Comm.*, November 1976.

378. Milstein, L. B., and Ragonetti, R. R. "Combination Sequences for Spread Spectrum Communications," *IEEE Trans. Comm.*, July 1977.

379. Glisic, S. G. "Power Density Spectrum of the Product of Two Time Displaced Versions of a Maximum Length Binary Pseudonoise Signal," *IEEE Trans. Comm.*, February 1983.

380. Pursley, M. B. "The Role of Coding in Multiple-Access Satellite Communication Systems," University of Illinois, Report R-724 April 1976.

381. Stark, W. E. "Performance of Concatenated Codes on Channels With Jamming," *IEEE I.C.C. Proc.*, 1982.

4. MATCHED FILTERS / CONVOLVERS

41. Armstrong, D. B. "Solid State Surface Acoustic Delay Lines," Litton Industries TN70-1 30, January 1970.

42. Collins, J. H., and Hagon, P. J. "Applying Surface Wave Acoustics," *Electronics*, November 10, 1969.

43. Cook, C. E. "Pulse Compression—Key to More Efficient Radar Transmission," *IRE Proc.*, March 1960.

44. Coquin, G. A., and Tsu, R. "Theory and Performance of Perpendicular Diffraction Delay Lines," *IEEE Proc.*, June 1965.

45. de Atley, E. "Surface Acoustic Waves Offer Cheap Signal Processing," *Electronic Design*, December 6, 1970.

46. Eveleth, J. H. "A Survey of Ultrasonic Delay Lines Operating Below 100 MC/S," *IEEE Proc.*, October 1965.

47. Eveleth, J. H. "Dispersive and Nondispersive Ultrasonic Delay Lines," Anderson Laboratories Tech. Note., undated.

48. Franklin, P. "Ultrasonics, Microwaves Join Forces," *Microwaves*, November 1968.

49. Franklin, P. "Echo Processor Offers 12 μs Delay with 50 dB Gain," *Microwaves*, October 1969.

410. Grasse, C. L., and Gandolfo, D. A. "400 MHz Acoustic Surface-Wave Pulse Expansion and Compression Filter," *IEEE Trans. Micro. Th. Tech.*, June 1971.

411. Haavind, R. C. "Praetersonics: Microwaves of the Future?" *Electronic Design*, July 18, 1968.

412. Klauder, J. R., Price, A. C., Darlington, S., and Albersheim, W. J. "The Theory and Design of Chirp Radars," *BSTJ*, July 1960.

413. Lean, E. G., and Broers, A. N. "Microwave Acoustic Delay Lines," *The Microwave Journal*, March 1970.

414. Newhouse, P. D. "Simplify EMC Design," *Microwaves*, May 1970.

415. Olson, F. A. "Today's Microwave Acoustic (Bulk Wave) Delay Lines," *The Microwave Journal*, March 1970.

416. Riezerman, M. J. "Lazer-Acoustic Line Reverses Time Functions," *Electronic Design*, June 7, 1969.

417. Gerard, H. M., Smith, W. R., Jones, W. R., and Harrington, J. B. "The Design and Applications of Highly Dispersive Acoustic Surface-Wave Filters," *IEEE Trans. MTT*, April 1973.

418. Martin, T. A. "The IMCON Pulse Compression Filter and Its Applications," *IEEE Trans. MTT*, April 1973.

419. Kino, G. S., Ludvik, S., Shaw, H. J., Shreve, W. R., White, J. M., and Winslow, D. K. "Signal Processing by Parametric Interactions in Delay-Line Devices," *IEEE Trans. MTT*, April 1973.

420. Bert, A. G., Epsztein, B., and Kantrowicz, B. "Signal Processing by Electron-Beam Interaction with Piezoelectric Surface Waves," *IEEE Trans. MTT*, April 1973.

421. Bell, Jr., D. T., Holmes, J. D., and Ridings, R. V. "Application of Acoustic Surface-Wave Technology to Spread Spectrum Communications," *IEEE Trans. MTT*, April 1973.

422. Burnsweig, J., and Wooldridge, J. "Ranging and Data Transmission Using Digital Encoded FM-"Chirp" Surface Acoustic Wave Filters," *IEEE Trans. MTT*, April 1973.

423. Grant, P. M., Collins, J. H., Darby, B. J., and Morgan, D. P. "Potential Applications of Acoustic Matched Filters to Air-Traffic Control Systems," *IEEE Trans. MTT*, April 1973.

424. Hagon, P. J., Micheletti, F. B., Seymour, R. N., and Wrigley, C. Y. "A Programmable Surface Acoustic Wave Matched Filter for Phase-Coded Spread Spectrum Waveforms," *IEEE Trans. MTT*, April 1973.

425. Morgan, D. P., and Sutherland, J. G. "Generation of Pscudonoise Sequences Using Surface Acoustic Waves," *IEEE Trans. Micro. Th. Tech.*, April 1973.

426. Berni, A. J., and Gregg, W. D. "On the Utility of Chirp Modulation for Digital Signalling," *IEEE Trans. Comm.*, July 1973.

427. Milstein, L. B., and Das, P. K. "A SAW Implemented Wideband Interference Rejection Spread Spectrum System," *IEEE I.C.C. Proc.*, 1982.

5. DIGITAL TRANSMISSION / EDAC

51. Birch, J. N. "Design Considerations for PCM and Delta Mod Systems," Magnavox ASAO Report TP 68-2174.

52. Cahn, C. R. "Performance of Digital Phase-Modulation Communication Systems," *IRE Trans. Comm. Syst.*, May 1959.

53. Cahn, C. R. "Application of Wagner Code to Digital Data Transmission System," Magnavox Research Laboratories STN-24, December 1965.

54. Cahn, C. R. "On Tansmitting Digital Data Over an FM System," Magnavox Research Laboratories MX-TM-3091-70, November 1970.

55. Creveling, C. J. "Comparison of the Performance of PCM and PFM Telemetry Systems," NASA GSFC 1965.

56. Landon, H. J. "Sampling, Data Transmission and the Nyquist Rate," *IEEE Proc.*, October 1967.

57. Lender, A. "Correlative Digital Communication Techniques," *IEEE Trans. Comm. Tech.*, December 1964.

58. Lugannani, R. "Intersymbol Interference and Probability of Error in Digital Systems," *IEEE Trans. Info. Th.*, November 1969.

59. Oliver, B. M., Pierce, J. R., and Shannon, C. E. "The Philosophy of PCM," *IRE Proc.*, November 1948.

510. Salz, J., and Koll, V. G. "Experimental Digital Multilevel FM Modem," *IEEE Trans. Comm. Tech.*, June 1966.

511. Sanders, R. W. "Communication Efficiency Comparison of Several Communication Systems," *IRE Proc.*, April 1960.

512. Tjhung, T. T. "Band Occupancy of Digital FM Signals," *IEEE Trans. Comm. Tech.*, December 1964.

513. Wall, M. E., and Kuhn, T. G. "Wideband Data Transmission: Which Medium Is Best?" *Microwaves*, April 1966.

514. Whelan, J. W. "Analog-FM vs. Digital-PSK Transmission," *IEEE Trans. Comm. Tech.*, June 1966.

515. Wintz, P. A., and Totty, R. E. "Principles of Digital Communications," *Electro-Technology*, February 1967.

516. "Hybrid PCM Improves Noise Performance When Bandwidth is Limited," *Communications Designer's Digest*, November 1969.

517. "Frame Sync Study Points Up Optimum Parameters for PCM Transmission," *Communications Designer's Digest*, February 1970.

518. Viterbi, A. J., and Jacobs, I. M. "Advances in Coding and Modulation for Noncoherent Channels Affected by Fading Partial Band, and Multiple Access Interference," in *Advances in Communication Systems*, Vol. 4, Academic, New York, 1975.

519. Yao, K. "Error Probability of Asynchronous Spread-Spectrum Multiple-Access Communication—Part I: System Analysis," *IEEE Trans. Comm.*, Vol. COM-25, pp. 803–809, August 1977.

520. Wu, K. T. "Average Error Probability for DS-SSMA Communications: The Gram-Charlier Expansion Approach," in *Proc. 19th Annu. Allerton Conf. Comm. Contr. Comput.*, September 1981, pp. 237–246.

521. Shaft, P. D. "Low-Rate Convolutional Code Applications in Spread-Spectrum Communications," *IEEE Trans. Comm.*, August 1977.

522. Omura, J. K., and Levitt, B. K. "Coded Error Probability Evaluation for Antijam Communication System," *IEEE Trans. Comm.*, May 1982.

523. Geraniotis, E. A. "Coding for PSK and QPSK SFH/SSMA Communications via Fading Channels," *IEEE Milcom. Conf. Proc.*, October 1981.

524. Chandler, E. W., and Cooper, G. R. "Error Correction Coding Performance Bounds for Spread Spectrum Systems," *IEEE Milcom. Conf. Proc.*, October 1981.

525. Shacham, N. "Performance of ARQ with Sequential Decoding Over One-Hop and Two-Hop Radio Links," *IEEE Milcom. Conf. Proc.*, October 1981.

526. Hanlon, R. C., and Gardner, C. S. "Error Performance of Direct Sequence Spread Spectrum Systems on Non-Selective Fading Channels," *IEEE Trans. Comm.*, November 1979.

527. Hopkins, P. M., and Simpson, R. S. "Probability of Error in Pseudonoise (PN)-Modulated Spread Spectrum Binary Communications Systems," *IEEE Trans. Comm.*, April 1975.

528. German, Jr., E. H. "A Comment on "Probability of Error in PN-Modulated Spread Spectrum Binary Communications Systems," *IEEE Trans. Comm.*, June 1978.

529. Hirt, W., and Pasupathy, S. "Continuous Phase Chirp (CPC) Signals for Binary Data Communication," *IEEE Trans. Comm.*, June 1981.

6. DIRECT SEQUENCE

61. Cahn, C. R. "Spectrum Reduction of Biphase Modulated (2-PSK) Carrier," Magnavox Research Laboratories MX-TM-3103-71.

62. Gilchreist, C. E. "Pseudonoise System Lock-In," JPL Research Summary No. 36-9.

63. Nilsen, P. W. "PN Receiver Carrier and Code Tracking Performance," Magnavox Research Laboratories MX-TM-8-674-3043-68.

64. Sage, G. F. "Serial Synchronization of Pseudonoise Systems," *IEEE Trans. Comm. Tech.*, December 1964.

65. Ward, R. B. "Acquisition of Pseudonoise Signals by Sequential Estimation," *IEEE Trans. Comm. Tech.*, December 1965.

66. Zegers, L. E. "Common Bandwidth Transmission of Information Signals and Pseudonoise Synchronization Waveforms," *IEEE Trans. Comm. Tech.*, December 1968.

67. Lewis, M. "PLLs Upconvert Chirp Radar Signals," *Microwaves*, June 1981.

68. Schilling, D. L., Milstein, L. B., Pickholtz, R. L., and Brown, R. W. "Optimization of the Processing Gain of an M-ary Direct Sequence Spread Spectrum Communication System," *IEEE Trans. Comm.*, August 1980.

69. Pursley, M. B. "Performance Evaluation for Phase-Coded Spread-Spectrum Multiple-Access Communications—Part I: System Analysis," *IEEE Trans. Comm.*, Vol. COM-25, pp. 795–799, August 1977.

610. Garber, F. D. "Analysis of Generalized Quadriphase Spread-Spectrum Communication," *Natl. Telec. Conf. Proc.*, November 1980.

611. Milstein, L. B., Pickholz, R. L., Schilling, D. L., and Brown, R. "Optimization of the Processing Gain of an M-ary Direct Sequence Spread Spectrum Communication System," *Int. Conf. Comm. Proc.*, June 1980.

612. Levitt, B. K. "On Direct Sequence Spread Spectrum Systems," *Natl. Telec. Conf. Proc.*, November 1980.

620. Low, J., and Waldstein, S. M. "A Direct Sequence Spread-Spectrum Modem for Wideband HF Channels," *IEEE Milcom. Proc. Conf.*, October 1981.

621. Schilling, D. L., Milstein, L. B., Pickholtz, R. L., and Brown, R. W. "Optimization of the Processing Gain of an M-ary Direct Sequence Spread Spectrum Communication System," *IEEE Trans. Comm.*, August 1980.

See also 16, 17, 110, 113, 114, 117, 119, 120, 121, 122, 26, 27, 28, 214, 215, 219, 220, 224, 225, 226, 229, 231, 232, 235, 236, 237, 238, 239, 33, 310, 318, 325, 331, 51, 58, 59.

7. FREQUENCY HOPPING

71. Abramson, N. "Bandwidth and Spectra of Phase- and Frequency-Modulated Waves," *IEEE Trans. Comm. Syst.*, December 1963.

72. Cahn, C. R. "Noncoherent Frequency Hop Sync Mode Performance," Magnavox Research Laboratories, STN-12, March 1964.

73. George, O. H. "Performance of Noncoherent M-ary FSK Systems With Diversity Under the Influence of Rician Fading," *IEEE Int. Conf. Comm.*, June 1968.

74. Huth, G. K. "Detailed Frequency Hopper Analysis," Magnavox Research Laboratories, STN-29, August 1966.

75. Kaplan, A. "Detection and Analysis of Frequency Hopping Radar Signals," *Sylvania Elect. Syst.* W.D.L. Mountain View, California.

76. Schreiber, H. H. "Self-Noise of Frequency Hopping Signals," *IEEE Trans. Comm. Tech.*, October 1969.

77. Splitt, F. G. "Combined Frequency and Time-Shift Keyed Transmission Systems," *IEEE Trans. Comm. Syst.*, December 1963.

78. Thomas, C. M. "A Matched Filter Concept for Frequency Hopping Synchronization," TRW 10C 7353.6-05.

79. Malm, R., and Schreder, K. "Fast Frequency Hopping Techniques," *Proc. Symp. Spr. Spec. Comm.*, March 1973.

710. Nossen, E. J. "Fast Frequency Hopping Synthesizer," *Proc. Symp. Sp. Spec. Comm.*, March 1973.

711. Henry, P. S. "Spectrum efficiency of a frequency-hopped-DPSK spread spectrum mobile radio system," *IEEE Trans. Veh. Tech.*, Vol. VT-28, November 1979.

712. Edell, J. D. "Wideband, Noncoherent, Frequency-Hopped Waveforms and their Hybrids in Low Probability-of-Intercept Communications," Naval Res. Lab., Washington, DC, NRL Rep. 8025, November 8, 1976.

713. Goodman, D. J., Henry, P. S., and Prabhu, V. K. "Frequency-Hopped Multilevel FSK for Mobile Radio," *Bell Syst. Tech. J.*, Vol. 59, pp. 1257–1275, September 1980.

714. Pawula, R. F., and Mathis, R. F. "A Spread Spectrum System with Frequency Hopping and Sequentially Balanced Modulation—Parts One and Two," *IEEE Trans. Comm.*, May 1980.

715. Simon, M. K., Huth, G. K., and Polydoros, A. "Differentially Coherent Detection of QASK for Frequency-Hopping Systems, Parts I and II," *IEEE Trans. Comm.*, January 1982.

716. Yue, O.-C. "Performance of Frequency Hopping Multiple-Access Multilevel FSK Systems with Hard-Limited and Linear Combining," *IEEE Trans. Comm.*, November 1981.

717. Simon, M. K., and Polydoros, A. "Coherent Detection of Frequency-Hopped Quadrature Modulations in the Presence of Jamming–Parts I and II," *IEEE Trans. Comm.*, November 1981.

718. Milstein, L. B., Pickholtz, R. L., and Schilling, D. L. "Optimization of the Processing Gain of an FSK-FH System," *IEEE Trans. Comm.*, Vol. COM-28, pp. 1062–1079, July 1980.

719. Omura, J. K., Levitt, B., and Stokey, R. "FH/MFSK Performance in a Partial Band Jamming Environment," *IEEE Trans. Comm.*, to be published.

720. Avidor, D. "Anti-Jam Analysis of Frequency Hopping M-ary FSK Communication Systems in HF Rayleigh Fading Channels," Doctoral dissertation, School Eng. Appl. Sci., University of California, Los Angeles, 1981.

721. Sarwate, D. V., and Pursley, M. B. "Hopping Patterns for Frequency-Hopped Multiple-Access Communication," in *Proc. 1978 IEEE Int. Conf. Comm.*, Vol. 1, pp. 7.4.1–7.4.3.

722. Henry, P. S. "Spectrum Efficiency of a Frequency-Hopped-DPSK Spread Spectrum Mobile Radio System," *IEEE Trans. Veh. Tech.*, Vol. VT-28, pp. 327–329, November 1979.

723. Yue, O. C. "Hard-Limited Versus Linear Combining for Frequency Hopping Multiple-Access Systems in a Rayleigh Fading Environment," *IEEE Trans. Veh. Tech.*, Vol. VT-30, pp. 10–14, February 1981.

724. Nettleton, R. W., and Cooper, G. R. "Performance of a Frequency-Hopped Differentially Modulated Spread-Spectrum Receiver in a Rayleigh Fading Channel," *IEEE Trans. Veh. Tech.*, Vol. VT-30, pp. 14–29, February 1981.

725. Geraniotis, E. A., and Pursley, M. B. "Error Probability Bounds for Slow Frequency Hopped Spread-Spectrum Multiple Access Communications over Fading Channels," in *Proc. 1981 IEEE Int. Conf. Comm.*

726. Budreau, A. J., Slobodnick, Jr., A. J., and Carr, P. H. "Fast Frequency Hopping Achieved With SAW Synthesizers," *Microwave Journal*, February 1982.

727. Ribchester, E. "Frequency Hopping Techniques Vary With Frequency," *Microwaves & RF*, March 1983.

728. Sussman, S. M., and Kotiveeriah, P. "Partial Processing Satellite Relays for Frequency Hop Antijam Communications," *IEEE Trans. Comm.*, August 1982.

729. Elhakeem, A. K. "Overall SNR Optimization of a FH/MFSK Pulse Code and Adaptive Data Modulation Systems in Mixed Jamming," *IEEE I.C.C. Proc.*, 1982.

730. Elnoubi, S. M. "Error Rate Performance of Frequency Hopped MSK Spread Spectrum Mobile Radio System With Differential Detection," *I.C.C. Proc.*, 1982.

731. Blanchard, J. E. "Performance of M-Ary FSK/FH Against Optimum Multitone Jamming," *I.C.C. Proc.*, 1982.

732. Niyonizeye, C., Lecours, M. and Huynh, H. T. "Address Assignment in a Multiple Access FH-FSK System," *I.C.C. Proc.*, 1982.

733. Muammar, R., and Gupta, S. C. "Performance of a Frequency-Hopped Multilevel FSK Spread Spectrum Receiver in a Rayleigh Fading and Log-Normal Shadowing Channel," *I.C.C. Proc.*, 1982.

734. Mizuno, M. "Randomization Effect of Errors by Means of Frequency Hopping Techniques in a Fading Channel," *IEEE Trans. Comm.*, May 1982.

735. Pursley, M. B., and Sarwate, D. V. "New Results on Frequency Hop, Spread-Spectrum, Multiple Access Communications," *Natl. Telec. Conf. Proc.*, November 1980.

736. Jacobs, I. M. "Dama-Frequency Hopping and Pre-Correction for a Processing Satellite," *Int. Conf. Comm. Proc.*, June 1980.

737. Lindsey, W. C., Beiderman, L., and Sherwin, R. P. "Coding and Modulation Tradeoffs for Frequency-Hopped Channels," *Int. Conf. Comm. Proc.*, June 1980.

738. Dixon, R. C. "Frequency Hopping Synthesizers Employing Conventional Commercially-Available Integrated Circuits, *ITC Conf. Proc.*, October 1981.

739. Haskell, B. G. "Computer Simulation Results on Frequency Hopped MFSK Mobile Radio-Noiseless Case," *Natl. Telec. Conf. Proc.*, November 1980.

See also 12, 14, 15, 16, 18, 110, 111, 113, 114, 117, 124, 29, 216, 217, 218, 219, 220, 226, 33, 310, 315, 323, 325, 332, 336, 510, 512, and Frequency Synthesis.

8. FREQUENCY SYNTHESIS

81. Barnum, J. L. "A Multioctave Microwave Synthesizer," *The Microwave Journal*, October 1970.

82. Burnell, R. W. "Phase-Lock Frequency Synthesizer," TRW 10C 7325.2-99, July 1968.

83. Blachowicz, L. F. "Dial Any Channel to 500 MHz," *Electronics*, May 2, 1966.

84. Davis, Jr., M. G., "Phase-Lock Frequency Synthesizer Equivalent to Lincoln Labs' Frequency Synthesizer," TRW 10C 7325-18, April 22, 1968.

85. Delaune, J. "MTTL and MECL Avionics Digital Frequency Synthesizers," Motorola AN-532, July 1970.

86. Hekemian, N. C. "Digital Frequency Synthesizers," *Frequency*, July/August 1967.

87. Illingworth, L. "Digital Methods Synthesize Frequency," *Electronics Design*, May 23, 1968.

88. Kahn, R. S. "Frequency Synthesizer Survey," Magnavox Research Laboratories MX-TM-8-672-3009-67, February 1967.

89. Koeper, B. "Shift Frequency Automatically Without Transients," *EDN*, November 11, 1968.

810. Lohrmann, D. R., and Sills, A. R. "Cut Synthesizer Current Consumption," *Electronic Design*, December 5, 1968.

811. Maag, R. A. "Spurious Output and Acquisition Time of an Indirect Frequency Synthesizer," TRW 10C 7352.1-20, February 1969.

812. Noordanus, J. "Frequency Synthesizers—A Survey of Techniques," *IEEE Trans. Comm. Tech.*, April 1969.

813. Renshler, E., and Welling, B. "An Integrated Circuit Phase-Locked Loop Digital Frequency Synthesizer," Motorola AN-463, March 1963.

814. Stone, Jr. R. R., and Hastings, H. F. "A Novel Approach to Frequency Synthesis," *Frequency*, September/October 1963.

815. "Modern Methods of Frequency Synthesis," *Antekna Tech. Bull.* 20-02.

See also Frequency Hopping.

9.   INFORMATION TRANSMISSION

91. T. Berger, "Optimum PAM Compared with Information-Theoretic Bounds," *NEREM Record*, 1966.

92. Cahn, C. R. "Theoretical Comparisons of Analog and Digital Modulations for Voice," Magnavox Research Laboratories STN-23, January 1966.

93. Cohen, W. "Signal-to-Noise Ratio in Modulated Systems," Parts 1 and 2, *Electro-Technology*, October 1965, January 1966.

94. Develet, J. A. "Coherent FDM/FM Telephone Communication," *IEEE Proc.*, September 1958.

95. Diaz, C. C., and Norvell, B. R. "Pulse Modulation Intelligibility," *Electro-Technology*, March 1966.

96. Gardenhire, L. W. "Selecting Sample Rates," *ISA Journal*, April 1964.

97. Hupert, J. J. "Frequency-Modulation Techniques," *Electro-Technology*, February 1965.

98. Hupert, J. J. "Distortion in Angle-Modulation Networks," *Electro-Technology*, September 1965.

99. Landau, H. J. "Sampling, Data Acquisition, and the Nyquist Rate," *IEEE Proc.*, October 1967.

910. Miller, G. A., and Licklider, J. C. R. "The Intelligibility of Interrupted Speech," *J. Acous. Soc. Amer.*, March 1950.

911. Porter, J. A. "Prediction of Speech Intelligiblity Over Voice Communications Systems," NASA EB-65-R2001.

912. Schuchman, L. "Dither Signals and Their Effect on Quantization Noise," *IEEE Trans. Comm. Tech.*, December 1964.

913. Walsh, C. P. "Frequency-Modulation Principles," *Electro-Technology*, January 1965.

914. Weiner, D. D., and Leon, B. J. "The Quasi-Stationary Response of Linear Systems to Modulated Waveforms," *IEEE Proc.*, June 1965.

915. Young, J. A. "Design Parameters Associated with Multilevel Transmission Systems," *NEREM Record*, 1966.

916. Jayant, N. S. "Characteristics of a Delta Modulator," *IEEE Proc.*, March 1971.

917. Hirt, W. and Pasupathy, S. "Continuous Phase Chirp (CPC) Signals for Binary Data Communication—Parts I and II," *IEEE Trans. Comm.*, June 1981.

918. Shimbo, O., and Celebiler, M. "The Probability of Error Due to the Intersymbol Interference and Gaussian Noise in Digital Communication Systems," *IEEE Trans. Comm.*, Vol. COM-19, pp. 113–119, April 1971.

919. Anderson, P. H., Hsu, F. M., and Sandler, M. M. "A New Adaptive Modem for Long Haul HF at Data Rates > 1BPS/Hz," *IEEE Milcom. Conf. Proc.*, October 1981.

920. Crozier, S., Lyons, R., and Tiedmann, K. "Wideband Adaptive Serial Data Modem for HF Communications," *IEEE Milcom. Conf. Proc.*, October 1981.

921. Das, P., and Shklarsky, D. "The Use of the Hilbert Transform to Double the Information Rate of a SAW Implement Direct Sequence Spread Spectrum System," *Natl. Telec. Conf. Proc.*, November 1980.

922. Dhar, S., and Perry, D. "Equalized Megahertz-Bandwidth HF Channels for Spread Spectrum Communications," *IEEE Milcom. Conf. Proc.*, October 1981.

923. Otto, J. "Chirping RPV Data Links for ECM Protection," *Microwaves Magazine*, December 1974.

924. Hammer, A., and Schaefer, D. J. "Performance Analysis of M-ary Code Shift Keying in Code-Division Multiple Access Systems," *IEEE I.C.C. Proc.*, 1982.

10. MODULATION / DEMODULATION

101. Adashko, J. G. "Design of Phase Sensitive Demodulators," *Electronic Design*, April 12, 1962.

102. Baghdady, E. J. "On the Noise Threshold of Conventional FM and PM Demodulators," *IEEE Proc.*, September 1963.

103. Cahn, C. R. "Phase Detector Characteristic's Effect on Signal-to-Noise Ratio, General Theory," STL 10C, April 1962.

104. Cahn, C. R. "Analysis of Correlation System with Receiver Filter," Magnavox Research Laboratories, STN-20, September 1965.

105. Cahn, C. R. "Error Bound for Non Optimally Demodulated Channel," Addendum to Magnavox Research Laboratories, STN-27, June 1966.

106. Chandler, J. P. "IC's Fill Need for Low-Drift Phase-Shift Keyed Detector," *EDN* January 15, 1969.

107. Channell, E. "The Semiconductor Ring Modulator," *EDN*, January 1964.

108. Cooper, P. W. "Correlation Functions for the Random Binary Wave," *IEEE Trans. Comm. Syst.*, December 1963.

109. Costas, J. P. "Synchronous Communications," *IRE Proc.*, December 1956.

1010. Datillo, J. A. "Incremental Phase Modulator," Magnavox Research Laboratories MX-TR-8-765-2002, March 1967.

1011. Develet, Jr., J. A., "An Analytic Approximation of Phase-Lock Receiver Threshold." *TRW* 9332.6-2, April 1962.

1012. Eckstrom, J. L. "Coherent Matched Filter Detection of Quadratically Phase-Distorted Carrier-Band Pulses, with Application to Transionospheric Signalling," *IEEE Trans. Space Elect. Telem.*, March 1964.

1013. Frankle, J. T. "Threshold Performance of Analog FM Demodulators," *RCA Review*, December 1964.

1014. Gross, T. A. O. "Increasing the Dynamic Range of AM Detectors," *IEEE*, November 1963.

1015. Hess, D. T. "Equivalence of FM Threshold Extension Receivers," *IEEE Trans. Comm. Tech.*, October 1968.

1016. Hirsch, S. R. "Convolution—A Graphical Interpretation," *EDN*, February 1967.

1017. Judge, W. J. "A Passive Correlator of Arbitrary Large T-W Product," Magnavox Research Laboratories R-502, April 1962.

1018. Katz, E. H., and Schreiber, H. H. "Design of Phase Discriminators," *Microwaves*, August 1965.

1019. Kurth, C. "Analysis and Synthesis of Diode Modulators," *Frequency*, January/February 1966.

1020. Langenthal, I. M. "Correlation and Probability Analysis," Signal Analysis Industries Corp., TB14.

1021. Levine, R. I. "Correlation—Theory and Practice," *Electronic Products*, November 1963.

1022. Mouw, R. B., and Fukuchi, S. M. "Broadband Double Balanced Mixer/Modulators," Parts I and II, *The Microwave Journal*, March/May 1969.

1023. Ogar, C. W. "Putting Diode Modulators to Work," *Electronic Industries*, July 1961.

1024. Polson, J. H. "Wideband Digital Data Link Quadriphase Demodulator," TRW 7322.05-202, December 1968.

1025. Stoll, A. W. "The Electronic Correlator," *Electronic Industries*, August 1965.

1026. Stone, M. S. "Analytical Considerations for Multiphase Modulation and Demodulation," TRW 7323.2-245, April 1968.

1027. Stone, R. B., and White, G. M. "Correlation Detector Detects Voice Fundamental," *Electronics*, November 22, 1963.

1028. Urbach, Jr., W. G., "Produce Fast and Clean RF Pulse," *Electronic Design*, September 1966.

1029. Vizkanta, V. Z. "I-Q Loop Receivers," TRW 10C 7323.2-289, September 1968.

1030. Voyce, A. M. "Ultimate Sensitivity of a Detector," *EEE*, January 1963.

1031. Kivett, J. A., and Bowers, G. F. "A Wideband Modem for Command and Control of Remote Vehicles," *Proc. Symp. Sp. Spec. Comm.*, March 1973.

1032. Lowry, R. B. "PSK-FSK Spread Spectrum Modulation/Demodulation," *Proc. Symp. Sp. Spec. Comm.*, March 1973.

1033. Schiff, M. L., and Dilley, D. M. "A Surface Acoustic Wave Spread Spectrum Modem," *Proc. Symp. Sp. Spec.*, March 1973.

1034. Bello, P. "Demodulation of a phase-modulated noise carrier," *IRE Trans. Info. Th.*, Vol. IT-7, pp. 19–27, January 1961.

1035. Simon, M. K. "Two-Channel Costas Loop Tracking Performance for UQPSK Signals with Arbitrary Data Formats," *IEEE Trans. Comm.*, September 1981.

1036. Simon, M. K., Huth, G. K., and Polydoros, A. "Differentially Coherent Detection of QASK for Frequency Hopping Systems—Parts I and II," *IEEE Trans. Comm.*, January 1982.

1037. Lindsey, W. C., and Simon, M. K. "Optimum Design for Performance of Costas Receivers Containing Soft Bandpass Limiters," *IEEE Trans. Comm.*, August 1977.

1038. Cahn, C., Leimer, D., Marsh, R., Huntowski, F., and LaRue, G. "Software Implementation of a PN Spread Spectrum Receiver To Accommodate Dynamics," *IEEE Trans. Comm.*, August 1977.

1039. Krasner, N. F. "Optimal Detection of Digitally Modulated Signals," *IEEE Trans. Comm.*, May 1982.

1040. Polydoros, A., and Weber, C. L. "Optimal Detection Considerations for Low Probability of Intercept," *IEEE Milcom. Conf. Proc.*, October 1981.

1041. Engeler, W., Baertsch, R. D., and Parks, H. G. "A Binary-Analog Correlator," *IEEE Milcom. Conf. Proc.*, October 1981.

1042. Glisic, S. G. "Noncoherent Demodulation of Spread Spectrum Signals Using Charge Coupled Devices," *IEEE Milcom. Conf. Proc.*, October 1981.

1043. Proakis, J. G., Anderson, P. H., and Ketchum, J. W. "Receiver Processing Techniques for DS Spread Spectrum in Dispersive Channels," *IEEE Milcom. Conf. Proc.*, October 1981.

1044. McDenzie, P. F., Coomber, D. B., and Hutchinson, B. H. "Design and Performance of a SAW-Based MFSK Demodulator," *IEEE Milcom. Conf. Proc.*, October 1981.

1045. Simon, M., and Lindsey, W. C. "Optimum Performance of Suppressed Carrier Receivers With Costas Loop Tracking," *IEEE Trans. Comm.*, February 1977.

1046. Messerschmitt, D. G. "Frequency Detectors for PLL Acquisition in Timing and Carrier Recovery," *IEEE Trans. Comm.*, September 1979.

1047. Holmes, J. K., and Biederman, L. "Delay-Lock-Loop Mean Time to Lose Lock," *IEEE Trans. Comm.*, November 1978.

1048. Amoroso, F. "Rotating Phase Demodulation," Hughes Aircraft Company 74/1461.20/40.

1049. Gronemeyer, S. A., and McBride, A. L. "MSK and Offset QPSK Modulation," *IEEE Trans. Comm.*, August 1976.

1050. Amoroso, F. "Pulse and Spectrum Manipulation in the Minimum (frequency) Shift Keying (MSK) Format," *IEEE Trans. Comm.*, March 1976.

1051. Pursley, M. B., Garber, F. D., and Lehnert, J. S. "Analysis of Generalized Quadriphase Spread-Spectrum Communications," *Proc. IIII Int. Conf. Comm.*, June 1980.

1052. La Rosa, R., Marynowski, T. J., and Henrich, K. J. "A Simple Modulator For Sinusoidal Frequency Shift Keying," *IEEE Trans. Comm.*, May 1982.

1053. Amoroso, F. "Enhancing Direct Sequence Pseudonoise (DSPN) Modulation by Increasing the Bandwidth per Chip Rate," *IEEE Milcom. Conf. Proc.*, October 1981.

1054. Stynes, P., Lovell, J., Sunkenberg, H., and Girodano, A., "Waveform Alternatives for a Frequency-Hopped Simulcast Radio Network," *IEEE Milcom. Conf. Proc.*, October 1981.

1055. Dixon, R. C. "Simplified Approach to Direct Sequence Carrier Modulation and its Selection," *IEEE Milcom. Conf. Proc.*, October 1981.

1056. Gorur, R., Payzin, A., and Oranc, H. S. "Multi-Subchannel Spread Spectrum Modulation," *IEEE Milcom. Conf. Proc.*, October 1981.

1057. Nettleton, R. W. "Addressing and Modulation for Frequency Hopping Multiple Access," *IEEE Milcom. Conf. Proc.*, October 1981.

1058. Bucher, T. T. N. "Spectrum Occupancy of Pulsed FSK," *IEEE Milcom. Conf. Proc.*, October 1981.

1059. Lindsey, W. C., Beiderman, L., and Sherwin, R. P. "Coding and Modulation Tradeoffs for Frequency-Hopped Channels," *Int. Conf. Comm. Proc.*, June 1980.

1060. Amoroso, F. "The Bandwidth of Spread Spectrum Signals," *ITC Conf. Proc.*, October 1981.

1061. Amoroso, F. "Enhancing Direct Sequence Pseudonoise (DSPN) Modulation by Increasing the Bandwidth Per Chip Rate," *IEEE Milcom. Conf. Proc.*, October 1981.

1062. Yazdani, H., Feher, K., and Steenaart, W. "Constant Envelope Bandlimited BPSK Signal," *IEEE Trans. Comm.*, June 1980.

1063. Bayless, J. W., and Pedersen, R. D. "Efficient Pulse Shaping Using MSK or PSK Modulation," *IEEE Trans. Comm.*, June 1979.

1064. Simon, M. K. "A Generalization of Minimum-Shift-Keying (MSK)-Type Signaling Based Upon Input Data Symbol Pulse Shaping," *IEEE Trans. Comm.*, August 1976.

1065. Rabzel, M., and Pasupathy, S. "Spectral Shaping in Minimum Shift Keying (MSK)-Type Signals," *IEEE Trans. Comm.*, January 1978.

1066. Amoroso, F. "The Use of Quasi-Bandlimited Pulses in MSK Transmission," *IEEE Trans. Comm.*, October 1979.

1067. Morais, D. H., and Feher, K. "The Effects of Filtering and Limiting on the Performance of QPSK, Offset QPSK, and MSK Systems," *IEEE Trans. Comm.*, December 1980.

11. MULTIPLE ACCESS

1101. Bedrosian, E., Feldman, N., Northrop, G., and Sollfrey, W. "Multiple Access Techniques for Communications Satellites: I. Survey of the Problem," Rand Report RM-4298-NASA, September 1964.

1102. Harman, W. H. "Multiple Access and Antijam Capabilities of Direct-Sequence Quadriphase on a Linear Channel," TRW 10C 7323.2-256, June 1968.

1103. Udalov, S. "Pseudo-Random Pulse Position Multiplexing for Random Access," Magnavox Research Laboratories STN-8, January 1964.

1104. Udalov, S. "Threshold Criterion for Virtual Carrier Multiplexing," Magnavox Research Laboratories STN-10, January 1964.

1105. Pursley, M. B., and Sarwate, D. V. "Performance Evaluation for Phase-Coded Spread Spectrum Multiple-Access Communication—Part II. Code Sequence Analysis," *IEEE Trans. Comm.*, Vol. COM-25, pp. 800–803, August 1977.

1106. Aein, J. M., and Turner, R. D. "Effect of Co-Channel Interference on CPSK Carriers," *IEEE Trans. Comm.*, Vol. COM-21, pp. 783–790, July 1973.

1107. Laforgia, D., Luvison, A., and Zingarelli, V. "Exact Bit Error Probability with Application to Spread-Spectrum Multiple Access Communications," in *Proc. IEEE Int. Conf. Comm.*, June 1981, Vol. 4, pp. 76.5.1–76.5.5.

1108. Raychaudhuri, D. "Performance Analysis of Random Access Packet-Switched Code Division Multiple Access Systems," *IEEE Trans. Comm.*, June 1981.

1109. Pursley, M. B. "Performance Evaluation for Phase-Coded Spread Spectrum Multiple-Access Communication—Part I: System Analysis," *IEEE Trans. Comm.*, Vol. COM-25, pp. 795–799, August 1977.

1110. Rowe, H. E. "Bounds on the Number of Users in Spread-Spectrum Systems," in *Proc. Natl. Telec. Conf.*, Nov. 30–Dec. 4, 1980, pp. 69.6.1–69.6.6.

1111. Borth, D., Pursley, M., Sarwate, D., and Stark, W. "Bounds on Error Probability for Direct Sequence Spread-Spectrum Multiple-Access Communications," in *Proc. MIDCON Conf.*, November 1979, paper 15/1, pp. 1–14.

1112. Laforgia, D., Luvison, A., and Zingarelli, V. "Exact Bit Error Probability with Application to Spread-Spectrum Multiple Access Communications," in *Proc. IEEE Int. Conf. Comm.*, June 1981, Vol. 4, pp. 76.5.1–76.5.5.

1113. Geraniotis, E. A., and Pursley, M. B. "Error Probability for Binary PSK Spread-Spectrum Multiple Access Communications," in *Proc. Conf. Info. Sci., Syst.*, The Johns Hopkins University, Baltimore, MD, March 1981, pp. 238–244.

1114. Pursley, M. B. "Spread Spectrum Multiple-Access Communications," in *Multi-User Communication Systems*, G. Longo, Ed., Springer-Verlag, New York, 1981, pp. 139–199.

1115. Pursley, M. B., Sarwate, D. V., and Stark, W. E. "On the Average Probability of Error for Direct-Sequence Spread-Spectrum Multiple-Access Systems," in *Proc. 1980 Conf. Info. Sci. Syst.*, Princeton University, Princeton, NJ March 1980, pp. 320–325.

1116. Wu, K.-T., and Neuhoff, D. L. "Average Error Probability for Direct Sequence Spread Spectrum Multiple Access Communication Systems," in *Proc. 18th Annu. Allerton Conf. Comm. Comput.*, October 1980, pp. 359–368.

1117. Yao, K. "Error Probability of Asynchronous Spread Spectrum Multiple Access Communication Systems," *IEEE Trans. Comm.*, Vol. COM-25, pp. 803–809, August 1977.

1118. Yao, K. "Performance of Offset Quadriphase Spread Spectrum Multiple Access Communication," *IEEE Trans. Comm.*, Vol. COM-29, pp. 305–314, March 1981.

1119. Wu, K.-T. "Direct Sequence Spread-Spectrum Communications: Applications to Multiple Access and Jamming Resistance," Ph.D. dissertation, University of Michigan, Ann Arbor, 1981.

1120. Pursley, M. B., Sarwate, D. V., and Stark, W. E. "Spread Spectrum Multiple-Access Communications—Par I: Upper and Lower Bounds," *IEEE Trans. Comm.*, May 1982.

1121. Anderson, D. R., and Wintz, P. A. "Analysis of a Spread-Spectrum Multiple Access System with a Hard Limiter," *IEEE Trans. Comm. Tech.*, Vol. COM-17, pp. 285–290, 1969.

See also 111, 115, 116, 121, 124, 26, 29, 214, 220, 218, 222, 31, 310, 313, 315, 515, 66, 94.

1122. Mohanty, N. C. "Spread Spectrum and Time Division Multiple Access Satellite Communications," *IEEE Trans. Comm.*, August 1977.

1123. Kochevar, H. J. "Spread Spectrum Multiple Access Communications Experiment Through a Satellite," *IEEE Trans. Comm.*, August 1977.

1124. Rowe, H. E. "Bounds on the Number of Signals with Restricted Cross Correlation," *IEEE Trans. Comm.*, May 1982.

1125. Geraniotis, E. A., and Pursley, M. B. "Error Probability for Direct-Sequence Spread-Spectrum Multiple Access Communications—Part II: Approximations," *IEEE Trans. Comm.*, May 1982.

1126. Geraniotis, E. A., and Pursley, M. B. "Error Probabilities for Slow Frequency Hopped Spread Spectrum Multiple-Access Communications Over Fading Channels," *IEEE Trans. Comm.*, May 1982.

1127. Huth, G., Batson, B., and Weber, C. "Performance of Code Division Multiple Access Systems," *Natl. Telec. Conf. Proc.*, November 1980.

1128. Pursley, M. B., and Geraniotis, E. A. "Performance Analysis for Fast Frequency Hopped Spread-Spectrum Multiple—Access Communications," *Natl. Telec. Conf. Proc.*, November 1980.

1129. Aein, J. M., and Pickholtz, R. L. "A Simplified Unified Phasor Analysis for PN Multiple Access Limiting Repeaters," *IEEE Trans. Comm.*, May 1982.

1130. Healy, T. J. "The Capacity Region of a Spread Spectrum Multiple-Access Channel," *IEEE Milcom. Conf. Proc.*, October 1981.

1131. Mayk, I., and Henderson, H. "Network Synchronization Initialization Performance of a TDMA System in the Ground Environment," *IEEE Milcom. Conf. Proc.*, October 1981.

1132. Weber, C. L., Huth, G., and Batson, B. "Performance Considerations of Code Division Multiple Access Systems," *IEEE Trans. Veh. Tech.*, February 1981.

1133. Garber, F. D., and Pursley, M. B. "Performance of Offset Quadriphase Spread-Spectrum Multiple-Access Communications," *IEEE Trans. Comm.*, March 1981.

1134. Musa S. A., and Wasylkiwskyj, W. "Co-Channel Interference of Spread Spectrum Systems in a Multiple User Environment," *IEEE Trans. Comm.*, October 1978.

1135. Aein, J. M. "Multiple Access to a Hard-Limiting Communication-Satellite Repeater," *IEEE Trans. Space Electron. Telem.*, Vol. SET-10, December 1964.

1136. Schwartz, J. W., Aein, J. M., and Kaiser, J. "Modulation Techniques for Multiple Access to a Hard-Limiting Satellite Repeater," *Proc. IEEE*, Vol. 54, May 1966.

1137. Roefs, H. F. A. "Binary Sequences for Spread Spectrum Multiple Access Communication," Ph.D. dissertation, Dept. Elec. Eng., University of Illinois, Urbana, and Coord. Sci. Lab., Rep. R-785, August 1977.

1138. Aein, J. M., and Pickholtz, R. L. "A Simple Unified Phasor Analysis for PN Multiple Access to Limiting Repeaters," this issue, pp. 1018–1026.

1139. Einarsson, G. "Address Assignment for a Time-Frequency-Coded, Spread-Spectrum System," *Bell Syst. Tech. J.*, Vol. 59, pp. 1241–1255, September 1980.

12. PHASE LOCK

1201. Davis, Jr., M. G. "On the Measurement of Phase Detector Scale Factor," TRW 10C 7325-17, April 1968.

1202. Develet, Jr., J. A. "Threshold Criterion for Phase-Lock Demodulation," *IEEE Proc.*, February 1963.

1203. Frankle, J. T., and Klappen, J. "Principles of Phase Lock and Frequency Feedback," *RCA*, October 1967.

1204. Frazier, J. P., and Page, J. "Phase-Lock Loop Frequency Acquisition Study," *IRE Trans. Space Elect. Telem.*, September 1962.

1205. Gruen, W. J. "Theory of AFC Synchronization," *IRE Proc.*, August 1953.

1206. Hoffman, L. A. "Receiver Design and the Phase-Lock Loop," *IEEE Elect. Space Exploration Lecture Series*, August 1963.

1207. Jaffe, R., and Rechtin, E. "Design and Performance of Phase-Lock Circuits Capable of Near-Optimum Performance Over a Wide Range of Input Signal and Noise Levels," *IRE Trans. Info. Th.*, March 1955.

1208. Janky, J. M. "Nomograms Simplify Phase-Lock-Loop Analysis," *Microwaves*, March 1970.

1209. Lohrmann, D. R. "Designing Sampling Phaselock Loops," *Electronic Design*, November 8, 1970.

1210. McAleer, H. T. "A New Look at the Phase-Locked Oscillator," *IRE Trans.*, June 1959.

1211. Nash, G. "Phase-Locked Loop Design Fundamentals," Motorola AN-535, October 1970.

1212. Preston, G. W., and Tellier, J. C. "The Lock-In Performance of an AFC Circuit," *IRE Proc.*, February 1953.

1213. Sanneman, R. W., and Rowbotham, J. R. "Unlock Characteristics of the Optimum-Type II Phase-Locked Loop," *IEEE Trans. Aerosp. Nav. Elect.*, March 1964.

1214. Tausworthe, R. C. "Theory and Practical Design of Phase-Locked Receivers," *JPL Tech. Report* 32-819, February 1966.

1215. Van Trees, H. L. "Functional Techniques for the Analysis of the Nonlinear Behavior of Phase-Locked Loops," *IEEE Proc.*, August 1964.

1216. Viterbi, A. J. "Acquisition and Tracking Behavior of Phase-Locked Loops," *JPL Ext. Pub.* 673, July 1959.

1217. Viterbi, A. J. "Phase-Locked Loop Dynamics in the Presence of Noise by Fokker-Planck Techniques," *IEEE Proc.*, December 1963.

1218. Young, W. D. "Receiver Lock Loop Theory and Application," *STL* 10*C* 9331.2-*258, January* 1962.

1219. "Non-Linear PLL's Flunk Test on Vulnerability to Noise and Interference," *Communications Designer's Digest*, November 1969.

1220. Lindsey, W. C., and Simon, M. K. "The Effect of Loop Stress on the Performance of Phase-Coherent Communication Systems," *IEEE Trans. Comm. Tech.*, October 1970.

1221. Didday, R. L., and Lindsey, W. C. "Subcarrier Tracking Methods and Communication System Design," *JPL Report* 32-1317, August 1968.

1222. Lindsey, W. C., and Simon, M. K. "The Performance of Suppressed Carrier Tracking Loops in the Presence of Frequency Detuning," *IEEE Proc.*, September 1970.

1223. Lindsey W. C., and Tausworthe, R. C. "A Bibliography of the Theory and Application of the Phase-Lock Principle," *JPL Report* 32-1581, April 1, 1973.

13.   NAVIGATION / RANGING

1301. Richardson, R. J. "Optimum Transponder Design for Pseudonoise-Coded Ranging Systems, Weak Signals," *IEEE Trans. Aerosp. Elect. Syst.*, January 1972.

1302. Rihaczek, A. W., and Golden, R. M. "Resolution Performance of Pulse Trains with Large Time-Bandwidth Products," *IEEE Trans. Aerosp. Elect. Syst.*, July 1971.

1303. Horton, B. M. "Noise Modulated Distance Measuring Systems," *IRE Proc.*, May 1959.

1304. Frank, R. L., and Zadoff, S. "Phase-Coded Hyperbolic Navigation System," U.S. Patent 3 099 835, July 30, 1963 (filed May 31, 1956).

1305. Guanella, G. "Direction Finding System," U.S. Patent 2 166991, July 25, 1939 (filed in U.S. November 24, 1937; in Switzerland, December 1, 1936).

1306. Torino, F. "NAVSTAR Global Positioning System Development," *Natl. Telec. Conf. Proc.*, November 1980.

1307. Sansom, R. E. "Design and Performance Considerations for Military Aircraft and Orbital Satellite GPS Applications," *Natl. Telec. Conf. Proc.*, November 1980.

1308. VanLeuwen, A. V. "Space Shuttle GPS Implementation," *Natl. Telec. Conf. Proc.*, November 1980.

1309. Thompson, T. "GPS Utilization for Trident Missile Testing," *Natl. Telec. Conf. Proc.*, November 1980.

1310. Stonestreet, B. "GPS Missile Accuracy Evaluator for the MX Program," *Natl. Telec. Conf. Proc.*, November 1980.

1311. Hui, P. S. P., and Neily, Jr., C. M. "Modular Architecture and Adaptive Interface Design for a GPS Orbital Terminal," *Natl. Telec. Conf. Proc.*, November 1980.

1312. Nilsen, P. W. "Threshold Performance of a GPS Receiver for Shuttle Applications," *Natl. Telec. Conf. Proc.*, November 1980.

1314. Helms, C. W., and Logsdon, T. "Comparison Between the Navigation Capabilities of the Navstar GPS and Competing Systems," *IEEE Eascon. Proc.*, November 1981.

1315. Feit, L., and Domanico, P. T. "Navigation Signal Pseudo Random Noise Generator Design and Performance Verification," *IEEE Eascon. Proc.*, November 1981.

1316. Esposito, R. J., and Sawtelle, E. M. "FAA Tests on Navstar Z Set," *IEEE Eascon. Proc.*, November 1981.

1317. Euler, W. "GPS Provides a Revolutionary New Military Capability," *IEEE Eascon. Proc.*, November 1981.

1318. Kruh, P. "Build-up and Replacement of NAVSTAR/GPS and the 18 Satellite Constellation," *ITC Conf. Proc.*, October 1981.

1319. Ward, P. "An Inside View of Pseudo Range and Delta Range Measurements in an NAVSTAR GPS Receiver," *ITT Conf. Proc.*, October 1981.

1320. Martin, E. "Applications of GPS Phase Coherency," *ITC Conf. Proc.*, October 1981.

1321. Cnossen, R., and Cardall, J. D. "Civil Air Applications of Differential GPS," *ITC Conf. Proc.*, October 1981.

1322. Painter, J., Noe, P., and Rhyne, T. "New Comparisons Between C/A and Other Marine Navaids," *ITC Conf. Proc.*, October 1981.

See also 215, 219, 232, 325, 330.

14. RF EFFECTS

1401. Aein, J. M. "On the Output Power Division in a Captured Hard-Limiting Repeater," *IEEE Trans. Comm. Tech.*, June 1966.

1402. Benoit, A. "Signal Attenuation Due to Neutral Oxygen and Water Vapour, Rain and Clouds," *The Microwave Journal*, November 1968.

1403. Blackman, N. M. "The Output Signal-to-Noise Ratio of a Bandpass Limiter," *IEEE Trans. Aerosp. Elect. Sys.*, July 1968.

1404. Bussgang, J. J., and Leiten, M. "Analysis of Phase-Shift Transmission Through Fading Channels," *NEREM Record*, 1965.

1405. Cahn, C. R. "Effects of Phase and Amplitude Distortion on PSK Signal Demodulation," *Magnavox Research Laboratories* MX-TM-3099-71, March 1971.

1406. Davenport, Jr., W. B. "Signal-to-Noise Ratios in Band-Pass Limiters," *J. Appl. Phys.*, June 1953.

1407. Doyle, W. "Elementary Derivation for Bandpass Limiter S/N," *IEEE Trans. Info. Th.*, March 1962.

1408. Eckstrom, J. L. "Coherent Matched Filter Detection of Quadratically Phase Distorted Carrier-Band Pulses, with Application to Trans-Ionospheric Signalling," *IEEE Trans.* SET-10, No. 1, March 1963.

1409. Elliott, R. S. "Pulse Waveform Degradation Due to Dispersion in Waveguide," *IRE Trans. Micro. Th. Tech.*, October 1957.

1410. Jacobs, I. "The Effects of Video Clipping on the Performance of an Active Satellite PSK Communication System," *IEEE Trans. Comm. Tech.*, June 1965.

1411. Jones, J. J. "Filter Distortion and Intersymbol Interference Effects on QPSK," *Proc. Hawaii Conf. Syst. Sci.*, January 1970.

1412. Karr, P. R. "The Effect of Phase Noise on the Performance of Bi-Phase Communications Systems," TRW 07791-6080-R000, September 1966.

1413. Kirlin, R. L. "Hard-Limiter Intermodulation with Low Input Signal-to-Noise Ratio," *IEEE Trans. Comm. Tech.*, August 1967.

1414. Kwan, R. K. "The Effects of Filtering and Limiting a Double-Binary PSK Signal," *IEEE Trans. Aerosp. Elect. Syst.*, July 1969.

1415. Sevy, J. L. "The Effect of Limiting a Biphase or Quadriphase Signal Plus Interference," *IEEE Trans. Aerosp. Elect. Syst.*, May 1969.

1416. Shaft, P. D. "Limiting of Several Signals and Its Effect on Communication System Performance," *IEEE Trans. Comm. Tech.*, December 1965.

1417. Staras, H. "The Propagation of Wideband Signals Through the Atmosphere," *IRE Proc.* **49**, No. 7, July, 1961.

1418. Sunde, E. D. "Pulse Transmission by AM, FM, and PM in the Presence of Phase Distortion," *BSTJ*, March 1961.

1419. Weidner, M. Y. "Analysis of Spread Spectrum Multiple Access Systems Experiencing Linear Distortion," TRW 10C 7323.4-117, May 1968.

1420. "Digital Transmission Study Rates Filtering Effects on Spread-Spectrum PSK," *Communications Designer's Digest*, August 1969.

1421. Speiser, J. M., and Whitehouse, H. J. "Signal Design for Synchronization and Multipath Resolution in Delay-Spread Channels," *Proc. Symp. Sp. Spec. Comm.*, March 1973.

1422. Cahn, C. R., and Moore, C. R. "Bandwidth Efficiency for Digital Comm Via a Hard Limiting Channel," *ITC Proc.*, **VIII**, 1972.

1423. Kwon, S. Y., and Simpson, R. S. "Effect of Hard Limiting on a Quadrature PSK Signal," *IEEE Trans. Comm.*, July 1973.

1424. Le Fande, R. A. "Effects of Phase Non Linearities on a Phase Shift Keyed Pseudonoise/Spread Spectrum System" *IEEE Trans. Comm. Tech.*, October 1970.

1425. Yue, O. C. "Hard-Limited Versus Linear Combining for Frequency Multiple-Access Systems in a Rayleigh Fading Environment," *IEEE Trans. Veh. Tech.*, Vol. VT-30, pp. 10–14, February 1981.

1426. Nettleton, R. W., and Cooper, G. R. "Performance of a Frequency-Hopped Differentially Modulated Spread-Spectrum Receiver in a Rayleigh Fading Channel," *IEEE Trans. Veh. Tech.*, Vol. VT-30, pp. 14–29, February 1981.

1427. Borth, D. E., and Pursley, M. B. "Analysis of Direct-Sequence Spread-Spectrum Multiple-Access Communication over Rician Fading Channels," *IEEE Trans. Comm.*, Vol. COM-27, pp. 1566–1577, October 1979.

1428. Geraniotis, E. A., and Pursley, M. B. "Error Probability Bounds for Slow Frequency-Hopped Spread Spectrum Multiple Access Communications over Fading Channels," in *Proc. 1981 Int. Conf. Comm.*

1429. Milstein, L. V., and Schilling, D. L. "Performance of a Spread Spectrum Communication System Operating over a Frequency-Selective Fading Channel in the Presence of Tone Interference," *IEEE Trans. Comm.*, Vol. COM-30, pp. 240–247, January 1982.

1430. Milstein, L. B., and Schilling, D. L. "The Effect of Frequency Selective Fading on a Noncoherent FH-FSK System Operating with Partial-Band Interference," *IEEE Trans. Comm.*, May 1982.

1431. Arnstein, D. S. "Power Division in Spread Spectrum Systems with Limiting" *IEEE Trans. Comm.*, Vol. COM-27, pp. 574–582, 1979.

1432. Monsen, P. "Fading Channel Communications," *IEEE Comm. Mag.*, Vol. 18, pp. 16–25, January 1980.

1433. George, S. S. "Antenna Multiplexers for VHF Frequency Hopping Radio," *IEEE Milcom. Conf. Proc.*, October 1981.

1434. Sass, P. F. "Propagation Measurements for Spread-Spectrum Communications," *IEEE Milcom. Conf. Proc.*, October 1981.

1435. Ochsner, H. "Analysis of Pseudo-Noise Spread-Spectrum Communication over Randomly Time-Variant Channels without Assuming Perfect Synchronization," *IEEE Milcom. Conf. Proc.*, October 1981.

1436. Kashian, H. C., Rogers, D. C., and Walker, J. R. "Propagation Measurements over Geographically Diverse Paths," *IEEE Milcom. Conf. Proc.*, October 1981.

1437. Garber, F. D., and Pursley, M. B. "Effects of Frequency-Selective Fading on Slow-Frequency Hopped DPSK Spread Spectrum Communications," *IEEE Milcom. Conf. Proc.*, October 1981.

1438. Tsui, E. T., and Ibaraki, R. "An Adaptive Binary DPSK Spread Spectrum Receiver for Multipath/Scatter Channels," *IEEE Milcom. Conf. Proc.*, October 1981.

1439. Lythe, A. C. "Signal Processing Distortion Loss in Spread-Spectrum Communication, Command, Control and Navigation Systems," *Natl. Telec. Conf. Proc.*, November 1980.

1440. Oetting, J. D. "An Analysis of Anti-Jam Communications Requirements in Fading Media," *Natl. Telec. Conf. Proc.*, November 1980.

1441. Kanayake, N. E. "The Effects of Transponder Non-Linearity on the Performance of MSK and Offset QPSK Signal Transmission," *Natl. Telec. Conf. Proc.*, November 1980.

1442. Alcuri, L., and Mamola, G., and Randazzo, E. "Signal-to-Noise Ratio in Bandpass Direct-Sequence Spread-Spectrum Modulation Systems," *IEEE Trans. Comm.*, March 1982.

1443. Weinberg, A. "The Effects of Repeater Hard-Limiting, Filter Distortion, and Noise on a Pseudo-Noise, Time-of-Arrival Estimation System" *IEEE Trans. Comm.*, September 1979.

1444. Price, R. "Notes on Ideal Receivers for Scatter Multipath," M.I.T. Lincoln Lab., Group Rep. 34-39, May 12, 1955 (AD 224559).

1445. Turin, G. L. "Communication Through Noisy, Random-Multipath Channels," in *IRE Conv. Rec.*, New York, NY, March 19–22, 1956, part 4, pp. 154–166.

1446. Price, R., and Green, Jr., P. E. "An Anti-Multipath Communication System," M.I.T. Lincoln Lab., Tech. Memo. 65, November 9, 1956 (not available from M.I.T.).

1447. Price, R., and Green, Jr., P. E. "A Communication Technique for Multipath Channels," *Proc. IRE*, Vol. 46, pp. 555–570, March 1958.

1448. Price, R., and Green, Jr., P. E. "Anti-Multipath Receiving System," U.S. Patent 2 982 853, May 2, 1961 (filed July 2, 1956).

1449. Sunstein, D. E., and Steinberg, B. "Communication Technique for Multipath Distortion," U.S. Patent 3 168 699, February 2, 1965 (filed June 10, 1959).

1450. Ehrich, W. G. "Common Channel Multipath Receiver," U.S. Patent 3 293 551, December 20, 1966 (filed December 24, 1963).

1451. Fano, R. M. "Anti-Multipath Communication System," U.S. Patent 2 982 852, May 2, 1961 (filed November 21, 1956).

1452. Bello, P. A., and Nelin, B. P. "The Effect of Frequency Selective Fading on the Binary Error Probabilities of Incoherent and Differentially Coherent Matched Filter Receivers," *IEEE Trans. Comm. Syst.*, Vol. CS-11, pp. 170–180, June 1963.

1453. Chang, J.-F. "The Effect of Multipath Interference on the Performance of a Digital Matched Filter," *IEEE Trans. Comm.*, April 1979.

1454. Huang, T. C., and Lindsey, W. C. "PN Spread Spectrum Signaling Through a Nonlinear Satellite Channel Disturbed by Interference and Noise," *IEEE Trans. Comm.*, May 1982.

1455. Gerionotis, E. A., and Pursley, M. B. "Effects of Specular Multipath Fading on Noncoherent Direct Sequence Spread-Spectrum Communications," *I.C.C. Proc.*, 1982.

15. SYNCHRONIZATION

1501. Cahn, C. R. "Synchronization Scheme with Sequential Detection," Magnavox Research Laboratories, MX-TM-3084-70, October 1970.

1502. Gagliardi, R. M. "A Geometrical Study of Transmitted Reference Communications Systems," *IEEE Trans. Comm. Tech.*, December 1964.

1503. Golomb, S. W., et al. "Synchronization" *IEEE Trans. Comm. Syst.*, December 1963.

1504. Kaneko, H. "A Statistical Analysis of the Synchronization of a Binary Receiver," *IEEE Trans. Comm. Syst.*, December 1963.

1505. Lindsey, W. C. *Synchronization Systems in Communication and Control*, Prentice-Hall, Englewood Cliffs, New Jersey, 1972.

1506. Mifflin, R. W., and Wheeler, J. P. "Transmitted Reference Synchronization System," U.S. Patent 3,641,433, February 8, 1972.

1507. "Coherent System Speeds Sync Acquisition in Military TDMA Systems," *Communications Designer's Digest*, January 1970.

1508. Carter, C. R., and Haykin, S. S. "A New Synchronization Technique for TDMA Satellite Systems," *Proc. Symp. Sp. Spec. Comm.*, March 1973.

1509. Gagliardi, R. M. "Rapid Acquisition Signal Design in a Multiple-Access Environment," *Proc. Symp. Sp. Spec. Comm.*, March 1973.

1510. La Rosa, R. "Switchable and Fixed-Code Surface Wave Matched Filters," *Proc. Symp. Sp. Spec. Comm.*, March 1973.

1511. Ramsey, J. L. "Effective Acquisition of FH TDMA Signals in Jamming," *Proc. Symp. Sp. Spec. Comm.*, March 1973.

1512. Cole, Jr., R. "Synchronization of a Frequency-Hopped Spread Spectrum Signal," *Proc. Symp. Sp. Spec. Comm.*, March 1973.

1513. De Couvreur, G. A. "Effect of Random Synchronization Errors in PN and PSK Systems," *IEEE Trans. Aerosp. Elect. Syst.*, January 1970.

1514. Sergo, J. R., and Hayes, J. F. "Analysis and Simulation of a PN Synchronization System," *IEEE Trans. Comm. Tech.*, October 1970.

16. TRACKING

1601. Gill, W. J. "A Comparison of Binary Delay-Lock Tracking-Loop Implementations," *IEEE Trans. Aerosp. Elect. Syst.*, July 1966.

1602. Haykim, S. S., and Thorsteinson, C. "A Quantized Delay-Lock Discriminator," *IEEE Proc.*, June 1968.

1603. Huff, R. J., and Reinhard, K. L. "A Sampled-Data Delay-Lock Loop for Synchronizing TDMA Space Communications Systems," *EASCON* 1968 Record.

1604. Spilker, Jr., J. J., "Delay-Lock Tracking of Binary Signals," *IEEE Trans. Sp. Elect. Telem.*, March 1963.

1605. Ward, R. B. "Application of Delay-Lock Radar Techniques to Deep-Space Tasks," *IEEE Trans. Sp. Elect. Telem.*, June 1964.

1606. Bruno, F. "Tracking Performance and Loss of a Carrier Loop Due to the Presence of a Spoofed Spread Spectrum Signal," *Proc. Symp. Sp. Spec. Comm.*, March 1973.

1607. Freeman, J. J. "The Action of Dither in a Polarity Coincidence Correlator," *IEEE Trans. Comm.*, June 1974.

1608. Hartman, H. P. "Analysis of a Dithering Loop for PN Code Tracking," *Proc. Symp. Sp. Spec. Comm.*, March 1973.

1609. Simon, M. K. "Non-Linear Analysis of an Absolute Value Type of an Early-Late Gate Bit Synchronizer," *IEEE Trans. Comm. Tech.*, October 1970.

1610. Grieco, D. M. "The Application of Charge-Coupled Devices to Spread Spectrum Systems," *IEEE Trans. Comm.*, September 1980.

1611. O'Clock, Jr., G. D., and Duffy, M. T. "Acoustic Surface Wave Properties of Epitaxially Grown Aluminum Nitride and Gallium Nitride on Sapphire," *Appl. Phys. Lett.*, 15 July 1973.

1612. O'Clock, Jr., G. D. "Acoustic-Wave Filters Improve Radar," *Electronic Design*, 5 July 1973.

1613. Kino, G. S., and Matthews, H. "Signal Processing in Acoustic Surface-Wave Devices," *IEEE Spectrum*, August 1971.

1614. Colvin, R. D. "Spread-Spectrum Devices Cater to New Systems," *Microwaves*, February 1981.

1615. Reible, S. "Acoustoelectric Convolver Technology for Spread Spectrum Communications," *IEEE Trans. MTT*, May 1981.

1616. Dostert, K., and Pandit, M. "Performance of a SAW Tapped Delay Line in an Improved Synchronization Circuit," *IEEE Trans. Comm.*, January 1982.

1617. Leung, V. C. M., and Donaldson, R. W. "Confidence Estimates for Acquisition Times and Hold-In Times for PN-SSMA Synchronizer Employing Envelope Correlation," *IEEE Trans. Comm.*, January 1982.

1618. Ward, R. B., and Yiu, K. P. "Acquisition of Pseudonoise Signals by Recursion-Aided Sequential Estimation," *IEEE Trans. Comm.*, Vol. COM-25, pp. 784–794, August 1977.

1619. Hopkins, P. M. "A Unified Analysis of Pseudonoise Synchronization by Envelope Correlation," *IEEE Trans. Comm.*, August 1977.

1620. Holmes, J. K., and Chen, C. C. "Acquisition Time Performance of PN Spread Spectrum Systems," *IEEE Trans. Comm.*, August 1977.

1621. Simon, M. K. "Noncoherent Pseudonoise Code Tracking Performance of Spread Spectrum Receivers," *IEEE Trans. Comm.*, March 1977.

1622. Yost, R. A., and Boyd, R. W. "A Modified PN Code Tracking Loop: Its Performance and Implementation Sensitivities," *Proc. Natl. Telec. Conf.*, December 1980.

1623. Baier, W. P., Grammuller, H., and Pandit, M. "Combined Acquisition and Fine Synchronization System for Spread Spectrum Receivers Using a Tapped Delay Line Correlator," *Proc AGARD Conf.*, 1977.

1624. Pandit, M. "The Mean Acquisition Time of Active and Passive Correlation Acquisition Systems for Spread Spectrum Communication Systems," *Proc. Inst. Elec. Eng.*, 1981.

1625. Baier, P. W., Dostert, K., and Pandit, M. "A Novel Spread-Spectrum Receiver Synchronization Scheme Using a SAW-Tapped Delay Line," *IEEE Trans. Comm.*, May 1982.

1626. Milstein, L. B., and Das, P. K. "Spread Spectrum Receiver Using Surface Acoustic Wave Technology," *IEEE Trans. Comm.*, August 1977.

1627. Maskara, S. L., "An Active RC Delay Line for Matched Filter Correlators," *IEEE Trans. Comm.*, August 1977.

1628. Munroe, S. C. "The Use of CCD Correlators in the SEEK COMM Spread Spectrum Receivers," *IEEE Milcom. Conf. Proc.*, October 1981.

1629. Annecke, K. H. "Adaptive Regeneration of Combination Spreading Functions by Nonlinear Control Loops," *IEEE Milcom. Conf. Proc.*, October 1981.

1630. Peterson, R., and Ziemer, R. "Multiple Dwell Serial Acquisition for Direct Sequence Spread Spectrum Systems," *IEEE Milcom Conf. Proc.*, October 1981.

1631. Woo, K. T. "Acquisition of Fast Frequency Hopping Spread Spectrum Systems," *IEEE Milcom Conf. Proc.*, October 1981.

1632. Davidovici, S., Milstein, L. B., and Schilling, D. L. "A Fast PN Synchronization Scheme," *IEEE Milcom Conf. Proc.*, October 1981.

1633. Dostert, K., and Pandit, M. "Synchronization of Spread Spectrum Binary Orthogonal Keyed (BOK) Burst Transmission Systems," *IEEE Milcom Conf. Proc.*, October 1981.

1634. Singh, R., and Kochar, S. "Acquisition and Network Response Times for Centrally Controlled SS Networks," *IEEE Milcom Conf. Proc.*, October 1981.

1635. O'Clock, G. D. "Comparison of Microwave and Optical Signal Processors for Spread Spectrum Correlation," *IEEE Trans. Comm.*, August 1977.

1636. LaRosa, R., Kerbel, S., Fowler, T., and Corcoran, P. "Bandwidth Partitioned Programmable Matched Filter for 54,000 Time-Bandwidth Product Direct Sequence," *IEEE Milcom Conf. Proc.*, October 1981.

1637. Baier, P. W., Meyer, J., and Waibel, H. "Power Level Adaptive Synchronization Circuit for Spread-Spectrum BOK Receivers in Burst Transmission Systems," *IEEE Milcom Conf. Proc.*, October 1981.

1638. Polydoros, A., and Weber, C. L. "Analysis and Optimization of Correlative Code Tracking Loop for Spread Spectrum Systems," *IEEE Milcom Conf. Proc.*, October 1981.

1639. Luecke, J. R., and Yost, R. A. "Decision Directed AFC for Noncoherent Detectors Experiencing Large Signal Dynamics," *IEEE Milcom Conf. Proc.*, October 1981.

1640. Putnam, C. A., Rappaport, S. S., and Schilling, D. L. "Tracking of Frequency Hopped Spread Spectrum Signals in Adverse Environments," *IEEE Milcom Conf. Proc.*, October 1981.

1641. Broome, N. K. "An All Digital Maximum A Posteriori Based Synchronizer for MSK," *IEEE Milcom Conf. Proc.*, October 1981.

1642. Judd, G. W., and Nelson, R. B. "Surface Acoustic Wave Device Applications to Communication Equipment," *Natl. Telec. Conf. Proc.*, November 1980.

1643. Rhodes, R. R., Hutchinson, W. K., and Hutchinson, B. H. "Correlators and Convolvers Used in Spread Spectrum Systems," *Natl. Telec. Conf. Proc.*, November 1980.

1644. Goll, J. H. "A SAW Convolver-Based Spread Spectrum Communications Subsystem," *Natl. Telec. Conf. Proc.*, November 1980.

1645. Bowles, W. M. "GPS Code Tracking and Acquisition Using Extended-Range Detection," *Natl. Telec. Conf. Proc.*, November 1980.

1646. Chang, J.-F. "The Performance of a Spread-Spectrum, Multiple-Access Communication System Using Digital Matched Filters," *Natl. Telec. Conf. Proc.*, November 1980.

1647. Ponnusamy, J. A., and Srinath, M. D. "Acquisition of FH PN Codes Using Partial Correlation Coefficients," *Natl. Telec. Conf. Proc.*, November 1980.

1648. Shreve, W. R. "Signal Processing Using Surface Acoustic Waves," *Hewlett-Packard Journal*, January 1982.

1649. Alem, W. K., Huth, G. K., Holmes, J. K., and Udalov, S. "Spread Spectrum Acquisition and Tracking Performance for Shuttle Communication Links," *IEEE Trans. Comm.*, November 1978.

1650. Rappaport, S. S., and Schilling, D. L. "A Two-Level Coarse Code Acquisition Scheme for Spread Spectrum Radio," *IEEE Trans. Comm.*, September 1980.

1651. Elhakeem, A. K., Takhar, G. S., and Gupta, S. C. "New Code Acquisition Techniques in Spread-Spectrum Communication," *IEEE Trans. Comm.*, February 1980.

1652. O'Clock, Jr., G. D., Grasse, C. L., and Gandolfo, D. A. "Switchable Acoustic Surface Wave Sequence Generator," *IEEE Proc.*, October 1971.

1653. Colvin, R. D. "Spread-Spectrum Devices Cater to New Systems," *Microwaves*, February 1981.

1654. Casasent, D. "A Review of Optical Signal Processing," *IEEE Comm. Mag.*, September 1981.

1655. Dostert, K., and Pandit, M. "Performance of a SAW Tapped Delay Line in an Improved Synchronization Circuit," *IEEE Trans. Comm.*, January 1982.

1656. Leung, V. C. M., and Donaldson, R. W. "Confidence Estimates for Acquisition Times and Hold-In Times for PN-SSMA Synchronizer Employing Envelope Correlation," *IEEE Trans. Comm.*, January 1982.

1657. Casasent, D. "Optical Signal Processing," *Electro-Optical Systems Design*, September 1981.

1658. Simon, M. K. "Two-Channel Costas Loop Tracking Performance for UQPSK Signals With Arbitrary Data Formats," *IEEE Trans. on Comm.*, September 1981.

1659. Kohlenberg, A., Sussman, S. M., and Van Meter, D. "Matched Filter Communication Systems," U.S. Patent 3 876 941, April 8, 1975 (filed June 23, 1961).

1660. Guillemin, E. "Matched Filter Communication Systems," U.S. Patent 3 936 749, February 3, 1976 (filed June 23, 1961). S. M. Sussman, "A Matched Filter Communication System for Multipath Channels," *IRE Trans. Info. Th.*, vol. IT-6, pp. 367–373, June 1960.

1661. Spilker, Jr., J. J., and Magill, D. T. "The Delay-Lock Discriminator—An Optimum Tracking Device," *Proc. IRE*, vol. 49, pp. 1403–1416, September 1961.

1662. Yost, R. A., and Boyd, R. W. "A Modified PN Code Tracking Loop: Its Performance and Implementation Sensitivities," in *Proc. Nat. Telecommun. Conf.*, Houston, TX, December 1980, pp. 61.5.1–61.5.5.

1663. Hunsinger, B. J. "Surface Acoustic Wave Devices and Applications—3. Spread Spectrum Processors," *Ultrasonics*, pp. 254–262, November 1973.

1664. Baier, W. P., Grammuller, H., and Pandit, M. "Combined Acquisition and Fine Synchronization System for Spread Spectrum Receivers Using a Tapped Delay Line Correlator," in *Proc. AGARD Conf.*, 1977, no. 230.

1665. Dostert, K., and Pandit, M. "Performance of a SAW Tapped Delay Line in a Synchronizing Circuit," *IEEE Trans. Comm.*, vol. COM-30, pp. 219–222, January 1982.

1666. Pandit, M. "The Mean Acquisition Time of Active and Passive Correlation Acquisition Systems for Spread Spectrum Communication Systems," *Proc. Inst. Elec. Eng.*, Part F, 1981.

1667. Putman, C. A., Rappaport, S. S., and Schilling, D. L. "A Comparison of Schemes for Coarse Acquisition of Frequency-Hopped Spread-Spectrum Signals," *IEEE Trans. Comm.*, Feb. 1983.

1668. DiCarlo, D. M., and Weber, C. L. "Multiple Dwell Serial Search: Performance and Application to Direct Sequence Code Acquisition," *IEEE Trans. Comm.*, May 1983.

1669. Meyr, H. "Delay-Lock Tracking of Stochastic Signals," *IEEE Trans. Comm.*, March 1976.

1670. Crofut, W. A. "SAWs Critical to Avionics," *MSN Magazine*, November 1980.

1671. Rajan, J. A. "Adaptive Acquisition of Multiple Access Codes," *I.C.C. Proc.*, 1982.

17. MISCELLANEOUS

1701. Burnell, R. W., and Ma, L. N. "Fourier Analysis of an Imperfect PSK Signal," TRW 10C 7322.14-7, October 1967.

1702. Cahn, C. R., and Moore, C. R. "Bandwidth Efficiency for Digital Comm. Via a Hard Limiting Channel," *ITC Proc.* **VIII**, 1972.

1703. Hancock, J. C. *An Introduction to the Principles of Communication Theory*, McGraw-Hill, New York, 1961.

1704. Klapper, J. "The Effect of the Integrator-Dump Circuit on PCM/FM Error Rates," *IEEE Trans. Comm. Tech.*, June 1966.

1705. Kwan, R. K. "The Effects of Filtering and Limiting a Double-Binary PSK Signal," *IEEE Trans. Aerosp. Elect. Sys.*, July 1969.

1706. Petrich, Mario, "On the Number of Orthogonal Signals which Can Be Placed in a WT-Product," *J. Soc. Indust. Appl. Math.*, December 1963.

1707. Richardson, R. J. "The Signal-Suppression Threshold of a Frequency Doubler," *IEEE Trans. Comm. Tech.*, December 1964.

1708. Sierra, H. M. "The Matched Filter Concept," *Electro-Technology*, August 1964.

1709. Solat, N. "Comparison of Power Spectra," Space General Corp. Interoffice Memo, October 1963.

1710. Stiffler, J. J. *Theory of Synchronous Communications*, Prentice-Hall, Englewood Cliffs, New Jersey, 1971.

1711. Turin, G. L. "An Introduction to Matched Filters," *IRE Trans. Info. Th.*, June 1960.

1712. Weaver, C. S. "An Adaptive Communications Filter," *IRE Proc.*, October 1961.

1713. Victor, W. K., and Brockman, M. H. "The Application of Linear Servo Theory to the Design of AGC Loops," *JPL Ext. Pub. 586*, December 1958.

1714. "Spectrum Analysis," *Hewlett-Packard AN*63, July 1964.

1715. "Digital Filter Studies Anticipate Potential of Spread Spectrum Designs," *Communications Designer's Digest*, July/August 1971.

1716. "Reference Data for Radio Engineers," H. W. Sams, 1968.

1717. Buss, D. D., Bailey, W. H., and Hite, L. R. "Spread-Spectrum Communications Using Charge Transfer Devices," *Proc. Symp. Sp. Spec. Comm.*, March 1973.

1718. Glazer, B. G. "Spread Spectrum Concepts—A Tutorial," *Proc. Symp. Sp. Spec. Comm.*, March 1973.

1719. Gooding, D. J. "Increasing the Utility of the Digital Matched Filter," *Proc. Symp. Sp. Spec. Comm.*, March 1973.

1720. Matthaei, G. L. "Acoustic Surface-Wave Transversal Filters," *IEEE Trans. Ckt. Th.*, September 1973.

1721. Mohanty, N. C. "Signal Design for Small Correlation," University of Southern California Report USCEE 438, February 1973.

1722. Ristenbatt, M. P. "Estimating Effectiveness of Covert Communications," *Proc. Symp. Sp. Spec. Comm.*, March 1973.

1723. Sussman, S. M., and Ferrari, E. J. "The Effects of Notch Filters on the Correlation Properties of a PN Signal," *IEEE Trans. Aerosp. Elect. Syst.*, May 1974.

1724. Torre, F. M. "A Unified Description and Design Format for Spread Spectrum Waveforms," *Proc. Symp. Sp. Spec. Comm.*, March 1973.

1725. Muraka, N. P. "Spread Spectrum Systems Using Noise Band Shift Keying," *IEEE Trans. Comm.*, July 1973.

1726. Cahn, C. R. "Performance of Digital Matched Filter Correlators with Unknown Interference," *IEEE Trans. Comm. Tech.*, December 1971.

1727. Hatsell, C. P., and Nolte, L. W. "Detectability of Burst-Like Signals," *IEEE Trans. Aerosp. Elect. Syst.*, March 1971.

1728. Millett, R. E. "A Matched-Filter Pulse-Compression System Using a Nonlinear FM Waveform," *IEEE Trans. Aerosp. Elect. Syst.*, January 1970.

1729. Lindsey, W. C., and Lewis, J. L. "Modeling, Characterization and Measurement of Oscillator Frequency Instability," *NRL Report* NRL-1.1351, June 14, 1974.

1730. Schiffer, J. "EW Systems Synthesizers Find a New Design Approach," *MSN*, December 1981.

1731. Casasent, D. "Optical Signal Processing," *Electro-Optical Systems Design*, September 1981.

1732. Johnston, S. L. "Radar ECCM History," in *Proc. NAECON*, May 1980.

1733. Viterbi, A. J., and Jacobs, I. M. "Advances in Coding and Modulation for Noncoherent Channels Affected by Fading, Partial Band, and Multiple Access Interference," *Advances in Communication Systems*, 1975.

1734. Henry, P. S. "Spectrum Efficiency of a Frequency-Hopped-DPSK Spread Spectrum Mobile Radio System," *IEEE Trans. Veh. Tech.*, November 1979.

1735. Holmes, J. K. *Coherent Spread Spectrum Systems*, Wiley, New York, 1982.

1736. Compton, Jr., R. T. "An Adaptive Array in a Spread Spectrum Communication System," *Proc. IEEE*, March 1978.

1737. Cooper, G. R., and Nettleton, R. W. "A Spread Spectrum Technique for High-Capacity Mobile Communications," *IEEE Trans. Veh. Tech.*, November 1978.

1738. Henry, P. S. "Spectrum Efficiency of a Frequency-Hopped-DPSK Spread Spectrum Mobile Radio System," *IEEE Trans. Veh. Tech.*, November 1979.

1739. Edell, J. D. "Wideband, Noncoherent, Frequency-Hopped Waveforms and their Hybrids in Low Probability-of-Intercept Communications," *Naval Res. Lab.*, Washington, D.C., November 8, 1976.

1740. Einarsson, G. "Address Assignment for a Time-Frequency-Coded Spread Spectrum System," *Bell Syst. Tech. J.*, September 1980.

1741. Cahn, C. R. "Spread Spectrum Applications and State-of-the-Art-Equipments," *Magnavox Comm. Navigation*, November 22, 1972.

1742. Viterbi, A. J. "Spread Spectrum Communications—Myths and Realities," *IEEE Comm. Soc. Mag.*, May 1979.

1743. Scholtz, R. A. "The Origins of Spread-Spectrum Communications," *IEEE Trans. Comm.*, May 1982.

1744. Schilling, D. L., Milstein, L. B., Pickholtz, R. L., and Brown, R. "Optimization of the Processing Gain of an M-ary Direct Sequence Spread Spectrum Communication System," *IEEE Trans. Comm.*, August 1980.

1745. Huth, G. K. "Optimization of Coded Spread Spectrum Systems Performance," *IEEE Trans. Comm.*, August 1977.

1746. Ananasso, F. "Matched Filters Up Power, Reduce Noise in Radar," *Microwaves*, April 1982.

1747. Milstein, L. B., Pickholtz, R. L., and Schilling, D. L. "Optimization of the Processing Gain of an FSK-FH System," *IEEE Trans. Comm.*, July 1980.

1748. Pickholtz, R. L., Schilling, D. L., and Milstein, L. B. "Theory of Spread-Spectrum Communications—A Tutorial," *IEEE Trans. Comm.*, May 1982.

1749. Winters, J. H. "Spread Spectrum in a Four-Phase Communication System Employing Adaptive Antennas," *IEEE Trans. Comm.*, May 1982.

1750. Smith, J. R. "Tradeoff Between Processing Gain and Interference Immunity in Co-Site Multichannel Spread Spectrum Communications," *IEEE Trans. Comm.*, May 1982.

1751. Scholtz, R. A. "The Spread Spectrum Concept," *IEEE Trans. Comm.*, August 1977.

1752. Ristenbatt, M. P., and Daws, Jr., J. L. "Performance Criteria for Spread Spectrum Communications," *IEEE Trans. Comm.*, August 1977.

1753. DeVore, W. J. "Computer-Aided Spread Spectrum Signal Design," *IEEE Trans. Comm.*, August 1977.

1754. Eckert, R. P., and Kelly, P. M. "Implementing Spread Spectrum Technology in Land Mobile Radio Services," *IEEE Trans. Comm.*, August 1977.

1755. Lukosevicius, A. V. "VHSIC Impact on Spread Spectrum Communications," *IEEE Milcom. Conf. Proc.*, October 1981.

1756. Major, P. A. "Spread Spectrum EMC: Measurement and Analysis," *IEEE Milcom. Conf. Proc.*, October 1981.

1757. El-Osmany, A., and Kader, Abdel "Interference Effects of Direct Sequence Spread Spectrum Signals on Voice AM and FM Communications Systems," *IEEE Milcom. Conf. Proc.*, October 1981.

1758. Budreau, A. J. "Compact Fast-Hopping Frequency Synthesizer for Spread-Spectrum Systems," *IEEE Milcom. Conf. Proc.*, October 1981.

1759. Budreau, A. J. "Frequency-Hopping Filters and Programmable Matched Filters for Spread-Spectrum Systems," *IEEE Milcom. Conf. Proc.*, October 1981.

1760. Bernardo, R. "SAW Delay Line Discriminators and Their Applications to Low Noise Agile UHF Frequency Sources," *IEEE Milcom. Conf. Proc.*, October 1981.

1761. Milstein, L. B., Das, P. K., and Gevargiz, J. "Processing Gain Advantage of Transform Domain Filtering DS-Spread Spectrum Systems," *IEEE Milcom. Conf. Proc.*, October 1981.

1762. Orr, R. S. "FH/MFSK Analyses," *IEEE Milcom. Conf. Proc.*, October 1981.

1763. Agustic, R., and Junyent, G. "Performance of a MH Multilevel FSK for Mobile Radio," *IEEE Milcom. Conf. Proc.*, October 1981.

1764. Kennedy, M. "Regulatory Aspects of Spread Spectrum," *IEEE Milcom. Conf. Proc.*, October 1981.

1765. Chesto, L. F. "Spread Spectrum and Radio Frequency Management," *IEEE Milcom. Conf. Proc.*, October 1981.

1766. Fienstein, H. "Spread Spectrum Experiments for Fixed and Land Mobile Applications," *IEEE Milcom. Conf. Proc.*, October 1981.

1767. Dodson, D., Huang, M., and Kagiwada, R. "Fast Hopping SAW Frequency Synthesizer," *IEEE Milcom. Conf. Proc.*, October 1981.

1768. Rhodes, R. R., Hutchinson, W. K., and Hutchinson, B. H. "Frequency Agile Phase-Locked Loop Synthesizer for a Communication Satellite," *Natl. Telec. Conf. Proc.*, November 1980.

1769. DePedro, H., and Wolf, J. K. "On the Design of Spread Spectrum Communication Networks," *Natl. Telec. Conf. Proc.*, November 1980.

1770. O'Clock, G. D., and Erickson, L. P. "Impact of Molecular Beam Epitaxy Technology on Spread Spectrum Systems," *ITC Conf. Proc.*, October 1981.

1771. Masenton, W. K. "Adaptive Interference Rejection Filters for Spread Spectrum Communication Systems," *ITC Conf. Proc.*, October 1981.

1772. Reilly, R. A., and Ward, C. R. "The Influence of Technology Advances on Integrated CNI Avionics," *Natl. Telec. Conf. Proc.*, November 1980.

1773. Moore, R. "Spread Spectrum Techniques," *Natl. Telec. Conf. Proc.*, November 1980.

1774. Haber, C. H., and Nossen, E. J. "Analog Versus Digital Antijam Video Transmission," *IEEE Trans. Comm.*, March 1977.

1775. Grieco, D. M. "Inherent Signal-to-Noise Ratio Limitations of Charge Coupled Device Pseudonoise Matched Filters," *IEEE Trans. Comm.*, May 1980.

1776. Ananasso, F. "Matched Filters Up Power, Reduce Noise In Radar," *Microwaves*, April 1982.

1787. Lewis, H. D. "Array Radars Solve Communication Jams," *Microwaves*, April 1982.

1778. Cuccia, C. L. "Spread Spectrum Systems Serve Nearly All C^3 Aspects," *MSN*, April 1982.

1779. Goggans, P. M., Jelinek, D. A., and Nielsen, W. F. "SECURICOM: An Antijam, Secure-Message, Security Force Radio Communication System. Technical Summary, Sandia Laboratories SAND80-0347, March 1981.

1780. Dixon, R. C. "SECURICOM: An Anti-Jam, Secure-Message, Security Force Radio Communication System. Design Feasibility Study," Sandia Laboratories SAND80-7049, March 1981.

1781. Jordan, G. B., et al. "Hybrid Spread Spectrum Systems: Theory, Advantages, and Limitations," Philco-Ford Report IRDP G692.FBLB.

1782. Spellman, M. "Spread-Spectrum Radios Thwart Hostile Jammers," *Microwaves*, September 1981.

1783. Cowan, C. F. N., and Grant, P. M. "Adaptive Filters Await Breakthroughs," *MSN*, August 1981.

1784. Lewis, M. "PLLs Upconvert Chirp Radar Signals," *Microwaves*, June 1981.

1785. Custance, K. "Hop, Skip, and Jump. Who Wins?" *Telecommunications*, August 1981.

1786. Bolen, N. E. "JTIDS Coordinates Tactical Elements," *Defense Electronics*, January 1982.

1787. Schiffer, J. "EW Systems Synthesizers Find a New Design Approach," *MSN*, December 1981.

1788. Melton, W. C. "Time-of-Arrival Measurements Possible With Use of Global Positioning System," *Defense Electronics*, December 1981.

1789. Ribchester, E. "Frequency-Hopping Radios Outwit 'Smart' Jammers" *Microwaves*, November 1979.

1790. Hirt, W., and Pasupathy, S. "Continuous Phase Chirp (CPC) Signals For Binary Data Communication—Parts I and II," *IEEE Trans. Comm.*, June 1981.

1791. Basore, B. L. "Noise-Like Signals and Their Detection by Correlation," M.I.T. Res. Lab. Electron. and Lincoln Lab., Tech. Rep. 7, May 26, 1952 (AD 004641).

1792. Pankowski, B. J. "Multiplexing a Radio Teletype System Using a Random Carrier and Correlation Detection," M.I.T. Res. Lab. Electron. and Lincoln Lab., Tech. Rep. 5, May 16, 1952 (ATI 168857; not available from M.I.T.).

1793. "Engineering Study and Experimental Investigation of Secure Directive Radio Communication Systems," Sylvania Elec. Products, Buffalo, NY, Interim Eng. Rep., Contr. AF-33(616)-167, Aug. 5–Nov. 5, 1952 (AD 005243).

1794. Dixon, R. C., (Ed.) *Spread Spectrum Techniques*, IEEE Press, New York, 1976.

1795. Costas, J. P. "Poisson, Shannon, and the Radio Amateur," *Proc. IRE*, Vol. 47, pp. 2058–2068, December 1959.

1796. Goldberg, B. "Applications of Statistical Communications Theory," presented at the Army Sci. Conf., West Point, NY, June 20–22, 1962 (AD 332048); republished in *IEEE Comm. Mag.*, Vol. 19, pp. 26–33, July 1981.

1797. Lindner, H. G. "Communication Security Method and System," U.S. Patent 4184 117, January 25, 1980 (filed April 16, 1956). J. W. Craig and R. Price, "A Secure Voice Communication System," *Trans. Electron. Warfare Symp.*, 1959.

1798. Schwartz, L. S. "Wide-Bandwidth Communications," *Space Aeronautics*, pp. 84–89, December 1963.

1799. Spilker, Jr., J. J., "Nonperiodic Energy Communication System Capable of Operating at Low Signal-to-Noise Ratios, U.S. Patent 3 638 121, January 25, 1972 (filed December 20, 1960).

17100. Lowrie, R. "A Secure Digital Command Link," *IRE Trans. Space Electron. Telem.*, Vol. SET-6, pp. 103–114, Sept.-Dec. 1960.

17101. Rechtin, E. "An Annotated History of CODORAC: 1953–1958," Jet Propulsion Lab., Pasadena, CA, Rep. 20-120, Contr. DA-04-495-Ord 18, August 4, 1958 (AD 301248).

17102. "Lincoln F9C-A Radio Teletype System," Sylvania Electron. Defense Lab., Mountain View, CA, Instruction Manual EDL-B8, December 21, 1956.

17103. Nicholson, Jr., M. G. "Time Delay Apparatus," U.S. Patent 2 401 094, May 28, 1946 (filed June 23, 1944).

17104. Deloraine, E. M., Busignies, H. G., and de Rosa, L. A. "Facsimile System," U.S. Patent 2 406 811, September 3, 1946 (filed December 15, 1942).

17105. Deloraine, E. M., Busignies, H. G., and deRosa, L. A. "Facsimile System and Method," U.S. Patent 2 406 812, September 3, 1946 (filed January 30, 1943). L. A. deRosa, "Random Impulse System," U.S. Patent 2 671 896, Mar. 9, 1954 (filed Dec. 13, 1942).

17106. Deloraine, E. M. "Protected Communication System," Fed. Radio Tel. Lab., New York, NY, Rep. 937-2, April 28, 1944 (from the National Archives, Record Group 227; this report was written to Division 15 of the National Defense Research Committee, Office of Scientific Research and Development, on Project RP-124).

17107. Busignies, H., Dodington, S. H., Herbst, J. A., and Clark, G. R. "Radio Communication System Protected Against Interference," Fed. Tel. Radio Corp., New York, NY, Final Rep. 937-3, July 12, 1945 (same source as [54]).

17108. Hoeppner, C. H. "Pulse Communication System," U.S. Patent 2 999 128, September 5, 1961 (filed November 4, 1945).

17109. Krause, E. H., and Cleeton, C. E. "Pulse Signalling System," U.S. Patent 4 005 818. February 1, 1977 (filed May 11, 1945).

17110. Pierce, J. R. "The Early Days of Information Theory," *IEEE Trans. Info. Th.*, Vol. IT-19, pp. 3–8, January 1973.

17111. "The WHYN Guidance System," Phys. Lab., Sylvania Elec. Products, Bayside, NY, Final Eng. Rep., Modulation Wave Form Study & F-M Exciter Develop., Contr. W28-099ac465, June 1949 (AD895816).

17112. "The WHYN Guidance System," Phys. Lab., Sylvania Elec. Products, Flushing, NY. Interim Eng. Rep. 5, Contr. W28-099ac465, October 1948 (ATI 44524).

17113. "The WHYN Guidance System," Phys. Lab., Sylvania Elec. Products, Bayside, NY, Final Eng. Rep., Equipment Develop. & East Coast Field Test, Contr. W28-099ac465, June 1950 (AD 895815).

17114. "Radio Countermeasures," Natl. Defense Res. Committee, Office Sci. Res. Develop., Washington, DC, Summary Tech. Rep. Div. 15, Vol. I, 1946 (AD 221601).

17115. Green, Jr., J. H., and Gordon, J. "Selective Calling System," U.S. Patent 3 069 657. December 18, 1962 (filed June 11, 1958).

17116. Oxley, V. C., and De Lisle, W. E. "Communications and Data Processing Equipment," U.S. Patnt 3 235 661. February 15, 1966 (fled July 11, 1962).

17117. deRosa, L. A., DiToro, M. J., and Fischer, L. G. "Signal Correlation Radio Receiver," U.S. Patent 2 718 638, September 20, 1955 (filed January 20, 1950).

17118. Frank, R. L. "Phase-Coded Communication System," U.S. Patent 3 099 795, July 30, 1963 (filed April 3, 1957).

17119. Shannon, C. E. "Communication in the Presence of Noise," *Proc. IRE*, Vol. 37, pp. 10–21, January 1949.

17120. _____, "A Mathematical Theory of Communication," *Bell Syst. Tech. J.*, Vol. 27, pp. 379–423, July, and 623–656, October 1948.

17121. Lee, Y. W., Wiesner, J. B., and Cheatham, Jr., T. P. "Apparatus for Computing Correlation Functions," U.S. Patent 2 643 819, June 30, 1953 (filed August 11, 1949).

17122. Lee, Y. W., Cheatham, Jr., T. P., and Weisner, J. B. "The Application of Correlation Functions in the Detection of Small Signals in Noise," M.I.T. Res. Lab. Electron., Tech. Rep. 141, October 13, 1949 (ATI 066538, PB 102361).

17123. Singleton, H. E. "A Digital Electronic Correlator," M.I.T. Res. Lab. Electron., Tech. Rep. 152, February 21, 1950.

17124. Harris, Jr., H. C., Leifer, M., and Cawood, D. W. "Modified Cross-Correlation Radio System and Method," U.S. Patent 2 941 202, June 14, 1960 (filed August 4, 1951).

17125. "The WHYN Guidance System," Phys. Lab., Sylvania Elec. Products, Bayside, NY, Final Eng. Rep., Equipment Syst. Lab. Tests & Anal., Contr. W28-099ac-465, June 1953 (AD 024044).

17126. Leifer, M., and Serniuk, W. "Long Range High Accuracy Guidance System," presented at the RDB Symp. Inform. Theory Appl. Guided Missile Problems, California Inst. Technol., Pasadena, February 2–3, 1953.

17127. Goldsmith, A. N. "Radio Signalling System," U.S. Patent 1 761 118, June 3, 1930 (filed November 6, 1924).

17128. Feldman, C. B. H. "Wobbled Radio Carrier Communication System," U.S. Patent 2 422 664, June 24, 1947 (filed July 12, 1944).

17129. Marchand, N., and Leifer, M. "Cross-Correlation in Periodic Radio Systems," presented at the IRE Conf. Airborne Electron., Dayton, OH, May 23–25, 1951.

17130. Budd, W. E. "Analysis of Correlation Distortion," M.E.E. thesis, Polytech. Inst. Brooklyn, Brooklyn, NY, May 1955.

17131. Kotowski, P., and Dannehl, K. "Method of Transmitting Secret Messages," U.S. Patent 2 211 132, August 13, 1940 (filed in U.S. May 6, 1936; in Germany May 9, 1935).

17132. Pursley, M. B., Garber, F. D., and Lehnert, J. S. "Analysis of Generalized Quadriphase Spread-Spectrum Communications," in *Proc. IEEE Int. Conf. Comm.*, June 1980, Vol. 1, pp. 15.3.1–15.3.6.

17133. Compton, Jr., R. T. "An Adaptive Array in a Spread-Spectrum Communication System," *Proc. IEEE*, Vol. 66, p. 289, March 1978.

17134. Cooper, G. R., and Nettleton, R. W. "Spectral Efficiency in Cellular Land-Mobile Communications: A Spread-Spectrum Approach," TR-EE 78-44, Final Rep., NSF Grant ENG 76-80536, October 31, 1978.

17135. Torrieri, D. J. *Principles of Military Communication Systems*, Artech House, Dedham, MA, 1981.

17136. Cahn, C. R. *et al. AGARD Lecture Series No. 58—Spread Spectrum Communications*, 1973.

17137. Dixon, R. C., and Gerhardt, L. A. (Eds.) Special Issue, *IEEE Trans. Comm.*, Vol. COM-25, August 1977.

17138. Cook, C. E., and Marsh, H. S. "An Introduction to Spread Spectrum," *IEEE Comm. Mag.*, March 1983.

17139. Campbell, M. R., Hoff, L. E., and Ziemer, R. E. "A Large Time-Bandwidth Product Signaling Technique for Nonwhite Noise Channels," *IEEE Trans. Comm.*, October 1976.

17140. Gottlieb, I. M. "From Smoke Signals to Spread Spectrum, Military Communications Decides Who Wins the War," *Military Electronics/Countermeasures*, December 1982.

17141. Moser, R. "Generation and Reception of Spread Spectrum Signals," *Microwave Journal*, May 1983.

17142. Moser, R., and Gross, G. J. "Spread Spectrum Techniques," *Microwave Journal*, October 1982.

17143. Oetting, J. "Spread Spectrum Communications," *Sea Technology*, May 1981.

17144. Moore, T. J., McKinley, R. L., Mortimer, V. D., and Nixon, C. W. "Evaluation of Word Intelligibility of Two Modulator/Demodulator Systems of a Spread Spectrum Communication System in Presence of Simulated Cockpit Noise," USAF Aerospace Medical Report AMRL-TR-78-54.

17145. Nielsen, D. L. "Microwave propagation and Noise Measurements for Mobile Digital Radio Applications," Stanford Research Institute Report for contract DAHC15-73-C-0187.

17146. Patti, J. J., and Roeder, A. W. "An Adaptive Spread Spectrum Correlation Receiver" Presented at Natl. Conf. on R.P.V's, June 1975.

17147. Cook, C. E., Ellersick, F. W., Milstein, L. B., and Schilling, D. L. (Eds.) Special Issue, *IEEE Trans. Comm.*, May 1982.

17148. Newhouse, P. "Procedures for Analyzing Interference Caused by Spread Spectrum Signals," IIT Research Inst. ESD-TR-77-003A, August 1978.

17149. Singh, R., Elkaheem, A. K., and Gupta, S. C. "Hybrid Diversities in a Spread Spectrum Mobile Communication System," *IEEE I.C.C. Proc.*, 1982.

APPENDIX 3

EQUIVALENCE OF SIMPLE AND MODULAR SEQUENCES

Assume a linear sequence $|n, p|_s$ and its alleged equivalent $|n, n - P|_m$. Writing equations and drawing diagrams for them, we get

SSRG

$$\lambda_s^0 = \lambda^n \oplus \lambda^p$$

$$= \lambda^0 D_n \oplus \lambda^0 D_p$$

MSRG

$$\lambda_m^0 = \left(\lambda^0 D_{n-p} \oplus \lambda^0 \right) L_p$$

$$= \lambda^0 D_{n-p} D_p \oplus \lambda^0 D_p$$

$$= \lambda^0 D_n \oplus \lambda^0 D_p$$

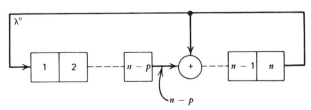

where $\lambda^{n-p} = \lambda^0 D_{n-p}$, and $D_{n-p} D_p = D_n$, because $(n-p) + P = n$. Therefore, because $\lambda_s^0 = \lambda_m^0$, the two are equivalent. Now, because this proof is so austere, let us try a more concrete example to help us to remember the equivalence of certain simple and modular sequence generator connections, at least for one specific case.

Simple generator for [5, 4, 3, 2]$_s$ code

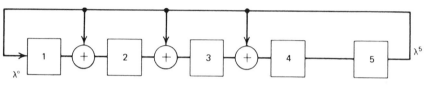

Modular SRG for [5, 3, 2, 1]$_m$ code

Given $[5, 4, 3, 2]$, and its equivalent $[5, 3, 2, 1]_m$, we find that the equation for the simple generator is just $\lambda^0 = \lambda^5 \oplus \lambda^4 \oplus \lambda^3 \oplus \lambda^2$.

Alternatively, we can write $\lambda^0 = ((\lambda^5 \oplus \lambda^4) \oplus \lambda^3) \oplus \lambda^2$. This is just the previous λ^0, once the brackets have been removed. We can write the equation for the modular sequence generator, noting that $\lambda^5 = \lambda^0$, as

$$\lambda^0 = \left(\left(\left(\lambda^0 D \oplus \lambda^0\right)D \oplus \lambda^0\right)D \oplus \lambda^0\right)D_2$$

$$= \left(\left(\lambda^0 D \oplus \lambda^0\right)D \oplus \lambda^0\right)D_3 \oplus \lambda^2$$

$$= \left(\lambda^0 D \oplus \lambda^0\right)D_4 \oplus \lambda^3 \oplus \lambda^2$$

$$= \lambda^5 \oplus \lambda^4 \oplus \lambda^3 \oplus \lambda^2,$$

and the modular generator is equivalent to the simple generator, as previously stated.

APPENDIX 4

MULTIPLICATION OF DIRECT SEQUENCE SIGNALS

Multiplication of a direct sequence signal is of interest from at least two aspects. One is in the context of Chapter 7, in which the possibility of multiplying a signal up to a desired transmitting frequency is discussed. The second is from the standpoint of a receiver who wishes to know if a DS signal is present. The example that follows simply shows that a biphase DS signal, after squaring, becomes a CW carrier (as does any symmetrically modulated signal). Though examples are not given here for all possible multiples, it should be obvious from the result that multiplication can be employed to detect a signal or, in the other extreme, should be avoided when the signal must be preserved. Table 7.1 shows the result of various multiplications of phase-modulated signals. The example given here is for squaring a biphase-modulated signal. The same result may be expected of squaring a quad-riphase-modulated signal twice.

Given a direct sequence biphase signal, $A \cos \omega_c t \pm 90$ (where the $90°$ term is a function of the code modulation). By squaring this signal (neglecting the amplitude coefficient A), we have

$$(\cos \omega_c t \pm 90)^2 = (\cos \omega_c t \cos 90 \pm \sin \omega_c t \sin 90)^2$$

$$= (\cos \omega_c t \cos 90)^2 \pm \sin \omega_c t \cos \omega_c t \sin 90 \cos 90$$

$$+ (\sin \omega_c t \sin 90)^2,$$

and, considering term by term,

$$(\cos \omega_c t \cos 90)^2 = \cos^2 \omega_c t = \tfrac{1}{2}(1 + \cos 2\, \omega_c t)$$

$$(\sin \omega_c t \sin 90)^2 = 0$$

$$\sin \omega_c t \cos \omega_c t \sin 90 \cos 90 = 0,$$

which leaves only a $\tfrac{1}{2}(1 + \cos 2\, \omega_c t)$ term, a CW signal at twice the original modulated carrier frequency. There is no modulated carrier at twice the input frequency. Therefore we see that squaring or doubling is not practical for use with a direct sequence signal.

APPENDIX 5

LINEAR CODE SEQUENCE VULNERABILITY

Linear* code sequences are often employed in communications systems for various uses, such as error correction, addressing, or spectrum spreading. Such codes have great utility, and in the case of the linear maximal codes (or m-sequences) can offer characteristics that are difficult to achieve with any other type of sequence.

It must be realized, however, that linear codes are limited by their vulnerability to simple analysis. That is, a system which uses a linear code for spectrum spreading can offer no advantage against the would-be interferor who can receive his or her transmitted signal, analyze it, and retransmit a replica of it to "spoof" the receiver. We will not consider the interferor's difficulty in receiving and detecting the code used. Suffice it to say that he or she must be able to procure any $2n + 1$ consecutive chips from the code sequence, and that given those $2n + 1$ chips, he or she can build a code generator that generates an identical code in its entirety. For example, if a $2^{31} - 1$ chip code is being employed, then any $(2 \times 31) + 1 = 63$ consecutive chips from the two-billion-chip-long code sequence can furnish all of the information needed to reproduce the entire code.

To be sure that the import of this note is clear—no linear code sequence can offer protection to information beyond the level of the casual observer.

---

*Any code sequence is considered to be linear if it is generated through the use of only those operations that are linear for the arithmetic field of interest (i.e., binary codes may be generated through modulo-2 addition, ternary codes through modulo-3 addition, etc.) Otherwise, the codes are nonlinear, and such codes are not considered here.

Any determined person who can procure $2n + 1$ consecutive chips has ready access to any information enbedded in the signal, as well as knowledge of all future code states. Furthermore, the process for determining this information is straightforward and simple, though perhaps somewhat cumbersome for those who must do the required operations by hand.

A tabular technique that produces code tap connections for an equivalent simple register code-sequence generator is given in the following pages, with an example to show its use. One assumption is made: that the analyst knows the length of the code-sequence generator that his or her sample comes from. This is a simplifying assumption, necessary only to make the example given more readily applicable to the question at issue—that of showing the vulnerability of linear codes.

For our first example, let us assume that the code generator in the transmitter is five stages long ($n = 5$). Typical code-sequence generators are much longer than this, but here we choose five stages because of the need for a simple example. The same principles apply, no matter what the register length. Now, let us suppose that we have just detected a subsequence 01110011011, 11 chips long. ($2n + 1$ chips $= 11$. Code length is $2^5 - 1 = 31$ chips, overall.)

First, we arrange the received subsequence in a column, with the last detected chip at the top. Then, shifting the same subsequence down by one row per column, generate more columns until the number of columns equals one more than the number of stages in the assumed generating register (five plus one).

The result should be as follows:

Q_5	Q_4	Q_3	Q_2	Q_1	λ
1					
1	1				
0	1	1			
1	0	1	1		
1	1	0	1	1	
0	1	1	0	1	1
0	0	1	1	0	1
1	0	0	1	1	0
1	1	0	0	1	1
1	1	1	0	0	1
0	1	1	1	0	0

This table is a listing of the state of shift register stages one through five, and the feedback (λ) necessary to feed into stage one, to produce the 11 chips that were received, using a five-stage shift register generator. This equivalent simple shift register generator is shown in Figure A5.1, which follows.

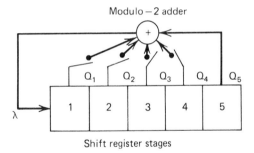

Figure A5.1   Five-stage simple shift register generator.

Now, noting two properties of shift register generators:

1. Code length is a function of the number of stages in the register, and the last stage, which determines the number, is always a member of the set in the feedback.
2. Stable, recurrent sequences do not exist for odd numbers of feedback taps (i.e., one, three, five, etc.).

Observe the feedback column, λ. A listing of shift register outputs, $Q_1, Q_2, \ldots$, can be generated that shows, for each register state, which stages may be added modulo-2 to the last stage output to give the desired λ for insertion to stage 1 in the next state. This produces the following result:

Q_5	Q_4	Q_3	Q_2	Q_1	λ	Possible feedback
1						
1	1					
0	1	1				
1	0	1	1			
1	1	0	1	1		
0	1	1	0	1	1	4, 3, 1
0	0	1	1	0	1	3, 2
1	0	0	1	1	0	2, 1
1	1	0	0	1	1	3, 2
1	1	1	0	0	1	2, 1
0	1	1	1	0	0	1

To be more specific: if the Q_5 output is modulo-2 added to Q_4, Q_3, or Q_1 outputs, then the result is the desired λ.

Now observe the possible feedback column. From this column we may determine the feedback taps that can be used to generate an m-sequence that contains the received 11-bit subsequence. The first entry in the λ

column, a one, would result from modulo-2 addition of Q_5 (a zero) with Q_4, Q_3, or Q_1. Similarly, the second λ would result from modulo-2 addition of Q_5 with either Q_3 or Q_2, and all other λ's would be generated by combination of Q_5 with the appropriate Q_n.

The last entry in the possible feedback column is a set with only one member, which means that for this state, only Q_1 would be added to Q_5 to produce the desired λ, a zero. Therefore, one of the feedback taps used to generate the code of which our 11-chip subsequence is a sample, is at the output of shift register stage 1.

Remembering that no code is possible with an odd number of feedback taps, and that stage five is always a member of the tap set, we see that there are four rows in the possible feedback column that are invalid: (3, 2) in rows two and four, and (2, 1) in rows three and five. Since taps at stage 2 and stage 1 cannot both be valid, and we know already that tap 1 is valid, the possibility of a tap at stage 2 can be discarded. We know, then, that feedback taps exist in the set we are seeking, at register stages 3 and 1, because 2 is also invalid in rows 2 and 4.

Again, we know that one more tap must exist because odd numbers of feedback taps are prohibited. Therefore, the remaining tap we seek must be at register stage 4, since it is the only one remaining, and it is contained in the top row of possible feedback taps. The code-sequence generator we are seeking is seen to be a five-stage register with feedback taps at stages 4, 3, and 1, as pictured in Figure A5.2.

Does this configuration generate a sequence containing the subset we started with? Indeed it does. The $2^5 - 1$ chip sequence produced by this generator is $\ldots 11111010001001010110000011100110\ldots$, in which our 11 chips start at chip 25 and run through chip 4. (The overall 31-chip sequence is cyclic.)

At this point, it is well that the rules used for evaluating possible feedback sets to be enumerated.

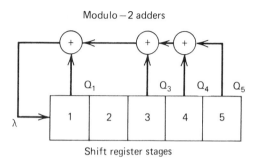

Figure A5.2 Five-stage shift register sequence generator derived from subsequence analysis.

1. Examine the possible feedback column for single member sets. Any single member is a feedback tap.
2. Determine if any one possible tap is found in all sets in the possible feedback column. If so, that tap is the only one used (i.e., the configuration is a single-tap generator).
3. If step 2 does not show that a single-tap generator is correct, then examine the possible feedback column for sets which contain two members, one of which is the tap found in step 1. Any two-member set that contains the tap already identified contains a second member which is not a valid tap, and that tap can be eliminated from further consideration.
4. Proceed to higher-order tap sets (3 to 4 to 5, etc.), eliminating invalid tap possibilities in higher-order sets through the same process as was used in step 3.
5. Assemble the results. All possible taps not eliminated by the process described are valid, and necessary to generating the desired code sequence.

The two properties of shift register sequence generators that were previously listed are all that were required to derive the taps used, and we have shown that a sequence generator can readily be constructed to simulate an unknown generator, if only its length is known.

Had it happened that the 11-chip input sequence was not a valid subsequence, then unresolvable ambiguities would have occurred in searching the possible feedback sets. It is obvious that for longer sequences, finding feedback taps by this handcrank method becomes more and more tedious and subject to error. The methods described are easily adaptable for computer solution, however, for any foreseeable code sequence length.

It has been assumed that the analyst knows the length of the register being used to generate the code. This is not too far fetched a notion, since the population of One/Zero runs in linear codes is fixed, and bounded in run length at n ones and $n - 1$ zeros. Analysis of the run length population in a received, coded signal could produce an estimate of the most likely code length, which in turn would support a set of trial solutions.

It is worthy of note that the analyst does not really care what the true length of the transmitted code is, or what the generator's feedback taps really are, as long as he or she can synthesize the code segment and project future states of the code.

With this thought in mind, we point out that a nonmaximal code produced by an n-stage register could just as well be a maximal length code from a shorter register. For instance, two of the nonmaximal codes produced by a $(5, 4)$ connected generator are identical to $(2, 1)$ and $(3, 1)$ connected m-sequences.

The analyst then would not care if a particular sequence was from a long register with nonmaximal connections or a shorter one with maximal connections. This view, in which the analyst does not really care how the sequence was generated, brings up a new point: Given any binary subsequence, some linear generator can be built up to reproduce it. Therefore, why bother to build nonlinear sequence generators?

The question is answered by recalling that the analyst's real interest is in finding out what the code sequence will do *in the future*. Even though a linear sequence or subsequence may reproduce a particular nonlinear subsequence perfectly, the future state of the nonlinear sequence cannot be linearly derived from the presently known subsequence. We have, on the other hand, shown that linear sequences can be reproduced by a simple process of analysis and generated by an equivalent simple shift register sequence generator.

APPENDIX 6

THE RELATION BETWEEN ω_N AND $\omega_{3\,dB}$

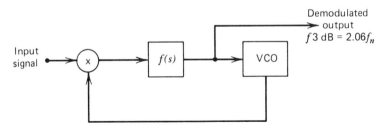

Figure A6.1 Phase-lock loop model.

(A) Given that it is desired that an FM-demodulating phase-lock loop be designed that has a bandpass $f_{3\,db}$. Hoffman[1206] gives a relation between B_N and ω_n:

$$B_N = \frac{4\zeta^2 + 1}{4\zeta}\omega_n \text{ cps}; \quad K \gg \omega_n/\zeta,$$

which for $\zeta = \sqrt{2}/2 = 0.7$ gives $B_N = 1.06\omega_n$ cps*.

In this same paper, Hoffman states that

$$f(3\,db) = 0.3088\,B_N \text{ cps } (\zeta = 0.7)$$

---

*Cycles per second, or Hertz.

Expressing B_N in Hertz, we get B_N in terms of f_n:

$$B_N = 2\pi \times 1.06 = 6.658 f_n \text{ cps.}$$

From the $f_0(-3\,\text{db}) = 3.088\, B_N$ equation, we get

$$B_N = \frac{f(3\,\text{db})}{0.3088}.$$

Substituting, then, we have

$$\frac{f(3\,\text{db})}{0.3088} = 6.658\, f_n$$

and $f(3\,\text{db}) = 2.055\, f_n \approx 2.06\, f_n$. A derivation confirming this relation follows.

(B) Derivation of the relationship between ω_n and $\omega_{3\,\text{DB}}$.

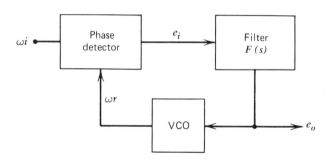

Definitions:

$$\omega_i = \omega_c + \omega_m,$$

$$K_1 = \text{multiplier gain in V/rad} = \frac{AB}{2}\sin\theta_m,$$

$$e_i = K_1(\phi_i - \phi_r)$$

$$= K_1\int(\omega_1 - \omega_r)\, dt,$$

$$e_{o(s)} = e_{i(s)}F_{(s)},$$

$$K_2 = d\omega_{\text{VCO}}/d_{eo} = \text{VCO gain in rad/V.}$$

$$\omega_r = \omega_c + K_2 e_o,$$

$$e_i = K_1(\omega_c + \omega_m - \omega_c - K_2 e)\, dt,$$

$$\omega_m = \dot{\omega}_m \sin \omega_m t = \text{modulating signal},$$

$$e_i = K_1 \int (\dot{\omega}_m \sin \omega_m t - K_2 e_o)\, dt,$$

$$e_{i(s)} = \frac{K_1}{s}\left[\dot{\omega}_m\left(\frac{s}{s^2 + \omega_m^2}\right) - K_2 e_{o(s)}\right],$$

$$e_{o(s)} = e_{i(s)}F_{(s)}\ (F_{(s)} = \text{loop filter response}),$$

$$e_{i(s)} = \frac{e_{o(s)}}{F_{(s)}}$$

Thus

$$\frac{e_{o(s)}}{F_{(s)}} = \frac{K_1}{s}\left[\frac{\dot{\omega}_m s}{s^2 + \omega_m^2} - K_2 e_{o(s)}\right]$$

and

$$e_{o(s)} = \frac{K_1 F_{(s)}}{s}\left[\frac{\dot{\omega}_m s}{s^2 + \omega_m^2} - K_2 e_{o(s)}\right]$$

$$= \frac{K_1 F_{(s)}\dot{\omega}_m}{s^2 + \omega_m^2} - \frac{K_1 K_2 e_{o(s)}F_{(s)}}{s}.$$

Gathering terms,

$$e_{o(s)} + \frac{K_1 K_2 e_{o(s)}F_{(s)}}{s} = \frac{K_1 F_{(s)}\dot{\omega}_m}{s^2 + \omega_m^2}$$

$$e_{o(s)} = \frac{s K_1 F_{(s)}\dot{\omega}_m}{(s_2 + \omega_m^2)(s + K_1 K_2 F_{(s)})}.$$

Now, since the loop filter is a lag-*RC* type, as is called for by most tracking phase lock loops,

$$F_{(s)} = \frac{1 + T_2 s}{1 + T_1 s}.$$

Substituting this value for $F_{(s)}$ into the above equation for $e_{o(s)}$,

$$e_{o(s)} = \frac{s K_1 \dot{\omega}_m (1 + T_2 s)}{(s^2 + \omega_m^2)s(1 + T_1 s) + K_1 K_2 (1 + T_2 s)\dot{\omega}_m},$$

rearranging terms

$$e_{o(s)} = \frac{K_1 s T_2 (1/T_2 + s) \dot{\omega}_m}{(s^2 + \omega_m^2)[(s T_1 (1/T_1 + s) + K_1 K_2 T_2 (1/T_2 + s) \dot{\omega}_m]},$$

and converting to the time domain

$$e_o(t) = \frac{K_1 T_2 \dot{\Delta}_\omega}{T_1} \left[\frac{\omega_m^2 + 1/T_1^2}{(K_1 K_2/T_1 - \omega_m)^2 + 4[(1 + K_1 K_2 T_2)/2T_1]^2 \omega_m^2} \right]^{1/2}$$

$$\sin(\omega_m' + \phi)$$

At $\omega_m = 0$

$$e_o(t) = \frac{\dot{\omega}_m}{K_2}.$$

At the point where $e_o(t)$ is down 3 db

$$e_o(t) = \frac{\sqrt{2} \, \dot{\omega}_m}{2K_2}.$$

Now, setting

$$e_o(t - 3\,\text{dB}) = \frac{\sqrt{2} \, \dot{\omega}_m}{2K_2}$$

$$= \frac{K_1 T_2 \dot{\omega}_m}{T_1} \left[\frac{(\omega_{3\,\text{dB}})^2 + (1/T_2)^2}{\left[K_1 K_2/T_1 - (\omega_{3\,\text{dB}})^2 \right]^2 + 4[(1 + K_1 K_2 T_2)/2T_1]^2 (\omega_{3\,\text{dB}})^2} \right]^{1/2}$$

$$\sin(\omega_s t + \phi),$$

defining

$$\omega_n = \sqrt{\frac{K_1 K_4}{R_1 C}}$$

$$\zeta = \frac{1 + K_1 K_2 T_2}{2T_1 \omega_n},$$

substituting the above, and squaring

$$1 = 2(\omega_n)^4 \frac{(\omega_{3\,dB})^2 T_2^2 + 1}{(\omega_3)^4 + (\omega_3)^2(4\omega_n^2\zeta^2 - 2\omega_n^2)\omega_n^4}$$

$$= \frac{2\left[(\omega_{3\,dB})^2 T_2^2 + 1\right]}{1 + (\omega_{3\,dB}^2/\omega_n^2)(4\zeta^2 - 2) + \omega_{3\,dB}^4/\omega_n^4}$$

and substituting

$$\zeta = 0.7, \qquad T_2 = \sqrt{2}/\omega_n,$$

$$\frac{\omega_{3\,dB}^4}{\omega_n^4} - \frac{4\omega_{3\,dB}^2}{\omega_n^2} - 1 = 0,$$

$$\omega_{3\,dB}^2 = \frac{4/\omega_n^2 \pm \sqrt{16/\omega_n^4 + 4/\omega_n^4}}{2/\omega_n^4}$$

$$= \frac{4 \pm 4.48}{2}\omega_n^2,$$

$$\omega_{3\,dB}^2 = 4.24\omega_n^2,$$

$$\omega_{3\,dB} = 2.06\omega_n.$$

This is the same result previously derived from Hoffman's values given for B_N as related to ω_n and $f_{(3\,dB)}$.

APPENDIX 7

GOLD CODE SELECTION

Any code can be represented by a polynomial, where the binary codes are represented by a polynomial of the form

$$1 + AX + BX^2 + CX^3 + \cdots + ZX^n$$

Here, each coefficient $(A, B, \ldots, Z)$ is either 0 or 1, each term of the polynomial (except for the first, 1) corresponds to a stage of a binary shift register, and there are n stages in the register. That is, each term in the polynomial containing an X corresponds one to one with a stage in a binary shift register.

The feedback connections in the code generator are defined by the terms in the polynomial whose coefficient is 1. For example, a code generator whose characteristic polynomial is $1 + 1X + 1X^2 + 1X^3 + 1X^7$ would have seven stages with feedback taken from its first, second, third, and seventh stages. Comparing this notation with that we have used previously, we see that the same code could be expressed as

$$[7, 3, 2, 1]_s = [7, 6, 5, 4]_m = 1 + X + X^2 + X^3 + X^7$$

Linear maximal codes, in which we have major interest, have characteristic polynomials that are primitive. That is, the primitive or nonfactorable polynomials each define a different linear maximal code. Fortunately, the tables of polynomials mentioned in Section 3.6 both define primitive polynomials and

TABLE A7.1   Performance of Preferred Pairs Compared with Worst Case Pairs

Degree	Period	Worst Case	Preferred Pair	Undesired Correlation Difference (DB)
5	31	11	9	1.7
6	63	23	15	3.7
7	127	41	17	7.6
8	255	95	31	9.7
9	511	113	33	10.7
10	1023	383	63	15.7
11	2047	287	65	12.9
12	4095	1407	127	20.9
13	8191	703	127	14.9

provide information that allows proper selection of pairs of codes for use in generating Gold codes.

Before going further, let us make it clear that generation of proper Gold codes requires that the codes need be properly chosen. Arbitrary selection of code pairs from the tables can result in very poor correlation performance, as is demonstrated by the result shown in Table A7.1. The "preferred" pairs of codes, as selected by the Gold-derived algorithm, always give undesired correlation that is bounded at $2^{(n+1)/2} + 1(n$ odd) and $2^{(n+2)/2} - 1(n$ even), however.

The Gold-derived algorithm for selection of preferred pairs requires the use of code tables that list the polynomial roots (as do Peterson's tables[326]).

TABLE A7.2   Description of Available Tables of Binary Polynomials

	Irreducible Polynomials	Number of Maximals	Degree	Polynomial Roots	Correlation Function of Sequences
Marsh (1957)	Yes	All	19	No	No
Peterson (1961)	Yes	All	16	Yes	No
		Partial	17–34		
Watson (1961)	No	1	100	No	No
Gold (1964)	No	All	13	Yes	Yes
Bradford (1965)	No	Partial	58	No	No

TABLE A7.3   Preferred Pairs for Gold Code Generation

Degree	Preferred Pair	Bound
5	[5, 2][5, 4, 3, 2]	9
6	NONE	
7	[7, 3][7, 3, 2, 1]	15
	[7, 3, 2, 1][7, 5, 4, 3, 2, 2]	
8	[8, 7, 6, 5, 2, 1][8, 7, 6, 1]	31
9	[9, 4][9, 6, 4, 3]	33
	[9, 6, 4, 3][9, 8, 4, 1]	
10	[10, 9, 8, 7, 6, 5, 4, 3][10, 9, 7, 6, 4, 1]	63
	[10, 8, 7, 6, 5, 4, 3][10, 9, 7, 6, 4, 1]	
	[10, 8, 5, 1][10, 7, 6, 4, 2, 1]	
	[10, 9, 8, 6, 5, 1][10, 5, 3, 2]	
	[10, 8, 4, 3][10, 9, 6, 5, 4, 3]	
	[10, 8, 7, 6, 2, 1][10, 9, 4, 2]	
	[10, 5, 2, 1][10, 9, 7, 6, 4, 3, 2, 1]	
11	[11, 2][11, 8, 5, 2]	65
	[11, 8, 5, 2][11, 10, 3, 2]	
	[11, 7, 3, 2][[11, 10, 9.7, 6, 4, 3, 2]	
12	[12, 11, 9.8, 7, 5, 2, 1][10, 10, 5, 4, 2, 1]	127
	[12, 9, 3, 2][12, 11, 8, 7, 6, 3, 2, 1]	
	[12, 11, 6, 4, 2, 1][12, 11, 8, 7, 5, 4, 3, 2]	
	[12, 11, 9.7, 6, 4][12, 6, 5, 3]	
	[12, 9, 8, 3, 2, 1][12, 11, 8, 7, 3, 1]	
	[12, 9, 8, 5, 4, 3][12, 7, 6, 4]	
	[12, 10, 7, 5, 3, 2][12, 11, 8, 4, 3, 1]	
	[12, 11, 7, 4, 2, 1][12, 10, 9.8, 4, 3]	
	[12, 10, 6, 5, 2, 1][12, 11, 10, 4]	
	[12, 11, 10, 9, 8, 7, 5, 4, 3, 2][12, 11, 7, 4]	
	[12, 9.7, 6, 3, 1][12, 11, 10, 7, 6, 5, 4, 1]	
	[12, 11, 10, 9.4, 2][12, 8, 4, 3, 2, 1]	
	[12, 9, 8, 5, 4, 3, 2, 1][12, 11, 7, 5, 4, 3]	
	[12, 10, 8, 7, 6, 2][12, 10, 2, 1]	
13	[13, 4, 3, 1][13, 10, 9, 7, 5, 4]	129
	[13, 10, 9.7, 5, 4][13, 12, 8, 7, 6, 5]	
	[13, 11, 8, 7, 4, 1][13, 11, 10, 5, 4, 3, 2, 1]	

The algorithm is used as follows:

1. Select a polynomial of the proper degree from the table (an n-stage shift register requires an nth-degree polynomial).

2. Read the number (k) in the polynomial roots column associated with the polynomial selected.

3. If the code generator has an odd number of stages, then calculate $2^k + 1$. If the number of stages is even, calculate $2^{(N+2)/2} + k$.

4. The number calculated in step 3 is the polynomial root of a second code that completes a preferred pair.

Use of any polynomial (code) with the polynomial root calculated in step 4 will produce Gold codes when combined with the original code that has properly bounded correlation with all members of the set.

Example: For 19-stage codes, suppose we select a code whose polynomial is 2000047, which converts to 111,001,000,000,000,000,010 or $1 + X + X^2 + X^5 + X^{19}$. The polynomial has 1 as its polynomial root.

With the algorithm from the foregoing definition (step 4), we calculate the second polynomial root required as $2^k + 1 = 2^1 + 1 = 3$. A second polynomial having polynomial root = 3 is 2020471 or $1 + X^3 + X^4 + X^5 + X^8 + X^{13} + X^{19}$.

This pair of codes would, when combined, produce Gold codes, every one of which would have cross-correlation bounded at $2^{(n+1)/2} + 1 = 1025$, which is $20 \log(2^{19}/1025) = 54 \, \text{dB}$ below the peak of autocorrelation. A listing of some readily available tables of primitive polynomials is given in Table A7.2. Note that of those given, however, only two list the polynomial roots.

A compilation of preferred pairs for generation of Gold codes is given in Table A7.3. This table lists all of the preferred pairs for degrees 5 through 13, which are the most commonly used Gold code lengths. These codes are most often used as synchronization preambles or for other not-so-critical applications. As code length becomes greater than $2^{13} - 1$ (8191 chips) more and more codes are available and the use of Gold codes may not be necessary.

APPENDIX 8

TYPICAL ERROR CORRECTION CODING INTEGRATED CIRCUITS

This appendix includes data sheets from Qualcomm Corporation for a convolutional encoder/Viterbi decoder, and from Space Research Technology, Inc., for a Golay encoder/decoder. In both cases, the material included here is excerpted from much larger and more comprehensive data booklets.

These are given here to illustrate the capability of current integrated circuit error correction subsystems, and represent only a small part of all that is available. However, these do represent well those error correction I.C.'s that are available readily on the open market, and that are easily adaptable for spread spectrum use.

This information is reproduced by permission of Qualcomm Corporation and Space Research Technology, Inc.

QUALCOMM

Q1650

k=7 MULTI-CODE RATE VITERBI DECODER

2.5, 10, 25 Mbps Data Rates

Technical Data Sheet

Q1650 Viterbi Decoder

FEATURES

- Stand-Alone Full-Duplex Rate 1/3, 1/2, 3/4, and 7/8 Encoder/Decoder
- On-Chip Bit Error Rate Monitor
- Three Pin-for-Pin Compatible Versions: 2.5, 10, and 25 Mbps Maximum Data Rates
- Processor Interface Simplifies Control and Status
- 3-Bit Soft-Decision Decoder Inputs
- 5.2 dB Coding Gain (Rate 1/2), 5.5 dB Coding Gain (Rate 1/3) at 10^{-5} BER
- Easy Implementation of Additional Code Rates
- Automatic Synchronization Capability
- Parallel or Serial Decoder Data Inputs
- On-Chip V.35 Data Scrambler/Descrambler
- Low Power CMOS Implementation
- Standard 84-Pin LCC Package
- Complies with INTELSAT IESS-308 and 309
- Commercial and MIL versions available

GENERAL DESCRIPTION

Forward Error Correction (FEC) techniques provide higher throughput data rates with improved bit error rate performance for power-limited (and in some cases bandwidth-limited) digital communication channels. Convolutional encoding of data combined with Viterbi decoding at the receiving node is the industry FEC standard for digital channels, especially those concerned with errors caused by the introduction of additive white Gaussian noise (AWGN), Satellite communication channels are examples of this noise environment. Figure 1 shows a typical application of FEC techniques in a communication system.

The Q1650 decoder provides convolutional encoding and Viterbi decoding of data channels while also incorporating powerful built-in features. Its capabilities have been optimized for modern digital communication channels. Figures 2 and 3 show block diagrams of the encoder and decoder functions of the Q1650 FEC system.

The Q1650 device processes data at one of four selectable code rates (1/3, 1/2, 3/4, and 7/8) using the industry standard constraint length (k) seven algorithms. It can operate at other code rates with minimal external circuitry. the Q1650 device provides built-in synchronization capability for standard BPSK (Binary Phase Shift Keying), QPSK (Quadrature), and OQPSK (Offset Quadrature) modems and operates with either 1-bit hard-decision data or 3-bit soft-decision encoded data.

The decoder is available in multiple speed grades with maximum data rates of 2.5, 10, and 25 Mbps to suit a variety of systems applications. All speed versions are pin-for-pin compatible; a single modem board design can be used for many of data rates.

The Q1650 device includes two powerful built-in techniques for monitoring synchronization status and for performing channel bit error rate measurements. In addition, the device includes a processor interface to facilitate control and status monitoring functions while keeping device pinout to a minimum.

The Q1650 decoder is packaged in an 84-pin plastic or ceramic LCC package and is implemented in fully static CMOS logic to reduce power consumption. The device uses fully parallel circuit architecture to negate the requirement for a higher speed computation clock as found in most Viterbi decoder implementations.

The Q1650 decoder is well suited to commercial satellite communication networks, including INTELSAT and INMARSAT, and to military and NASA communication systems. The low cost and high performance of the Q1650 product line make it ideal for FEC requirements in systems like DBS, VSAT, digital modems, and digital cellular telephone applications.

QUALCOMM Incorporated, VLSI Products Division · 10555 Sorrento Valley Road · San Diego, California 92121-1617 USA
TEL (619) 597-5005 · Data may change without notice · FAX (619) 452-9096

Typical Application of FEC in a Communication System

Q1650 Encoder Block Diagram

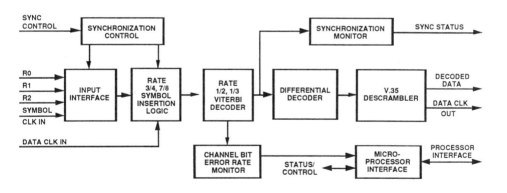

Q1650 Decoder Block Diagram

Figure A8.1   Application, encoder, and decoder block diagrams for Q1650.

Q1650 Coding Performance

Figure A8.2 Q1650 performance: BER vs. Eb/No.

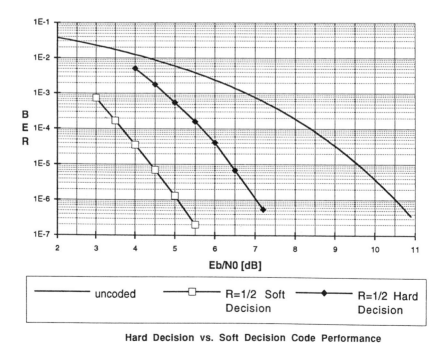

Hard Decision vs. Soft Decision Code Performance

Figure A8.3 Hard vs. soft decision performance: BER vs. Eb/No.

HYPER-FEC III CMOS HARD DECISION FORWARD ERROR CORRECTION ENCODER/DECODER 241203

FEATURES

◊ Information data bit rates from DC to 20 Mbits per second

◊ Decoder corrects up to 3 bit errors per 24-bit codeword and detects 4 errors

◊ Approximately 3dB of net coding gain and 6dB of transfer gain at one error per million

◊ Automatically synchronizes using error free data with no loss of data. Short sync acquisition time.

◊ Decoder delay of only 35 information bit periods

◊ Decoder outputs number of errors in each received codeword

◊ Single +5V supply

◊ Loopback and bypass modes

◊ Low power CMOS design - compatible with both TTL and CMOS logic

PIN CONFIGURATION (Top View)

```
EDO  1 •      24  UDI
EDC  2        23  ECK
VDD  3        22  ESI
ESO  4        21  CDI
EBP  5        20  LPB
     6        19
     7        18  WT0
DBP  8        17  WT1
GND  9        16  ASE
WT2  10       15  DCK
DDO  11       14  DSI
DDC  12       13  DSO
```

PIN NAMES

UDI - UNCODED DATA INPUT	EDO - ENCODER DATA OUTPUT
ECK - ENCODER CLOCK	EDC - ENCODER DATA CLOCK
ESI - ENCODER SYNC INPUT	ESO - ENCODER SYNC OUTPUT
CDI - CODED DATA INPUT	EBP - ENCODER BYPASS
LPB - LOOPBACK ENABLE	DBP - DECODER BYPASS
WT0 - ERROR WEIGHT BIT 0	WT2 - ERROR WEIGHT BIT 2
WT1 - ERROR WEIGHT BIT 1	DDO - DECODER DATA OUTPUT
ASE - AUTO-SYNC ENABLE	DDC - DECODER DATA CLOCK
DCK - DECODER CLOCK	VDD - POWER
DSI - DECODER SYNC INPUT	GND - GROUND
DSO - DECODER SYNC OUTPUT	

FIGURE 1. PIN CONFIGURATION.

DESCRIPTION

The 241203 HYPER-FEC III Encoder/Decoder consists of a Hard Decision encoder and decoder designed to detect and correct errors in digital transmission channels. The Encoder accepts a 12-bit serial data word, and by adding check bits, converts it to a 24-bit serial codeword. The Decoder receives the serial codeword and performs the decoding algorithm to correct up to three and detect any four random errors occuring anywhere in the received codeword. The corrected 12-bit data word is then extracted from the 24-bit codeword and is outputted serially. The Encoder and Decoder are functionally independent, enabling full duplex operation at the same or different data rates. The HYPER-FEC III is based on a one-half rate (24,12) extended Golay short block code.

The only inputs required by the Encoder (Decoder) are uncoded data (coded data) and clock. The user can provide an optional encoder sync to specify the 12-bit data word boundaries. An encoder sync output is provided to help synchronize the coded output data with other equipment.

The Decoder contains an auto-synchronization circuit which locks in after only two consecutive error-free words have been received. The auto-sync circuit is constantly monitoring itself to prevent false sync acquisition and to reacquire sync, if sync is lost. The auto-sync circuit can be overridden, if desired, by applying an external synchronization signal. A sync output is provided to synchronize the decoded output data with other equipment. The decoder measures the bit-error-rate by counting and outputting the number of errors detected in each input codeword.

The HYPER-FEC III is offered in a 24-pin, 600 mil, dual-in-line plastic package and is guaranteed to operate from 0°C to 70°C. Industrial and military grade parts are available by special request.

CODE SPECIFICATION

CODE TYPE:	Hard Decision (24,12) Extended Golay	CODE RATE:	1/2
NET CODING GAIN:	3dB @ BER of 1×10^{-6}	TRANSFER CODING GAIN:	6dB @ BER of 1×10^{-6}
EFFICIENCY:	Theoretical upper bound, i.e., an optimal quasi-perfect code	THROUGHPUT RATE:	DC to 20 MBits per second of user data

PIN DESCRIPTIONS (ENCODER)

ESI ENCODER SYNC INPUT - An optional input used to synchronize the encoder timing. The sync signal should occur during the last bit period (bit 12) of a data word to establish the timing for all subsequent data bits until the next sync pulse occurs. ESI is clocked into the encoder with the falling edge of ECK.

ECK ENCODER CLOCK - This input provides the timing for the encoder and shouldbe run at the coded data rate, i.e. twice the rate of the uncoded input data (UDI).

UDI UNCODED DATA INPUT - The serial input data to be encoded. One data word consists of 12 bits of data. The input data bit rate is one-half the rate of the encoder clock (ECK).

ESO ENCODER SYNC OUTPUT - This output establishes the encoder data output (EDO) codeword boundaries. The pulse width of ESO is equal to one output bit time with the falling edge occuring at the midpoint of the the 24th bit of the codeword.

EDC ENCODER DATA CLOCK - This output is a clock signal running at the rate of the input data, i.e. one-half the rate of the encoder clock (ECK). EDC is used to clock in the uncoded data input (UDI).

EDO ENCODER DATA OUTPUT - This output is the serially coded data output running at the rate of the encoder clock (ECK). It is formatted into 24-bit codewords.

EBP ENCODER BYPASS - This input signal allows the user to bypass the encoding process. When enabled (HIGH), the uncoded data input (UDI) is connected directly to the encoder data output (EDO). EBP must be disabled (LOW) for encoder operation.

PIN DESCRIPTIONS (DECODER)

CDI CODED DATA INPUT - This input is the serial input coded data for decoding. One codeword consists of 24 bits (12 data plus 12 parity check bits). This input runs at the rate of the decoder clock (DDC).

DCK DECODER CLOCK - This input clock provides the timing for the decoder and runs at the coded data rate. Codeword data bits are clocked into the decoder on the rising edge of DCK.

DSI DECODER SYNC INPUT - This optional input sync signal establishes codeword synchronization. DSI should occur during the last bit period (bit 24) of a codeword to establish the timing for all subsequent codewords until the next sync pulse occurs.

DDO DECODER DATA OUTPUT - This output is the serially decoded data output running at half the rate of the decoder clock (DCK). It should be considered as an error free data word unless 4 errors are detected (WT2 = HIGH).

DDC DECODER DATA CLOCK - This output clock signal runs at the rate of the decoder data output, i.e. one-half the rate of the decoder clock (DCK). DDC clocks out the decoder data output (DDO) on the falling edge.

DSO DECODER SYNC OUTPUT - This output signal is synchronous with the last bit of each output data word. DSO can be used to synchronize the transfer of output data words to external equipment or circuitry.

ASE AUTO SYNC ENABLE - This input signal is used to enable (HIGH) or disable (LOW) the decoder's internal auto sync circuitry. ASE has an internal pullup so that the auto sync is enabled with no pin connection.

LPB LOOPBACK ENABLE - This input signal enables (HIGH) or disables (LOW) the connection of the encoder output to the decoder input to facilitate loopback testing.

DBP DECODER BYPASS - This input signal allows the user to bypass the decoding process. When enabled (HIGH), the decoder data input (CDI) is connected directly to the decoder data output (DDO). DBP must be disabled (LOW) for encoder operation.

SPACE RESEARCH TECHNOLOGY, INC.
2323 S. VOSS RD., SUITE 123
HOUSTON. TEXAS 77057
PHONE: (818) 991–0693

BER VS. SNR FOR A CHANNEL WITH AND WITHOUT (24,12) EXTENDED GOLAY
CODING. ASSUMES AN AWGN* CHANNEL AND BPSK OR QPSK MODULATION.

*AWGN = ADDITIVE WHITE GAUSSIAN NOISE

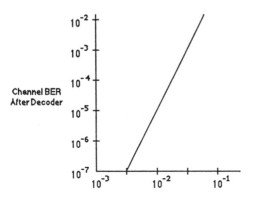

CHANNEL BER BEFORE DECODER VS. BER AFTER DECODER FOR (24,12) EXTENDED
GOLAY CODING. ASSUMES AN AWGN* CHANNEL AND BPSK OR QPSK MODULATION.

*AWGN = ADDITIVE WHITE GAUSSIAN NOISE

Figure A8.4   Performance of SRT Golay encoder-decoder.

APPENDIX 9

TYPICAL INTEGRATED-CIRCUIT FREQUENCY SYNTHESIZERS

This appendix includes excerpts from the data sheets for three typical integrated-circuit frequency synthesizers, all of the indirect (phase-lock) type. No integrated circuit of the direct type capable of generating hundreds to thousands of frequencies is known to this author. Of the three integrated circuits shown, the first is from Motorola. This I.C., the MC145152, is one of the very few available that has parallel frequency control, which makes it readily adaptable to accepting the output from multiple stages of a code generator's shift register. That is, a code generator's output can be fed directly to this synthesizer chip, which translates the code state to an output frequency that holds until the next code clock occurs. This is a very convenient configuration for a frequency-hopping synthesizer.

The other two synthesizers for which excerpts are given here are from Fujitsu: the MB1507 and MB1508. Both of these synthesizers have the advantage that they include high-speed dual-modulus prescalers cable of counting to 2.0 and 2.5 GHz, which means that both are capable of generating output signals to 2.0 GHz or greater with very little added circuitry except for a voltage-controlled oscillator. Both of these I.C.'s, however, require serial control through a proprietary three-wire bus, which makes them much more difficult to use in frequency-hopping systems.

512

MOTOROLA

MC145152-1

HIGH-PERFORMANCE
CMOS
LOW-POWER COMPLEMENTARY MOS
SILICON-GATE

**PARALLEL INPUT PLL
FREQUENCY SYNTHESIZER**

Advance Information

PARALLEL INPUT PLL FREQUENCY SYNTHESIZER

The MC145152-1 is programmed by sixteen parallel inputs. The device features consist of a reference oscillator, selectable-reference divider, two output phase detectors, 10-bit programmable divide-by-N counter and 6-bit programmable $\div$ A counter. When combined with a loop filter and VCO, the MC145152-1 can provide all the remaining functions for a PLL frequency synthesizer operating up to the device's frequency limit. For higher VCO frequency operation, a down mixer or a dual modulus prescaler can be used between the VCO and MC145152-1.

The MC145152-1 offers improved performance over the MC145152. Modulus Control output drive has been increased and the ac characteristics have been improved. Input current requirements have also been changed.

- General Purpose Applications:
  CATV   TV Tuning
  AM/FM Radios   Scanning Receivers
  Two-Way Radios   Amateur Radio
- Low Power Consumption
- 3.0 to 9.0 V Supply Range
- On- or Off-Chip Reference Oscillator Operation
- Lock Detect Signal
- Dual Modulus/Parallel Programming
- 8 User-Selectable $\div$ R Values — 8, 64, 128, 256, 512, 1024, 1160, 2048
- $\div$ N Range = 3 to 1023, $\div$ A Range = 0 to 63
- Chip Complexity: 8000 FETs or 2000 Equivalent Gates

MC145152L1
CERAMIC PACKAGE
CASE 733

MC145152P1
PLASTIC PACKAGE
CASE 710

PIN ASSIGNMENT

1	f_{in}	LD	28
2	V_{SS}	OSC_{in}	27
3	V_{DD}	OSC_{out}	26
4	RA0	A4	25
5	RA1	A3	24
6	RA2	A0	23
7	ϕ_R	A2	22
8	ϕ_V	A1	21
9	Mod Control	N9	20
10	A5	N8	19
11	N0	N7	18
12	N1	N6	17
13	N2	N5	16
14	N3	N4	15

BLOCK DIAGRAM

Note: N0 through N9, A0 through A5 and RA0 through RA2 have pullup resistors not shown.

V_{DD} = Pin 3
V_{SS} = Pin 2

This document contains information on a new product. Specifications and information herein are subject to change without notice.

FIGURE 8 — AIRCRAFT NAV RECEIVER SYNTHESIZER DEMONSTRATES
A LOW COST DUAL MODULUS SYSTEM EMPLOYING THE MC145152-1

NOTES:
1. $f_R = 50$ kHz, $\div R = 64$; 22.0 MHz low side injection; $N_{TOTAL} = 1720 \rightarrow 1919$.
2. Using 22.0 MHz for the receiver I.F. demonstrates how the choice of I.F. value can sometimes reduce the number of $\div N$ bits that must be programmed. Using the more common 21.4 MHz I.F. would require six rather than four $\div N$ programming inputs.

FIGURE 9 — SYNTHESIZER FOR LAND MOBILE RADIO UHF BAND
COVERAGE DEMONSTRATES USE OF THE MC145152-1
IN SYSTEMS OPERATING TO SEVERAL HUNDRED MHz

NOTES:
1. $N_{TOTAL} = N \cdot 64 + A = 32480$ to 37600; N = 507 to 587; A = 0 to 63.
2. $f_R = 12.5$ kHz, $\div R = 1024$ (code 101).
3. The prescaling approach can be chosen for the application to enhance economy e.g., single chip MC3393P to approximately 100 MHz.
MC12011 or MC12013 with dual flip flop to approximately 250 MHz.
MC12011 or MC12013 with MC10178 to over 500 MHz.

March 1991
Edition 3.0

FUJITSU

===== DATA SHEET =====

MB1507
SERIAL INPUT PLL FREQUENCY SYNTHESIZER

SERIAL INPUT PLL FREQUENCY SYNTHESIZER
WITH 2.0GHz PRESCALER

The Fujitsu MB1507 is a single chip serial input PLL frequency synthesizer designed for BS tuner and cellular telephone applications.

It contains a 2.0 GHz dual modulus prescaler which enables pulse swallow function, and an analog switch to speed up lock up time.

It operates supply voltage of 5.0V typ. and dissipates 18mA typ. of current realized through the use of Fujitsu's unique U-ESBIC Bi-CMOS technology.

- High operating frequency: $f_{IN\ MAX}$=2.0GHz ($V_{IN\ MIN}$=−4dBm)
- Pulse swallow function: 128/129 or 256/257
- Low supply current: I_{CC}=18mA typ.
- Serial input 19-bit programmable divider consisting of:
  - Binary 8-bit swallow counter: 0 to 255
  - Binary 11-bit programmable counter: 16 to 2047
- Serial input 15-bit programmable reference divider consisting of:
  - Binary 14-bit programmable reference counter: 8 to 16383
  - 1-bit switch counter (SW) sets divide ration of prescaler
- On-chip analog switch achieves fast lock up time
- Two types of phase detector output
  - On-chip charge pump (Bipolar type)
  - Output for external charge pump
- Wide operating temperature: −40°C to +85°C
- 16-pin Plastic Flat Package (Suffix: −PF)

PLASTIC PACKAGE
FPT-16P-M06

PIN ASSIGNMENT

OSC$_{IN}$	1	16	øR
OSC$_{OUT}$	2	15	øP
V$_P$	3	14	I$_{OUT}$
V$_{CC}$	4	13	BISW
D$_O$	5	12	FC
GND	6	11	LE
LD	7	10	Data
I$_{IN}$	8	9	Clock

TOP VIEW

ABSOLUTE MAXIMUM RATINGS (see NOTE)

Rating	Symbol	Value	Unit
Power Supply Voltage	V$_{CC}$	−0.5 to +7.0	V
	V$_P$	V$_{CC}$ to 10.0	V
Output Voltage	V$_{OUT}$	−0.5 to V$_{CC}$ +0.5	V
Open-drain Voltage	V$_{OOP}$	−0.5 to 8.0	V
Output Current	I$_{OUT}$	+10	mA
Storage Temperature	T$_{STG}$	−55 to +125	°C

NOTE: Permanent device damage may occur if the above **Absolute Maximum Ratings** are exceeded. Functional operation should be restricted to the conditions as detailed in the operational sections of this data sheet. Exposure to absolute maximum rating conditions for extended periods may affect device reliability.

This device contains circuitry to protect the inputs against damage due to high static voltages or electric fields. However, it is advised that normal precautions be taken to avoid application of any voltage higher than maximum rated voltages to this high impedance circuit.

Copyright© 1991 FUJITSU LIMITED

MB1507

TYPICAL APPLICATION EXAMPLE

V_p, V_{PX} : 8V max.
C_1, C_2 : Depends on crystal oscillator
LE,FC : With pull up resistor
øP : Open drain output

April 1991
Edition 2.0

DATA SHEET

MB1508
SERIAL INPUT PLL FREQUENCY SYNTHESIZER

SERIAL INPUT PLL FREQUENCY SYNTHESIZER
ON CHIP 2.5GHz PRESCALER

The Fujitsu MB1508 on chip 2.5 GHz dual modulus prescaler is a serial input PLL (Phase Locked Loop) frequency synthesizer with pulse swallow function. It is well suited for BS tuner, CATV system, and TV tuner applications.

It operates supply voltage of 5.0V typ. and dissipates 16mA typ. of current realized through the use of Fujitsu's unique U-ESBIC Bi-CMOS technology.

- Power supply voltage: V_{cc} = 4.5 to 5.5V

- High operating frequency: f_{in} = 2.5GHz (V_{in} = –4dBm)

- 2.5GHz dual modulus prescaler: P = 256/272, 512/528

- Low power supply current: I_{cc} = 16mA typ.

- Programmable reference divider consisting of:
  Binary 2–bit programmable reference counter (R = 256, 512, 1024, 2048)

- Programmable divider consisting of:
  Binary 5–bit swallow counter (A = 0 to 31)
  Binary 12-bit programmable counter (N = 32 to 4095)

- Wide operating temperature: T_A = –40 to +85°C

- Plastic 20-pin flat package (Suffix: –PF)

PLASTIC PACKAGE
FPT-20P-M01

PIN ASSIGNMENT

	TOP VIEW	
FC 1		20 LD
LE 2		19 f_{OUT}
Data 3		18 V_{CC2}
Clock 4		17 f_{in}
V_{CC1} 5		16 GND2
OSC_{IN} 6		15 $\overline{f_{in}}$
OSC_{OUT} 7		14 BC1
GND 8		13 BC2
D_{O1} 9		12 BC3
D_{O2} 10		11 BC4

ABSOLUTE MAXIMUM RATINGS (see NOTE)

Rating	Symbol	Value	Unit
Power Supply Voltage	V_{cc}	–0.5 to 7.0	V
Output Voltage	V_o	0.5 to V_{cc} +0.5	V
Output Current	I_o	±10	mA
Storage Temperature	T_{STG}	–55 to +125	°C

NOTE: Permanent device damage may occur if the above **Absolute Maximum Ratings** are exceeded. Functional operation should be restricted to the conditions as detailed in the operational sections of this data sheet. Exposure to absolute maximum rating conditions for extended periods may affect device reliability.

This device contains circuitry to protect the inputs against damage due to high static voltages or electric fields. However, it is advised that normal precautions be taken to avoid application of any voltage higher than maximum rated voltages to this high impedance circuit.

Copyright © 1991 by FUJITSU LIMITED

MB1508

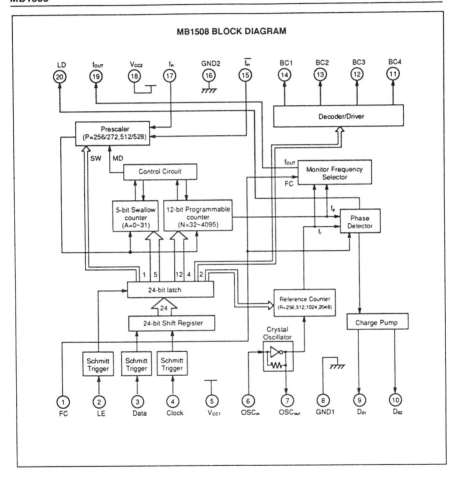

MB1508 BLOCK DIAGRAM

MB1508 APPLICATION CIRCUIT

C₁, C₂ : depends on the crystal oscillator.
FC : with internal pull up resistor.

APPENDIX 10

REPORT ON SPREAD SPECTRUM AND OTHER INTERFERENCE WITH RESPECT TO ITS EFFECT ON STANDARD MICROWAVE COMMUNICATIONS LINKS

This report has been included in its entirety because of the importance of spectrum sharing to new commercial spread spectrum personal communications services that have been proposed and are currently being considered by the Federal Communications Commission. Included in the report are both laboratory and field test results, carried out on operational microwave links, that show that the degree of interference from a spread spectrum signal to a microwave link is considerably less than the interference from a narrowband signal to the same microwave link. In fact, according to this report, any required exclusion zones (to prevent interfering with a typical microwave link) are "likely to be 100 times larger for narrowband systems" (than for spread spectrum systems).

FDM-FM Microwave Link

Propagation and

Interference Test Results

Interim Report

June 25, 1992

Omnipoint Communications, Inc.

1.0 INTRODUCTION

Responsible spectrum sharing with incumbent OFS microwave users requires a clear understanding of how these systems react to different types of interfering signals as a function of range, bandwidth and format. Depending on modulation format and bandwidth, OFS microwave receivers exhibit markedly different responses. In particular, OFS receivers are as much as 20 dB more susceptible to narrowband interference than Omnipoint's 10-MHz signals. As a consequence of this, narrowband systems will require large exclusion zones and have greater difficulty in sharing with incumbent OFS users. Required exclusion areas are likely to be up to 100 times larger for narrowband systems. Complicating the issue, propagation conditions into the victim OFS receiver are highly variable. The Hata[1] propagation models used in most filings present an overly simplistic view of the interference potential from PCS. Based on field measurements taken in the *1962–1965* time frame by Okumura;[2] the Hata models represent median propagation path loss in quasi smooth terrain. As such, they do not address the "rogue" handset problem. Compared with Hata, Large city Urban predictions, a single handset, located in a region where near free space propagation conditions prevail, can appear one hundred thousand times stronger to an OFS receiver. We have observed this situation in our field testing.[3]

2.0 MICROWAVE RECEIVER SENSITIVITY TO PCS INTERFERENCE

Examining the principles of operation for FDM-FM receivers, we have concluded that narrowband interference is more damaging to OFS microwave receivers. Specifically, the narrowband FDM-FM modulation format used by 83% of all licensed users is more susceptible to narrowband interference effects because each baseband voice channel's spectral occupancy is concentrated in two, 3.1-kHz bands. A series of tests; both in the lab and in the field, were undertaken to validate our theoretical models.

The principle of operation is essentially the same among all OFS analog FDM-FM equipment. Figure 2.0-1, taken from the operations manual, shows specifications for the Motorola Starpoint transmitter/receiver pair used in our testing. The equipment used for testing operates in the common carrier band at a 2178-MHz center frequency. We used a 96-channel model (P65 option) with a per channel RMS deviation of 47 kHz, Table 2.0-1, taken from the TSB10-E identifies an OFS receiver with identical parameters, except

---

[1] M. Hata, Empirical Formula for Propagation Loss in Land Mobile Radio Services, *IEEE Trans. Vehic. Tech.*, August 1980.

[2] Y. Okumura, Field Strength and Its Variability in UHF and VHF Land-Mobile Radio Services, Tokyo Rev. Elec. Commun. Lab. vol. 16, 1968. Reprinted in *Land-Mobile Communications Engineering*, IEEE Press, 1984.

[3] Propagation results are contained in section 4.0 of this report.

2100–2300 MHz PERFORMANCE CHARACTERISTICS

GENERAL

Model Series:	K06RBF1100, K06RBF2100, K16RBF1100, and K16RBF2100
Frequency Range:	2100–2300 MHz
Channel Capacity:	48 Standard; options for 24 and 96 channels
FCC Transmitter Designation	CC6001
Modulation:	FM
Input Voltage:	-24 V dc
Baseband Frequency Response:	± 0.5 dB (.3 kHz–408 kHz)
Baseband Impedance	75 ohm unbalanced
Baseband Return Loss:	26 dB (12 kHz to 408 kHz)
Emphasis:	CCIR: Standard-Optionally Deleted
T/R Separation: to 120 MHz	50 MHz Standard (40 MHz to 65 MHz and 75 MHz available)
Baseband Connection:	BNC
Antenna Port Connection:	Type "N"
Operating Temperature Range:	$-30°$ C to $+55°$ C
Guaranteed Performance Range:	$0°$ C to $+50°$ C
Humidity:	95% @ $+40°$ C

SYSTEM GAIN SPECIFICATIONS (measured at antenna port)

				System Gain (dB) 1 Watt Models			
				Non-Standby Models		Standby Models	
Model/Option*	Channel Capacity	Deviation (kHz RMS)	Pilot Freq. kHz	Standard	Option Without Emphasis	Standard	Option Without Emphasis
P87 Option*	24	41.9	119	126.5	123.0	124.0	122.0
Standard Models*	48	26.2	331	117.0	113.5	116.0	112.5
P37 Option**	48	79.8	331	126.5	123.0	125.5	122.0
P65 Option**	96	46.9	607	116.0	112.5	115.0	111.5

*Emission Designator 800F9
**Emission Designator 1600F9   NOTE: For 100 mW models subtract 10 dB from all measurements

NOISE POWER RATIO (measured back-to-back at receiver level of -30 dBm)

		FCC Loading NPR (db)/dBrnc0		CCIR Loading NPR(dB)/dBrnc0	
Model/Option	Channel Capacity	Std.	Without Pre/ De-Emphasis	Std.	Without Pre/ De-Emphasis
P87 Option	24	56/22.5	56/22.5	56/21.6	56/21.6
Standard Models	48	51/25.0	52/24.0	51/24.8	52/23.8
P37 Option	48	56/20.0	53/23.0	56/19.8	53/22.8
P65 Option	96	52/21.9	52/21.9	52/21.9	52/21.9

SPACE AND POWER REQUIREMENTS

	Depth		Rack Units	Rack Space		Weight		Power-24 V dc Requirements
Configuration	Inches	mm		Inches	mm	lb.	kg.	
Non-Standby	18	457.2	3 RU	5.25	133.4	42	19.1	24 W
Standby	18	475.2	7.5 RU	13.125	333.4	91	41.3	48 W

2100–2300 MHz PERFORMANCE CHARACTERISTICS *(Continued)*

TRANSMITTER

Power Out Measured at Antenna Port (K06RBF Series)	Non-Standby + 20 dBm Standby + 19.5 dBm
Power Out Measured at Antenna ort (K16RBF Series)	Non-Standby + 30 dBm Standby + 29.5 dBm
Frequency Stability	$\pm 0.001\%$ ($-30°$ C to $+55°$ C)
Carrier Generation	Crystal controlled direct launch
Metering	Built-in metering for many critical functions

Transmit Baseband Levels (Non-Standby)
 	
(with emphasis)
	Input 1: -13.5 dBm to -45 dBm adjustable
	  (without emphasis)
	        -8.5 to -40 dBm adjustable
	
	Input 2: -30 dBm

Transmit Baseband Levels (Standby)	Input 1: -7.5 dBm to -39 dBm adjustable
	  (without emphasis)
	        -2.5 dBm to -34 dBm adjustable
	  (with emphasis)
	Input 2: -30 dBm

RECEIVER

Sensitivity (dBM) @30 dB S/N	Model/Option	With Emphasis	Without Emphasis
Measured at Antenna Port (Non-Standby)	P87	-96.5	-93.0
	Standard	-87.0	-83.5
	P37	-96.5	-93.0
	P65	-86.0	-82.5

Sensitivity (dBm) @30 dB S/N Measured at Antenna Port	Model/Option	With Emphasis	Without Emphasis
(Standby)	P87	-94.5	-92.5
	Standard	-86.5	-83.0
	P37	-96.0	-92.5
	P65	-85.5	-82.0

Noise Figure	6.5 dB max.
Conversion	Single
Receiver I-F	35 MHz
Image Rejection	70 dB min.
RF Bandwidth	20 MHz
I-F Bandwidth	48 and 96 Channel 2.5 MHz 24 Channel 1.5 MHz
Max. Receiver Input Power	-30 dBm
Receive Baseband Levels	Output 1: -15 dBm to -46.5 dBm, adjustable (with or without emphasis) Output 2: -15 dBm or -30 dBm

SPECIFICATIONS SUBJECT TO CHANGE WITHOUT NOTICE

(M), Motorola and STARPOINT are trademarks of Motorola, Inc. ■ © 1980 Motorola Inc. ■ Printed in U.S.A.

Figure 2.0-1 Microwave Equipment Characteristics, P65 Option, 96 Channel.

TABLE 2.0-1 Modulation Parameters

Channels	Per Channel RMS Deviation (KHz)	Average Channel Power (dBmQ)	Total rms Deviation F (KHz)	Baseband Frequencies (KHz) Lowest (f_l)	Highest (f_n)	Modulation Index (M)	Bandwidth (MHz)
			1.9/6.7 GHz Bands				
780	140	−19.6	409	60	3284	0.125	10
600	140	−15.0	610	60	2540	0.240	10
480	200	−15.0	779	60	2044	0.380	10
300	200	−15.0	616	60	1300	0.470	10
120	200	−15.0	464	60	552	0.840	10
120	200	−15.0	464	60	552	0.840	5
300	175	−19.6	319	60	1300	0.245	5
			"Grandfathered" 1.9 GHz Band				
600	88	−15.0	383	60	2540	0.15	.8
420	160	−15.0	583	60	1800	0.32	8
		2130 MHz–2150 MHz and 2180 MHz 2200 MHz Band					
132	47	−19.6	66	12	552	0.12	1.6
96	47	−15.0	104	12	408	0.25	1.6
72	60	−15.0	126	12	300	0.42	1.6
48	25	−15.0	50	12	204	0.24	0.8
24	42	−15.0	78	12	108	0.72	0.8
			12.5/13.0 GHz Band				
1200	140	−15.0	866	564	5772	0.15	20/25
600	200	−15.0	871	60	2540	0.34	20/25
300	200	−15.0	616	60	1300	0.47	20/25
300	200	−15.0	616	60	1300	0.47	10/12.5

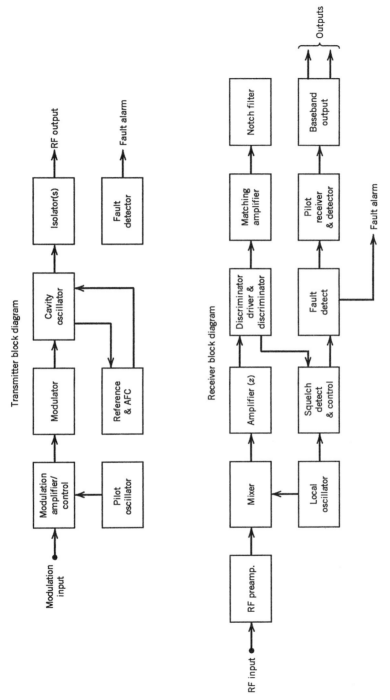

Transmitter block diagram

Receiver block diagram

Figure 2.0-2 Nonstandby station functional diagram (simplified, see Motorola no. PEPs-28.825-0, 10/1/79)

527

operating in the 2130–2150 MHz and 2180–2200 MHz POFS allocations. Of particular importance, the modulation index is small, around 0.25. These systems are in actuality using a narrowband FM signaling format, even though the bandwidths are large. Figure 2.0-2 provides a system block diagram of the transmitter and receiver.

2.1 BENCH TEST RESULTS

In quantifying an OFS receiver's sensitivity to interference, both loaded and unloaded operation must be considered. Unloaded operation refers to situations where essentially none of the voice channels are in use while loaded operation considers the case where essentially all voice channels are in use. In the unloaded situation; the OFS receiver's sensitivity is set mainly by noise Figure. As voice channels become active, intermodulation distortion becomes important in setting the apparent noise floor of the OFS receiver. This is because nonlinearies in the OFS Transmit/Receive process introduce inter-modulation products between voice channels that map into other voice channels.

2.1.1 Test Procedure (Bench)

Figures 2.1.1-1a and -1b diagram the baseline bench test setup used. Referring to the first Figure; a Marconi noise generator and a Krohn-Hite programmable filter is used to simulate baseband voice channel loading in accordance with CCIR recommendations. Switchable narrowband bandstop filters at 14-kHz, 245-kHz, and 342-kHz center frequencies can notch out selected frequencies to simulate unused voice channels and observe inter-modulation distortion. In unloaded testing, the noise generator is switched "off" so there is no input to the Motorola FM transmitter.

The Motorola Starpoint transmitter has a nominal output level of 1 Watt (+30 dBm). To simulate typical path loss, 90 dB of attenuation is inserted between the transmitter and receiver. This corresponds to a nominal fade margin of 22.5 dB; considerably less than most OFS systems.[4] In consequence, our interference studies are conservative in their estimates of how much power is needed to interfere with the OFS operator.

The Motorola Starpoint receiver used has been specially modified to include several monitoring functions in a Rx/Tx drawer test set. These modifications do not affect receiver operations but allow monitoring receiver internals not normally output. Specifically, it allows us to determine whether or not the 607-kHz pilot tone is being received correctly.

---

[4]One study conducted by COMSEARCH, considering 107 links in the Houston area, found that the average fade margin is 54.4 dB.

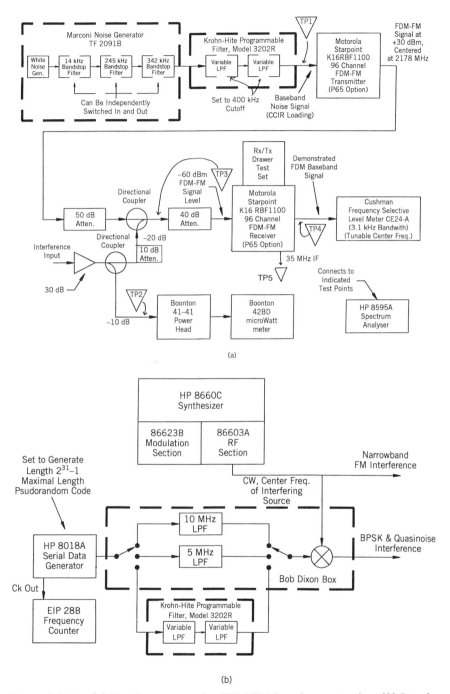

Figure 2.1.1-1 (*a*) Bench test setup for FDM-FM interference testing; (*b*) Interference source test setup

The demodulated baseband FDM signal is then input into a Cushman, frequency selective level meter. Selecting a bandwidth of 3.1 kHz, this meter allows direct observation of the power level in any voice channel since the center frequency is tunable.

Interference signal input is also shown in Figure 2.1.1-1a. After 30 dB of amplification, the interfering source is passed through a -10-dB directional coupler into a Boonton 41-41 Power Head where it is envelope detected. A Boonton 42BD microwatt meter then gives a calibrated reading of the interference source power. The main feedthrough output of the directional coupler is input into the -20 dB port of another directional coupler to inject interference into the simulated OFS link.

Five test points are indicated in Figure 2.1.1-1a: TP1 through TP5. An HP 8595A Spectrum analyzer can be connected at each of these test points to obtain a picture of the composite spectrum.

One of the key objectives in our testing is to quantify the effects of different types of interference sources. Referring to Figure 2.1.1-1b, we have constructed a specialized item of test equipment, the Bob Dixon Box, to generate a variety of interfering sources. An HP-8018A serial data generator is set to generate length $2^{31} - 1 = 2,147,483,647$ maximal length (ML) pseudo random code.[5] An attached frequency counter monitors the serial bit rate generated.

Serial data from the HP 8018A are then filtered by one of three filters: a 10-MHz LPF, a 5-MHz LPF, or a Krohn-Hite programmable filter. The resultant is then mixed with an appropriate CW signal from the HP 8660C synthesizer to translate to RF. Depending on the selected filter bandwidths and serial data chipping rate, a wide variety of interference source signals can be provided. As an example, if the code generator is run at a high rate and a narrow filter is used, the output will be a quasinoise source with essentially Gaussian characteristics.[6] Slowing the chip rate, the output of the mixer takes on a BPSK, constant envelope characteristic, typical of many narrow-band modulation formats and direct sequence spread spectrum signals. Figures 2.1.1-2a and 2b, measured at TP2, illustrate this effect.

In the first figure, a nominal 1-MHz chipping rate signal is fed into a 50-kHz LPF[7] and then translated to a 2,178,014,000-Hz center frequency to produce a noise like spectrum. In the second, the chipping rate is cut to 100 kHz while other settings remain constant. Note the slight residual sin x/x sidelobe present.

Figures 2.1.1-2c and 2d show spectrums of 0.2 Mhz and 0.5 MHz band-width quasinoise interference source signals used in this study. Although they

---

[5]An ML code was selected because it has a flat line spectrum, with equal power in each spectral line.

[6]By the central limit theorem, when the LPF "adds" together a large set of identical distributed random variables, the resultant is Gaussian.

[7]Implemented using two sections of the Krohn-Hite filter.

Figure 2.1.1-2 (*a*) 200 kHz quasinoise (peakhold) 2 kHz; (*b*) 100 kbps BPSK (filtered); (*c*) 200 kHz quasinoise (peakhold) 2 kHz; (*d*) 500 kHz quasinoise (peakhold) 2 kHz; (*e*) 1.25 MHz quasinoise (peakhold) 2 kHz (1.25 MHz filtered BPSK); (*f*) 5.0 MHz quasinoise (peakhold) bob box (5 MHz filtered BPSK); (*g*) 10.0 MHz quasinoise (peakhold) bob box (10 MHz filtered BPSK); (*h*) 100 kHz FSK (peakhold, 1 kHz baseband sinewave)

21: 49: 54 JUN 05, 1992

MKR Δ −101 kHz
−5.72 dB

REF −51.3 dBm #ATTEN 0 dB
PEAK
LOG
10
dB/

MARKER Δ
−101 kHz
−5.72 dB

MA SB
SC FC
CORR
TH
−131.2
dBm

CENTER 2.178002 GHz
RES BW 10 kHz   #VBW 10 kHz

SPAN 1000.0 kHz
SWP 30 msec

(c)

21: 55: 08 JUN 05, 1992

MKR Δ −250 kHz
−4.13 dB

REF −51.3 dBm #ATTEN 0 dB
PEAK
LOG
10
dB/

MARKER Δ
−250 kHz
−4.13 dB

MA SB
SC FC
CORR
TH
−131.2
dBm

CENTER 2.178002 GHz
#RES BW 30 kHz   #VBW 10 kHz

SPAN 2.500 MHz
SWP 25 msec

(d)

Figure 2.1.1-2   *Continued*

(e)

(f)

Figure 2.1.1-2   *Continued*

22: 24: 29 JUN 05, 1992

MKR Δ −50.0 kHz
−4.48 dB

REF −51.3 dBm #ATTEN 0 dB
PEAK
LOG
10
dB/

MARKER Δ
−5.00 MHz
−4.48 dB

MA SB
SC FC
CORR

CENTER 2.17800 GHz
#RES BW 300 kHz
#VBW 10 kHz

SPAN 50.00 kHz
SWP 50 msec

(g)

20: 44: 03 JUN 05, 1992

MKR Δ −50.0 kHz
−20.30 dB

REF −51.3 dBm #ATTEN 0 dB
PEAK
LOG
10
dB/

SWEEPTIME
100 msec

MA SB
SC FC
CORR
TH
−131.2
dBm

CENTER 2.1780020 GHz
RES BW 3.0 kHz
#VBW 10 kHz

SPAN 250.0 kHz
SWP 100 msec

(h)

Figure 2.1.1-2 *Continued*

appear identical, they are not; the spectrum analyzer span (indicated in the lower right corner of the figure) is different between the two figures. Figures 2.1.1-2e through -2g depict 1.25-, 5.0- and 10.0-MHz bandwidth filtered BPSK interference source signals. The 5- and 10-MHz cases are representative of Omnipoint signals while the 1.25-MHz signal is representative of other proposed CDMA systems.

We also tested using the HP 8660C's FM modulation section. Figure 2.1.1-2h shows the spectrum of the 100-kHz bandwidth FM signal used in our tests. The baseband modulating waveform was a 1-kHz sine wave and the modulation index was adjusted to give the desired bandwidth.

2.1.2 Unloaded Interference Sensitivity (Bench)

In the unloaded test configuration, the Marconi noise generator is turned *off* resulting in zero input to the Motorola Starpoint transmitter. Figure 2.1.2-1 depicts the transmitted signal spectrum, measured at TP3, Nominal FDM-FM signal power at this point is -20 dBm. The spectrum consists of two main components: the carrier and the pilot channel, located at a 607-kHz offset. The pilot channel is generated in the transmitter by adding a 607-kHz sine wave to the baseband signal. Since its frequency is above the top 400-kHz voice channel, it does not interfere with voice-channel information. The pilot signal serves several purposes: It is used to set gain on the baseband signal, it

(i)

Figure 2.1.2-1 TP3 signal spectrum—no modulation input (improved cable)

14: 09: 00 JUN 07, 1992

REF –37.2 dBm #ATTEN 0 dB

MKR Δ 23.7 kHz
–41.43 dB

PEAK
LOG
10
dB/

VIDEO BW
30 Hz

WA SB
SC FS
CORR

START 1.0 kHz
#RES BW 1.0 kHz

#VBW 30 kHz

STOP 500.0 kHz
SWP 500 msec

Figure 2.1.2-2 TP4 baseband output (unloaded) (no baseband input to Tx)

provides link quality assessment, and it serves as a BFO reference for the SSB-FDM signal.[8]

Figure 2.1.2-2, measured at TP4, shows the output baseband signal spectrum under unloaded conditions. As expected, with preemphasis/deemphasis turned off, the noise level rises with frequency. The spiky signals in the lower portion of the spectrum are of undetermined origin but are generated within the receiver itself. Although they appear large, they are in actuality quite small in comparison with an active voice signal.[9]

In Figure 2.1.2-3a, we have added in a 100 kbps, filtered BPSK[10] signal at a frequency of 2178.2 MHz, 200 kHz above the Motorola Starpoint's nominal center frequency. The interfering source's power level is − 80 dBm, referenced to the Starpoint receiver's input. The resultant baseband signal spectrum is shown in Figure 2.1.2-3b. Note how the power bump has formed at an offset of 200 kHz. FDM voice channels of 3.1 kHz operating in the neighborhood of this bump would experience significantly higher interference compared with voice channels at far removed frequencies.

---

[8]Because the baseband FDM signal is a series of SSB signals, slight frequency offsets in the SSB demodulation process result in the "Donald Duck" syndrome. The pilot signal provides a reference signal to measure any frequency shifts introduced by the FM transmitter/receiver pair.

[9]See Sect. 2.1.3 for examples of loaded voice circuits.

[10]See Fig. 2.1.1-2b in section 2.1.1.

Figure 2.1.2-3 (*a*) TP3 RF signal (−80 dBm, 100 kbps filtered BPSK at 2178.2 MHz); (*b*) TP4 Baseband output (−80 dBm 100 kbps, filtered BPSK at 2178.2 MKz) unloaded

Repeating the experiment except using a 10-MHz filtered BPSK signal[11] typical of Omnipoint's system, we obtain the results of Figures 2.1.2-4a and -4b. Comparing Figure 2.1.2-4a with Figure 2.1.2-1 (no interference case), we see that the wideband Omnipoint signal appears much like an elevated noise floor going into the FM receiver except for the spur at a 200 kHz offset. This spur is due to LO leakage through the double-balanced mixer in Bob Dixon's Box.[12] In practice we can eliminate this leakage term using a proprietary technique, but for test purposes we decided to leave it in since it makes a handy marker for the interfering-source center frequency.

Examining Figure 2.1.2-4b, we note that the baseband FDM signal does not exhibit the pronounced power bump seen in Figure 2.1.2-3b. In fact, comparing the two cases, we see that the power bump is 20 dB higher than corresponding frequencies for the Omnipoint spread spectrum signal. *This is because the narrowband FM* [13] *demodulation process used in FDM-FM receivers is essentially a spectrum-preserving process. A narrowband signal going into the FM discriminator will come out a narrowband signal.* The spread spectrum signal spreads its energy over a wider range of frequencies; hence, it has a low-spectral density going into the FM discriminator. This low-spectral density is preserved on output from the FM discriminator.

In Figures 2.1.2-5a and -5b, we consider an intermediate case, using the 1.25-MHz filtered BPSK signal[14] as an interfering source. This is representative of the direct-sequence QPSK CDMA signals proposed by several companies in their pioneers preference filings. In the RF spectrum plot, we note some spectrum rolloff, but because the span is only 2 MHz, we don't observe the full spectrum of the interfering signal. Examining the baseband spectrum plot (Fig. 2.1.2-5b), we see results that at first glance, appear very similar to the results obtained with a 10-MHz, spreading bandwidth. *They are not!* If we superimpose the two plots, the Omnipoint signal results in a factor of 9-dB-less interference because it has a 9-dB-lower spectral density.

Up to this point, we have been examining spectral plots to show how the interfering source's bandwidth affects the baseband signal. To make quantitative statements about how much interfering power can be tolerated as a function of bandwidth, we use the Cushman frequency-selective level meter to measure power in a 3.1-kHz voice-channel bandwidth. FDM voice channels centered at 14 kHz, 245 kHz, and, 342 kHz were taken as representative and designated Low, Middle, and High respectively. We then introduced each of the eight types of interference identified in Figures 2.1.1-2a through -2h and measured the required interference power necessary to cause 1 dB, 3 dB, 6 dB, and 10 dB of degradation in the selected voice channel. In each

---

[11]The spectrum of the test signal is shown in Figure 2.1.1-2g. In actual systems, the small sidelobes are eliminated using pulse shaping techniques not included in our test setup.

[12]Diagrammed in Figure 2.1.1-1b.

[13]Modulation index of 0.25.

[14]Depicted in Figure 2.1.1-2e.

Figure 2.1.2-4 (*a*) TP3 RF signal (− 80 dBm 10 MHz BPSK at 2178.2 MHz); (*b*) TP4 baseband output (− 80 dBm 10 MHz BPSK at 2178.2 MHz)

15: 30: 44 JUN 07, 1992

MKR 2.177990 GHz
−22.15 dBm

REF −20.0 dBm #ATTEN 10 dB

CENTER 2.178000 GHz
#RES BW 10 kHz   #VBW 1 kHz   SPAN 2.000 MHz
SWP 600 msec

(a)

15: 07: 20 JUN 07, 1992

MKR 248.0 kHz
−68.93 dB

REF −37.2 dBm #ATTEN 0 dB

BPSK mixer
leak through

× Corresponding power
level for 10 MHz chip rate

START 1.0 kHz
#RES BW 1.0 kHz   #VBW 30 kHz   STOP 500.0 kHz
SWP 50 msec

(b)

Figure 2.1.2-5 (*a*) TP3 RF signal (−80 dBm 1.25 MHz BPSK at 2178.2 MHz); (*b*) TP4 baseband signal (unloaded) (−80 dBm, 1.25 MHz BPSK at 2178.2 MHz)

TABLE 2.1.2-1 "Unloaded" Bench Test Interference Results

LAB INTERFERENCE DATA by JOHN SANELLI 07JUN92

Notes:
Shading denotes squelch bypassed.
Italics indicates Mini-circuits amplifier is in line.

	CKT	NOISE BAND	HOP UNLOAD BASEBAND RE	HOP UNLOADED RF POWER FOR 1 dB	HOP UNLOADED RF POWER FOR 3 dB	HOP UNLOADED RF POWER FOR 6 dB	HOP UNLOADED RF POWER FOR 10 dB	HOP UNLOADED RF POWER FOR DEGRADATI 20 dB
Low	L	0.1 BPSK	-99.2	-105.7	-99.9	-95.6	-93.3	-80.7
	L	0.1 FM	-99.2	-103.4	-98.6	-93.9	-89.2	-79.1
	L	0.1 QUASI NO	-99.2	-105.6	-100	-95.4	-90.7	-80.4
	L	0.2	-99.2	-102.4	-97.8	-92.8	-88.1	-77.9
	L	0.5	-99.2	-99.8	-95.2	-90	-85.3	-75.1
	L	1.25	-99.2	-95.3	-91	-86.1	-81	-70.3
	L	5	-99.2	-89.4	-84.2	-79.6	-75	-66.3
	L	10	-99.2	-85.5	-81.2	-76.6	-72.1	-64.1
Medium	M	0.1 BPSK	-82.8	-113.4	-109	-104.5	-99.8	-88.9
	M	0.1 FM	-82.8	-110	-105	-100.4	-95.6	-85.3
	M	0.1 QUASI NO	-82.8	-112.5	-107	-102.2	-97.3	-86.7
	M	0.2	-82.8	-104.1	-102	-98.3	-94.1	-84
	M	0.5	-82.8	-103	-99	-95.6	-91.2	-80.7
	M	1.25	-82.8	-102	-97	-93.5	-88.7	-78.3
	M	5	-82.8	-97.7	-93	-88.3	-83.4	-72.6
	M	10	-82.8	-95.3	-90	-85.1	-80.1	-69.3
High	H	0.1 BPSK	-79.6	-111.5	-108	-102.7	-97.8	-87.5
	H	0.1 FM	-79.6	-109.3	-106	-100.9	-96.1	-85.7
	H	0.1 QUASI NO	-79.6	-111.6	-107	-102.5	-97.6	-87.4
	H	0.2	-79.6	-109.7	-105	-99.7	-94.9	-84.6
	H	0.5	-79.6	-106.2	-101	-96	-91.2	-80.6
	H	1.25	-79.6	-102.5	-97.3	-92.2	-87.2	-76.5
	H	5	-79.6	-98.6	-93.3	-88.1	-83.1	-72.2
	H	10	-79.6	-95	-90.1	-85.1	-80	-69.2

case, the interferer was centered on the target voice channel's baseband FDM frequency offset. As an example, if the Medium channel (245 kHz) voice channel is to be targeted, the interfering source was tuned to a center frequency of 2178 MHz + 245 kHz = 2178.245 MHz. Wideband interference sources were moved over an additional 30 kHz to avoid biasing the results with mixer LO leakage.

Table 2.1.2-1 summarizes the results of this study. For each interference type, we first measure the base-channel noise power with no interference present. Then, we gradually increase interference power until the voice-channel power increases by 1 dB as measured by the appropriately tuned Cushman meter. We then note the required interference power, as measured by the Boonton microwattmeter and reference it to the FM receiver antenna post (subtract 60 dB from the measured value). This procedure is then repeated for 3-dB voice channel degradation, and so forth. In examining these results and the plotted results seen in Figure 2.1.2-6, we again confirm our expectation that FDM-FM systems are more susceptible to narrowband interference. As expected, the relative susceptibility goes as the ratio of the bandwidths. Comparing results for a 100-kHz interfering source with those for a 10-MHz source, we see that the FDM-FM receiver is

$$10 \log(10 \text{ MHz} / 100 \text{ kHz}) = 20 \text{ dB}$$

more vulnerable to the narrowband interferer. *With regards to spectrum sharing, the implications of this are enormous. For a given power level, we can be a factor of ten times closer to on OFS receiver before we cause the same level of interference. This means that our exclusion zones[15] will be 100 times smaller in terms of area.*

2.1.3   Loaded Interference Sensitivity (Bench)

In assessing the potential for interfering with FDM-FM OFS links, we also need to consider the loaded case where the link is carrying voice traffic. Rather than try and find 96 telephones and hook them up to the transmitter with 96 people on the line, the CCIR recommends using a noise source to simulate voice traffic. In the test setup of Figure 2.1.1-1a, the Marconi noise generator and Krohn-Hite filter are used to generate simulated voice traffic.

Figure 2.1.3-1a shows the resultant simulated traffic power spectrum measured at TP1. Feeding this signal into the FM transmitter results in the RF signal spectrum shown in Figure 2.1.3-b. Note how the pilot and carrier signals are no longer visible. They are still there except with significantly lower amplitudes, energy is being "robbed" to carry the voice traffic. After demodulation in the FM receiver, we obtain the signal spectrum shown in

[15]Regions where inband PCS operation is prohibited since it will violate OFS interference criteria.

Figure 2.1.3-1 (*a*) Input baseband signal (no notches); (*b*) TP3 signal spectrum (no notches, noise loaded); (*c*) TP4 baseband output (−60 dBm input FM signal) no jamming

14: 00: 09 JUN 07, 1992

Figure 2.1.3-1   *Continued*

Figure 2.1.3-1c, measured at TP4. It looks about the same as the input baseband signal measured at TP1.

To measure the effects of interference on the FM receiver, we notch out selected portions of the input baseband spectrum as shown in Figure 2.1.3-2a. Here, the bandstop notch filters shown as part of the Marconi noise generator are switched in to simulate empty voice channels. This signal, measured at TP1, is then fed into the FM transmitter/receiver chain. After demodulation by the FM receiver, we obtain the results of Figure 2.1.3-2b, measured at TP4. Of particular importance, note how the notch depth has been filled in. This is because of intermodulation distortion inherent to the FM radio chain.

Nonlinearities in the transmitter receiver chain cause elements of the baseband signal to mix together, producing new and undesired frequencies. In a sense, the FDM-FM system is jamming itself.

In Figure 2.1.3-3a, we have added in a -80 dBm (referenced to the receiver port), 100-kbps, filtered, BPSK, interfering source centered at 2178.27 MHz. As expected, the notch filled in with interference. Comparing with Figure 2.1.3-2b,[16] the 245 kHz notch depth has degraded from 35.77 dB to 11.93 dB. Voice channels operating in the neighborhood of 245 kHz have had SNR degraded by 35.77 − 11.93 = 23.84 dB.

---

[16]This figure was measured about three hours earlier than Figure 2.1.3-3a through -3c. Apparently the Krohn-Hite filter had drifted over this interval, hence the shift in level and shape.

Figure 2.1.3-2 (*a*) TP1 baseband input; (*b*) TP4 baseband output (− 60 dBm input RF signal)

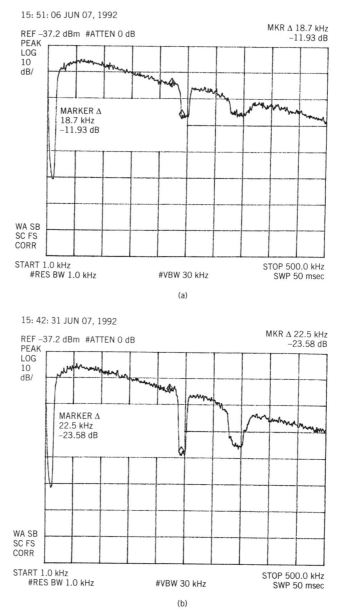

15: 51: 06 JUN 07, 1992

REF –37.2 dBm #ATTEN 0 dB

MKR Δ 18.7 kHz
–11.93 dB

PEAK
LOG
10
dB/

MARKER Δ
18.7 kHz
–11.93 dB

WA SB
SC FS
CORR

START 1.0 kHz
#RES BW 1.0 kHz #VBW 30 kHz

STOP 500.0 kHz
SWP 50 msec

(a)

15: 42: 31 JUN 07, 1992

REF –37.2 dBm #ATTEN 0 dB

MKR Δ 22.5 kHz
–23.58 dB

PEAK
LOG
10
dB/

MARKER Δ
22.5 kHz
–23.58 dB

WA SB
SC FS
CORR

START 1.0 kHz
#RES BW 1.0 kHz #VBW 30 kHz

STOP 500.0 kHz
SWP 50 msec

(b)

Figure 2.1.3-3 (*a*) TP4 baseband signal (loaded) (−80 dBm jamming noise: 100 Kbps filtered BPSK at 2178.2451 MHz); (*b*) TP4 baseband signal (loaded) (−80 dBm jamming noise: 10 MHz BPSK at 2178.27 MHz); (*c*) TP4 baseband signal (loaded) (−80 dBm jamming noise: 1.25 MHz BPSK at 2178.27 MHz)

15: 38: 58 JUN 07, 1992

Figure 2.1.3-3 *Continued*

Figure 2.1.3-3b repeats the experiment with an Omnipoint 10-MHz signal. Notch depth improves 11.5 dB, demonstrating the reduced interference potential of the Omnipoint waveform. In Figure 2.1.3-3c, we consider the intermediate case of a 1.25-MHz, filtered, BPSK, interfering source. Notch depth degrades by about 6 dB. Since the lab results are based on rather low FDM-FM received levels, we expect that redoing the series for a higher FDM-FM signal level, we will again see results more along the lines of the unloaded case.

Figures 2.1.3-4a and 4b summarize S/N degradation versus interference level, with type of interference and bandwidth as parameters.

2.1.4   Bench Test Conclusions

Our bench testing has conclusively shown that under both unloaded and loaded conditions, FDM-FM receivers are significantly more susceptible to narrowband interference as we had claimed in our pioneers preference filing. Specifically, we have demonstrated that for a given power level, FDM-FM receivers are 100 times more susceptible to 100-kHz narrowband interference than 10-MHz Omnipoint type signals in the unloaded case. As a consequence of this, we expect that our exclusion zones will be 100 times smaller in terms of area than comparable narrowband systems.

Figure 2.1.4-1, taken from our filing, reiterates this fact, showing required exclusion radius as a function of transmit EIRP for both narrowband and wideband signaling formats. In all cases, free space propagation is assumed to provide an upper bound on required exclusion radius.

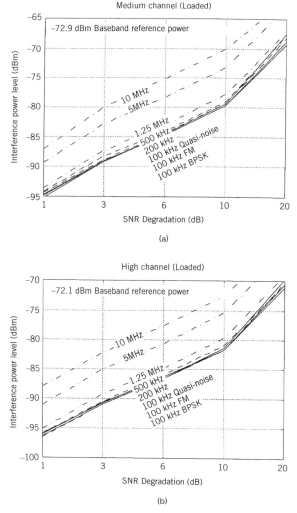

Figure 2.1.3-4 (*a*) medium channel (loaded); (*b*) high channel (loaded)

2.2 FIELD TEST RESULTS (INTERFERENCE)

As part of our testing, we put together a mobile van test setup to experiment with jamming an active point to point microwave link. The link was operating at normal power levels with a fade margin of 35 dB.

The microwave link test setup used essentially the same equipment as in the bench test except actual dish antennas were used. The test site[17] transmits either unloaded or loaded FDM-FM signals as described in

---

[17]Tower base located at 38 44′ 45.74″ N, 104 50′ 34.69″ W, Elevation 6925′ AMSL. Antenna height of 55′ AGL.

Figure 2.1.4-1   Direct sequence spread spectrum signal provides more compatible coexistence with OFS analog formats (85% of all systems)

section 2.1. An additional 10 dB of attenuation can be switched in to model fading effects.

The victim FDM-FM receiver was located in downtown Pueblo Colorado[18] on top of a 130-foot tall building. As described in section 2.1, a Cushman Frequency Selective Level Meter was used to measure interference effects.

The mobile van unit carried the complement of equipment indicated in Figures 2.2-1b and -1c. This is fundamentally the same equipment used in the lab except with the interference source driving a linear power amplifier[19] and a discone antenna located 13-feet AGL. The Boonton power meter taps into the transmission path through a net 30 dB of attenuation to provide a calibrated measure of the transmitted interference power.

When using the Bob Dixon Box (BDB), maximum EIRP from the van is around 500 mW because of insertion losses associated with the double-balanced mixer. When using a CW interfering source from the HP 8660 synthesizer, the BDB is bypassed, and maximum EIRP increases to 2.4 W.

A continuous wave (CW) signal is indicative of a homogeneous 100-kHz signal (a string of 1's or 0's) as is common in some systems during an "idle" state. Table 2.2-1a (Unloaded Filed Test Results [Interference]) clearly indicates that a CW signal is very likely to interfere with an OFS user even when the CW source is not in the main beam. At nearly 14 miles, 1.6 mW of

---

[18]Tower base located at 38 17' 21.33" N, 104 35' 46.65" W, Elevation 4742.7' AMSL. Antenna height of 144' AGL.

[19]Nominal 1-W maximum power output.

TABLE 2.2-1a   **Unloaded Field Test Results (Interference) (Full Link Power)**

Distance from Main Beam (Miles)	Offset from Main Beam (Deg.)	Interference Type (MHz)	EIRP (mW) for Degradation of*		Max Van EIRP (mW)
			1 dB	3 dB	
13.8	−30	CW	1.58	5.62	2455
13.8	−30	100 kHz BPSK	44.67	173.78	501
13.8	−30	10 MHz	n	n	501
7.5	−34	CW	0.15	0.95	2455
7.5	−34	100 kHz BPSK	8.32	38.90	501
7.5	−34	10 MHz	n	n	501
6.3	−45	100 kHz BPSK	7.08	23.44	501
6.3	−45	10 MHz	478.63	n	501
5.9	−60	100 kHz BPSK	9.12	43.65	501
5.9	−60	10 MHz	n	n	501
4.7	−63	100 kHz BPSK	61.66	489.78	501
4.7	−63	10 MHz	n	n	501
3.8	−57	100 kHz BPSK	13.80	54.95	501
3.8	−57	10 MHz	n	n	501

*n denotes insufficient power available in van high voice circuit.

CW signal 30° outside of the mainbeam caused the maximum allowable TSB10-E degradation, that is, 1 dB. Similarly, one seventh of a milliwatt of CW 7.5 miles away and 34° offset from the main beam causes 1 dB of degradation.

Personnel at the van and the test sites were linked together via cellular phone conference calls permitting a test procedure similar to that used in the lab. At each test site, we performed a series of interference tests using the narrowband, 100 kbps BPSK signal and the 10 MHz wideband spread spectrum signal. In general, the results agreed with our bench tests; the narrowband signal proved approximately 20 dB more damaging for a given power level in the unloaded case and 10 to 15 dB more damaging in the loaded case. Required exclusion zones will be significantly larger for narrowband equipment.

Tables 2.2-1a and -1b show field test results for unloaded cases respectively. In several instances, we did not have sufficient interference source transmit power to achieve the desired interference level into the FDM-FM victim receiver. These cases are noted with an "n" in the tables. Blanks indicate tests not performed.

In these tables, the first column shows interference source range from the Pueblo FDM-FM receiving station. The second column indicates how many degrees off boresight the interfering source is, measured in a clockwise direction. Zero degrees is the mainbeam direction where the dish antenna is pointed. The interference type is then indicated. Quiescent baseband noise

TABLE 2.2-1b   Unloaded Field Test Results (Interference)
(Link Power Reduced by 10 dB)

Distance from Main Beam (Miles)	Offset from Main Beam (Deg.)	Interference Type (MHz)	EIRP (mW) for Degradation of*		Max Van EIRP (mW)
			1 dB	3 dB	
6.9	−89	100 kHz BPSK	44.7	141.3	501
6.9	−89	10 MHz	n	n	501
4.7	−63	100 kHz BPSK	77.6	n	501
4.7	−63	10 MHz	n	n	501
3.8	−57	100 kHz BPSK	44.7	131.8	501
3.8	−57	10 MHz	n	n	501

*n denotes insufficient power available in van high voice circuit.

level (no interference) is measured prior to each interference test, providing a reference level for measuring voice channel SNR degradation. We also experimented with the switchable 10-dB attenuator at the transmit site to simulate moderate fading conditions.

3.0   FIELD TEST RESULTS (PROPAGATION)

In reading the various filings and propagation test reports, we have come to the conclusion that the use of Hata propagation models may be unrealistic for establishing exclusion zones. This is true in the context of mobile users since a single handset can potentially obtain near free space propagation to that microwave tower and jam it. In our opinion, several reports submitted to the FCC paint overly optimistic results on sharing by using Hata Urban models to derive exclusion ranges.

In obtaining some propagation data of our own, we mounted a discone antenna on the Pueblo test site microwave antenna at a height of 138 feet above ground level (AGL) and transmitted at 2178 MHz with an EIRP of 32 dBm. On the van, we mounted two discone antennas; one on the roof at 6 feet AGL and another on a mast at 13 feet AGL. We equipped the van with a Trimble GPS navigation unit to measure position. We then drove around in the Pueblo area measuring signal-received signal strength.

Results from this testing are shown in Figures 3.0-1a through -1c. In each of these figures, expected results for the various[20] Hata models are included as well as the free space prediction. In particular, we note that near free space propagation can be obtained at a range of 12 miles. *Signal levels from a PCS handset at this location into the microwave tower would be* 100,000 *times stronger than predicted by Hata Urban models. A single handset at this location*

[20]Large City models with a base station height of 138 feet and a mobile antenna height of 6 feet.

Figure 3.0-1 (*a*) Pueblo test site results (first look data); (*b*) received power vs. range with 90% bars; (*c*) received power vs. range with 90% bars; piecewise linear

would be equivalent to 100,000 *handsets at an average location.* Rogue handsets present a significant threat to the incumbent OFS user unless properly accounted for in the propagation modeling.

Figures 3.0-2a and -2b[21] show that this is not just a pathological case; significant deviations from the Hata predictions are observed in both cases.

---

[21]Taken from Motorola's July 24, 1991, filing with the FCC.

Figure 3.0-1 *Continued*

This is in fact the main weakness in using the Hata median propagation loss formulas for developing path loss predictions; they only predict medians. In developing responsible criteria for sharing with incumbent OFS users, it is important to not gloss over this fact.

4.0   CONCLUSIONS AND PLANNED ADDITIONAL INTERFERENCE TESTING

Bench testing with a representative FDM-FM receiver has conclusively shown that interference susceptibility is a function of interfering source bandwidth with narrowband formats doing the most damage. As a consequence of this, narrowband systems will require large exclusion zones and have greater difficulty in sharing with incumbent OFS users. Required exclusion areas are likely to be 100 times larger for narrowband systems. Outdoor interference testing indicated that the bench test results will hold in actual practice; we were able to interfere with the FDM-FM receiver at much longer ranges when using narrowband signals.

Digital radios are the other major type of equipment used by OFS users although they only constitute a small minority of systems. We have obtained the use of a Faranon-Harris digital radio link capable of supporting a DS2 data rate[22] and plan to conduct another series of interferences tests with this equipment. A BER degradation criteria along the lines of TSB10-E will be

[22]6.176 MBPS or four times the T1 data rate of 1.544 MBPS.

Figure 3.0-2 (*a*) Schaumburg-Gretnalink; (*b*) Link 24, Ill. Central RR

**Even Outside the OFS Beampath a Single
100mW Narrowband Handset can violate the 1db
Threshold for Interference well over 20 miles away**

Figure 4.0-1   Illustration of interference by narrowband PCS.

Figure 4.0-2   (*a*) Freespace loss model exclusion zones; (*b*) Hata loss model exclusion zones.

Omnipoint's Exclusion Zones Are Dramatically
Smaller For The Same Propogation Assumptions

HATA LOSS MODEL
(median city suburban assumption)

9 miles

(RF Horizon)

2.2 miles

NARROWBAND EXCLUSION ZONE AREA IS UP TO 15 TIMES
THE SIZE OF OMNIPOINT'S EXCLUSION ZONE AREA OUTSIDE
OF THE MAIN BEAMPATH.

OMNIPOINT'S EXCLUSION ZONE AREA IS ROUGHLY 7%
OF THE NARROWBAND EXCLUSION ZONE AREA OUTSIDE
OF THE MAIN BEAMPATH.

Assumptions: Both PCS devices are transmitting at 100 mW using a 0 dBd gain,
omnidirectional antenna at a height of 30 feet
100 KHz narrowband signal; 10 MHz Omnipoint spread spectrum signal
2 GHz microwave transmitter, directional antenna at a height of 200 feet

Figure 4.0-2   *Continued*

used to evaluate interference effects. We expect to see differences in suscep-tibility between narrowband and wideband cases based on published T/I[23] criteria as a function of frequency.

Finally, we have measured propagation path losses as a function of range to a microwave tower and found that it is highly variable. Results are consistent with Motorola's published data on this topic. In some cases, the propagation path losses were 100,000 times smaller than predicted by the Hata Urban models used by some authors in their reports on spectrum sharing. Even in large cities, we expect that free space propagation can prevail into a microwave tower if the handset is located in a rogue position (e.g., balcony, rooftop, etc.). Responsible sharing must address this problem. Omnipoint's approach has been to develop signaling formats that are inher-ently less damaging to OFS receivers and then assume very conservative propagation path loss models.

Figure 4.0-1 illustrates the case of interference from a narrowband, uncontrolled radio. Figures 4.0-2a and 2b show exclusion zones for freespace and Hata model propagation.

---

[23]Threshold to Interference.

ANSWERS TO PROBLEMS

CHAPTER ONE

1. $10 \log 250 = $ **24 dB.**
2. $M_J + S/N + L_{sys} = $ **34 dB** + system losses.
3. $BW_{RF} \geq 6 \times 2.5 \times 10^6 = $ **15 MHz.** (Remember to allow for the 10 dB output S/N.)
4. $\dfrac{1.5 \times 10^7}{2} = 7.5 \times 10^6$ cps.
5. $\geq$ **2500**$(\log^{-1} 3.4 = 2500)$.
6. $TW = 20$ dB or 100; $T = (10{,}000)^{-1} = 1 \times 10^{-4}$; $W = (1 \times 10^2)/(1 \times 10^{-4}) = 1 \times 10^6 = 1$ MHz.
7. $G_p = 10 \log(2 \times 10^6)/(1 \times 10^5) = 13$ dB. If output S/N must be 6 to 10 dB and losses are 1 to 2 dB, then M_J is a maximum of 6 dB. Probably not worth the trouble. Decide for yourself.
8. With 100-khop/sec rate minimum channel spacing should be 100 kHz. In 5 MHz this would allow for 50 channels. This number of channels would permit only a 17-dB process gain and high vulnerability to partial band interference.
9. The chip rate is a function of information rate, but the sweep bandwidth is usually the maximum of which the generator is capable.
10. The transmission bandwidth is a function of the code sequence used in a direct sequence system.
11. The frequency hop transmission bandwidth is a function of the number of channels employed. Channel spacing may or may not be a function of the hop rate.
12. The simple time-hopping system transmits at a constant center frequency and thus does not avoid a single-frequency interferer. Reducing duty cycle does increase potential process gain but often forces an increase in transmitted power to maintain E_b/N_0 for the demodulator.

CHAPTER TWO

1. Primarily to suppress the RF carrier while providing a constant signal envelope.

2. (a) Let $0.5 = (\sin^2 x)/x^2$; 0.707 $x = \sin x/x$; 0.707 times $2\pi/16 \sim \sin 7\pi/16$; bandwidth $\approx 0.44 \times$ clock rate (one-sided).

3. Approximately 90%.

4. (a) $\dfrac{P_T}{2 \times 10^7}$ W/Hz.

   (b) $\dfrac{P_T}{2 \times 10^7} \dfrac{(\sin \pi/4)^2}{(\pi/4)^2}$ W/Hz.

5. $\log^{-1} 30/10 = 1000$; the number of jammed channels $= 1000$: $1000/0.01 = $ **100,000 channels**.

6. 3 kHz $\times 100,000 = $ **300 MHz**.

7. $S/N@1 \times 10^{-2}$ error rate ≈ 7 dB (allow 2 dB for implementation loss): $30 + 7 + 2 = 39$ dB; $\log^{-1} 39/10 = 8000$; $1000 \times 8000 = 8$ MHz.

8. $10 \log 100 = 20$ dB. $10 \log(4 \times 10^7)/(3 \times 10^3) = 41$ dB. Process gain $= 20 + 41 = $ **61 dB**.

9. A coherent frequency hopping receiver maps the received signal to a single phase coherent frequency, pulsed at the data rate. The noncoherent hopper performs the same frequency mapping but the carrier pulses corresponding to data received are at random phases with respect to one another, that is, the noncoherent receiver receives a carrier signal that is both pulse- and random-phase modulated. The demodulator must be approximately twice as wide as accept the same signal when the carrier pulses are not coherent, so that a 3-dB penalty is paid for noncoherent reception.

10. The three-chip-per-bit demodulator sees an effective chip rate three times that of the 1-chip-per-bit system and its demodulator bandwidth must be three times as great. The resulting 4.7-dB loss due to increased noise bandwidth must be considered when calculating interference rejection.

11. In a chip-matched filter a particular delay is associated with each increment of frequency input. If each frequency in a particular sweep is offset by R^* Hz, then each increment at the filter output is delayed a different amount. Thus the compressed output pulse is offset by an amount $R\Delta t/\Delta f$(seconds). If the slope of Δf is positive, then positive Doppler causes increased delay. For Δf negative positive Doppler causes decreased delay. Upchirp-downchirp delays can therefore be combined to develop a signal unaffected by Doppler shift.

$^*R$ is the mean frequency offset due to Doppler for a particular frequency sweep.

12. Multipath signals consist of components arriving from two or more paths —the direct path and one or more indirect paths. Signals arriving with greater than one code chip time are uncorrelated in either the direct sequence or frequency hopping receivers and are treated the same as any other interference. If delay is less than a code chip time (a hop interval for frequency hoppers), the signals do interfere directly with one another and the receiver will track either one or a combination of the signals that appear. If the receiver is a noncoherent frequency hopper, it accepts the vector sum of its inputs and makes its decisions on that basis.

13. Time-frequency hopping implies that from time slot to time slot different frequencies are used, and an alternating set of transmitters takes turns making use of a common channel. Because only one is on the air at any one time, there is no possibility that more powerful transmitters will overpower the weaker ones.

14. Direct path = 100 miles. Indirect path = 103.2 miles. For a propagation delay of 6.1 μ sec per nautical mile, differential path delay = 3.2 × 6.1 × 10^{-6} = 19.52 μ sec. The hopper should change frequencies before the repeater can recognize the signal and repeat it to the receiver. Maximum chip time is 19.52 μ sec. The minimum hop rate is therefore (19.52 μsec)$^{-1}$ = 51,229 khps.

15. The power spectrum of a chirp signal is an equal-level set of lines separated by an amount corresponding to the frequency related to the sweep period (i.e., frequency separation = $1/\Delta t$). Therefore the power level in one line is $P/\Delta f \Delta t$ and the average power density is $P/\Delta f$ for a linear chirp signal. FH signals place all their transmitted power in each channel, as it is chosen. The power density is $P/BW_{channel}$. DS signals have power density that varies according to $(\sin x)/x)^2$. Power density varies as a function of distance from center frequency, with respect to code rate. On the basis of approximating the $(\sin x/x)^2$ main lobe with a triangular spectrum, the power density at any point is approximately $(PR_c - f_d)/R_c$, where R_c = code clock rate and f_d = distance from the center frequency (Hz).

CHAPTER THREE

1. $2^{23} - 1$ = **8,388,607.**
2. $\phi(2'' - 1)$ = 131070; 131070/17 = **7,710.**
3.

T	D_1	D_2	D_3	D_4	D_4	$\oplus$	D_2
1	1	1	0	1		0	
2	0	1	1	0		1	
3	1	0	1	1		1	
4	1	1	0	1		0	

The sequence is **101.**

4. $2^{11} - 1 = \mathbf{2047}$.

5. The sequence is

$$\ldots 1111100110100100001010111011000\ldots$$

Therefore the sequence is maximal $((2^5 - 1) = 31)$ and $\psi_{max} = \mathbf{31}$; $\psi_{min} = -\mathbf{1}$.

6. For initial conditions = 1111111:

$\ldots 111111100110010110100011101011000100101000010000001010$

$10111011100100111101001101111000110000\ldots$ (length = 93

chips).

For initial conditions = 1100110:

$\ldots 0110011100001101010010001011111\ldots$ (length = 31 chips).

For initial conditions = 1101101:

$\ldots 101\ldots$ (length = 3 chips).

7. $[9, 6, 4, 3]_s$:

$[9, 8, 4, 1]_s$:

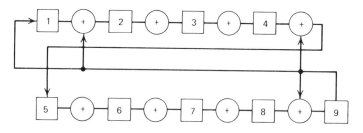

8. $[5, 3]_s$:

$$\ldots 1111100110100100001010111011000\ldots$$

$[5, 4, 3, 2]_s$:

$$\dots 11111001001100001011010100011110\dots$$

Shift	A − D	Shift	A − D
0	7	16	7
1	−1	17	−3
2	−1	18	−11
3	7	19	−9
4	7	20	7
5	−1	21	−1
6	7	22	5
7	−1	23	7
8	−1	24	−1
9	−9	25	−1
10	−5	26	−1
11	−1	27	−9
12	−1	28	7
13	−3	29	−9
14	−9	30	7
15	−1		

$A - D_{max} = 7 \quad A - D_{min} = -11$.

9. **0.0131072 sec; 76.293 Hz.**

10. $2^{n-(p+2)} \Big| \begin{matrix} n = 13 \\ p = 3 \end{matrix}$
$$= 2^{13-(3+2)} = 2^8 = 256.$$

11. **256.**

12. Any modulo-2 combination of two or more 12-stage, m-sequence generators with selected sequences.

13. Let $[\varphi(2^n - 1)]/n \geq 1000$. It is known that there are only 630 sequences generated by a 13-stage register in the m-sequence configuration (8191 chips). For $n = 15$, $2^n - 1 = 32{,}767$; 32,767 has prime factors 7, 31, and 151. $[(7 - 1) \times (31 - 1) \times (151 - 1)]/15 = 1800$. A 15-stage register has enough m-sequence available.

14. Using $5, 3_s$, the sequence produced is

$$\dots 11111000110111010100001001011000\dots$$

Selecting a pair of arbitrary phases and adding them to the above,

(a)
$$\begin{array}{l} 11111000110111010100001001011000 \\ 00101100111110001101110101000001 \\ \hline 11010100001001011001111110001101 \end{array}$$

(b)
$$\begin{array}{l} 11111000110111010100001001011000 \\ 10010110011111000110111010100000 \\ \hline 01101110101000010010110011111100 \end{array}$$

Both (a) and (b) are just new phase shifts of the same sequence. The same case holds for all possible phase shifts. Therefore for the operation of addition the set of phase of an *m*-sequence connected generator is closed.

15. Given a code sequence $\lambda^0 = \lambda^n \oplus \lambda^{n-m}$, where $\lambda'' \oplus \lambda^{n-m}$ is a maximal register connection, let there be n zeros in the register: $\lambda''(0) \oplus \lambda^{n-m}(0)$ $= 0 = \lambda^0(0)$. At t_1, with λ'' and λ^{n-m} both containing zeros, the feedback input to the first register stage (λ^0) is zero. For all $t_n(t_n = t_1, t_2, \ldots, t_n)\lambda^0$ $= 0$. Because register input is always zero, the register never leaves the all-zeros state and no *m*-sequence is generated.

16. Assume a code sequence $\lambda^0 = \lambda^n \oplus \lambda^{n-m} \oplus \lambda^{n-p}$. If the sequence is maximal, it must contain the all-ones state at some point in its progression. When all-ones are inserted in the *n*-stage register, λ^0 is always a one. Therefore the generator can never leave the all-ones state and an *m*-sequence cannot be generated.

17. Because each state occurs once and only once in each sequence, sensing any state and triggering an oscilloscope when that state occurs allows the oscilloscope to be synchronized so that the code can be observed.

CHAPTER FOUR

1. $BW_{BPSK} = 2\ BW_{QPSK}$.
2. Let $16^x \geq 8000$, $16^3 = 4096$, $16^4 = 65,536$. **Four.**
3.

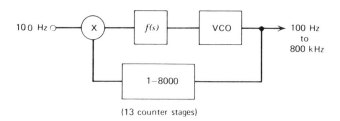

(13 counter stages)

4. Assume that minimum sampling rate is two samples per Hertz: Data rate $= 2 \times 3 \times 3 \times 10^3 = $ **18 Kbps.** 40 decibels correspond to 10,000 times RF-to-baseband ratio. Therefore the bandwidth is $18 \times 10 \times 10^6$ $= 180$ MHz.

5.

	Direct	Indirect
Hop rate	Determined by inter-stage filters	Loop must relock at each hop; loop BW determines rate

	Direct	Indirect
Frequency stability		
Long term	Set by reference	Same
Short term	Set by reference	Depends on loop jitter
Spurious	Many spurious signals generated by mixing processes; interstage filters determine output	Usually close-in noise caused by loop jitter and internal noise

6. For 64 frequencies use four add-and-divide modules: $K^A = 4^4 = 64$. Four base frequencies are needed. Let $M = 4$; $f_0 = 70$ MHz (given); $f_m = f_0 - f_0/M = 70 - 70/4 = 52.5$ MHz. Base separation $= K^{A-1} \times 10$ kHz $= 4^3 \times 10^4 = 640$ kHz.

$$
\begin{array}{ccccc}
| & | & | & | & | \\
f_1 & f_2 & f_m & f_3 & f_4 \\
 & & | & &
\end{array}
$$

$$f_2 \text{ and } f_3 = f_m \mp \frac{\text{base separation}}{2}$$

$$= 52.5 \times 10^6 \mp \frac{6.4 \times 10^5}{2} = 52.5 \text{ MHz} \mp 320 \text{ kHz}$$

$$f_2 = 52.18 \text{ MHz},$$

$$f_3 = 52.82 \text{ MHz},$$

$$f_1 = 52.18 - 0.64 = 51.54 \text{ MHz},$$

$$f_4 = 52.82 + 0.64 = 53.46 \text{ MHz}.$$

7. Because squaring the direct sequence signal produces a twice-frequency carrier (see Appendix 4), the modulated carrier could be detected by a single squaring process.
8. $\log_2 2000 = $ **10.966.** 11 binary bits are required.
9. $K^A \geq $ **256** is desired. How many modules (A) with K base frequencies should be used to make the simplest synthesizer? Solving $\log_K X$ for A as

the next larger integer:

K	A
2	8
3	6
4	4
5	3
$\vdots$	$\vdots$
15	3
16	2

$K = 4$ or 6 appears to be optimum, because an even number of frequencies balances the spectrum around f_m and f_0 and little is gained by reducing K once 5 or more base frequencies are used.

10. Flip-flop output is identical to the suppressed-clock PDM signal (i.e., the carrier clock may be subtracted by simple division by two).

CHAPTER FIVE

1. $30 - 20 = $ **10 dB.**

2.

3. **0.3.**

4.

5. The m-ary FH system has n/m frequency cells available, compared with n cells for a binary FH system. The binary system must hop faster to convey information at the same rate.

6. $\log_2 m$ bit PCM or m-level analog (for the base band information).

7. If signal 1 has a structure $A\cos(\omega_{c1} \pm \omega_{m1})t$ and this is multiplied with $B\cos(\omega_{c2} \pm \omega_{m2})t$, the output bandwidth produced corresponds to the maxima of ω_{m1} and ω_{m2}. (BW $= g(\omega_{m1}, \omega_{m2})$). The output signal is $AB/2\cos(\omega_{c1} \pm \omega_{c2} \pm \omega_{v1} \pm \omega_{m2})t$. Therefore at either $\omega_{c1} + \omega_{c2}$ or $\omega_{c1} - \omega_{c2}$ the modulation bandwidth is $2(\omega_{m1} + \omega_{m2})$ maximum. The multiplier output due to a similar cochannel signal would be twice as wide as the local reference bandwidth.

8. $\left(\dfrac{S + I}{I}\right)_{out} = (SG_p + I)/I(1000 + 10)/10 = 1010/10 = 101;$
   $10\log 101 = \textbf{20.04 dB.}$

9. The code repetition would be at a $1 \times 10^6/4095 = 244.2$ repetitions per second rare. Line spacing would be 244.2 Hz.

10. $\sin = \omega_{offset}/K = (2 \times 10^4)/(1 \times 10^5) = 0.628; \delta = \textbf{38.9}°.$

11. Subcarrier FM or FSK modulation for direct sequence signals is a DSB suppressed-carrier technique. Therefore the output of a correlator when such a signal is being sent is a pair of mirror-image modulated signals, separated by twice the subcarrier frequency, around the output IF center frequency. Costas or squaring loops are designed to combine coherently such DSB signals and regenerate the missing carrier. A phase-lock loop can track only one of the pair of sidebands and is therefore at a 3-dB disadvantage. Two independent phase-lock loops, each tuned to one of the two sidebands (the upper and lower subcarrier-produced frequencies), can be employed, with summing of their demodulated outputs. (This alternate does provide for recovering the lost 3 dB.)

12. Assuming that the frequency shift of the carrier is the same as that applied to the subcarrier, we can generate the following matrix:

Shift (f_c, f_s)	$++$	$+-$	$-+$	$--$
f_c	δ	δ	$-\delta$	$-\delta$
f_{us}	2δ	0	0	-2δ
f_{1s}	0	2δ	-2δ	0

where f_c = carrier,

f_s = subcarrier,

$+$ signifies a positive shift (increase in frequency),

$-$ signifies a negative shift (decrease in frequency),

δ is the amount shifted.

If the demodulator is such that it can accommodate the unsymmetrical signal, the two can be separately demodulated. A Costas demodulator, for instance, can track carrier and DSB subcarrier shifts separately. Two separate baseband channels can be accommodated with minimum mutual interference.

CHAPTER SIX

1. Drift per day (from Figure 6.1) = 1.1×10^2 chips; $1.1 \times 10^2 \times 2.5 = 2.75 \times 10^2$ chips. Total uncertainty is $\pm$**275 chips.**

2. Drift per day = 1.1 chips. Total uncertainty = $1.1 \times 2.5 = \pm$**2.75 chips**.

3. Allowing 2-dB implementation loss, BW_{RF}/BW_{info} corresponding to 33 dB is 2×10^3. The information bandwidth then is $(2 \times 10^7)/(2 \times 10^3)$ = 10 kHz. This is also assumed to be the synch recognition bandwidth. Recognition rise time is approximately $0.35/(1 \times 10^4) = 0.35 \times 10^4 = 35$ μsec. To search 4095 chips, at $\frac{1}{2}$ chip per search, increment would require at least $8190 \times 35 \times 10^{-6} =$ **0.286 sec.**

4. $2.5 \times 10^{-2} =$ **0.025** (from Figure 6.15).

5. The search offset is ± 0.25 chip. Therefore the maximum loss corresponds to a 0.25-chip offset. The correlation loss is also 0.25 or 1.2 dB. The average loss should be much less.

6. $10 \times 1.48 \times 10^{-3} \times 9.8 \times 10^2 =$ **14.5**. Travel is away from us and Doppler shift is negative. Apparent clock rate is 9,999,985.5 Hz.

7. $10 \times 1.48 \times 10^{-3} \times 5.8 \times 10^2 =$ **8.58**. He is approaching. Therefore Doppler shift is +8.58 Hz. Apparent clock rate is now 10,000,008.5 Hz.

8. Case 1: $3.75 \times 10^2 \times 1.48 \times 10^{-3} \times 9.8 \times 10^2 =$ **543.9**. Shift is **−543.9 Hz**.
   Case 2: $3.75 \times 10^2 \times 1.48 \times 10^{-3} \times 5.8 \times 10^2 = 321.9$. Shift is +321.9.

9. The period of each hop is $\frac{1}{10}^5 = 10^{-5}$ sec. Propagation time from transmitter is $6.1 \times 10^{-6} \times 1.12 \times 10^2 = 683.2$ μsec. The receiver must search $(683.2 \times 10^{-6})/(1 \times 10^5) = 68.32$ chips (assuming that $R_{hop} = R_{clock}$).

10. Twice as far (two times round trip delay) or 136.64 chips, without reset. With reset, 68.32 chips.

11. Code period = $(5 \times 10^6)^{-1} = 2 \times 10^{-7}$ sec. $1023 \times 2 \times 10^{-7} = 2.046 \times 10^{-4}$. Line length is 0.2046 msec. Tap are 200 nsec apart.

12. **11.5 dB** or greater (see Figure 6.15).

13. Each chirp signal, or pulse, is usually independent of all others and synchronized on the basis over one chirp interval.

CHAPTER SEVEN

1. Noise power in 10 MHz = 3.5×10^{-14} W (from Figure 7.3). Noise power in 50 kHz = 1.8×10^{-16} W: $10 \log(3.6 \times 10^{-14})/(1.8 \times 10^{-16}) = 10 \log 2 \times 10^2 = 20$ dB. Assuming 10-dB output S/N ratios, 30% average modulation, 6-dB noise figures, and 2-dB DS system implementation loss, AM sensitivity = 32 dB above 1.8×10^{-6} W, or 2.88×10^{-13} W = 2.05 μV. DS sensitivity = 8 dB below 3.6×10^{-14} W, or 5.76×10^{-15} W = 0.411 μV.

2. 20 dB corresponds to $J/S = 100$; 10 dB corresponds to $J/S = 10$. Total worst case $J/S = 110$. Difference in overall interference due to the second interference source is 10 percent, or approximately 0.4 dB.

3. For -20-dB signal input, a 20-dB jamming signal reduces the output S/N ratio by 3 dB (see Figure 7.2).

4. Pulsed effects are enhanced.

5. (a) 400-MHz center frequency and 40-kHz spacing. (b) 400 MHz CW carrier only.

6. Biphase signal power loss $= 3.4$ dB. Quadriphase signal power loss $= 0$ dB (if two 2.5 Mbps data streams are used for modulation). Same as biphase if two 5.0 Mbps data streams are used.

7. If there are 100 link users, then $S/N_{RF} = -20$ dB. An additional interfering signal, 15 dB larger than the desired signal, appearing at a receiver would be less effective than the 20-dB friendly interference level already existing there. 20 dB $= 100$ times the desired signal power; 15 dB $= 31$ times the desired signal power. Total interference $= 131$ times the desired signal; $10 \log 131 = $ **21.17 dB.** The signal would be degraded by only 1.7 dB by the 15-dB jammer, beyond the loss already existing due to code division multiplex.

8. (See Figure 7.3) Noise power $= 3.6 \times 10^{-14}$ W. NF $= 10$ dB. Minimum signal is 20 dB below 3.6×10^{-13} W, or 3.6×10^{-15} W.

9. Noise power increases to **3.6×10^{-13} W.** M_J increases to **30 dB.** Minimum signal is 30 dB below 3.6×10^{-12} W, or **3.6×10^{-15} W.** (Note that sensitivity is not improved.)

10. Modulation is $\sin(\omega_c \pm \omega_m)t$, where $\omega_m t = 0.90, 180,$ and $270°$.

$$\omega_m t = 0 \quad 90 \quad 180 \quad 270$$

$$64 \times \sin(\omega_c \pm \omega_m)t = 0 \quad 0 \quad 0 \quad 0$$

There would be only a $\times 64$ carrier remaining.

CHAPTER EIGHT

1. 10-Mcps period is 100 nsec. The worst-case resolution then is ± 100 nsec.

2. 20 Mcps corresponds to a 50-nsec period. Range resolution is approximately 1 nsec/ft. 350 nautical miles 2.1×10^6 ft; $(2.1 \times 10^6/50 = 42,000$. Therefore code length must be at least **42,000 chips.**

3. $25 \times 0.161875 \times 10^6 = $ **4.046875 Mcps.**

4. $\cos \theta = (c_1 - c_2)/(a_1 - a_2) = (192.05 - 192.03)/01 = 0.2; \theta = $ **78.5°.**

5. Assuming a $\pm\frac{1}{10}$ chip resolution accuracy, 0.5 mile corresponds to approximately 150 chips at 50 Mcps. Directional increments = 150/0.1 = 1500/quadrant. Directional resolution = 90/1500 = **±0.06.**

6. With simple ranging by code position, resolution would be to within $\frac{1}{10}\frac{5}{}= 10^{-5}$ sec, or $10^{-5}/(6.1 \times 10^{-6}) =$ **1.64 miles.**

7. The carrier-frequency offset does not directly effect range accuracy. The clock rate offset appears as a direct range inaccuracy to the extent that transmitting and receiving code sequences drift apart. Doppler and oscillator inaccuracies have the same effect.

8. Given that the estimate

$$\cos \theta = \frac{\text{antenna separation in chips}}{\text{distance difference in chips}}$$

is employed, errors of 1 and 0.1% produce the following angle measurements:

True $\theta°$	$\cos \theta$	θ est*	$\theta(0.01)$	$\theta(0.001)$
5	0.996	5.126	0	4.442
10	0.984	10.26	5.926	9.94
15	0.966	14.98	12.66	14.76
30	0.866	30.002	28.99	29.9
45	0.707	45.008	44.43	44.95
60	0.5	60.0	59.67	59.96
75	0.259	74.98	74.84	74.97
89	0.017	89.02	88.98	89.02

*θ est is given only to show the amount of error generated by rounding the cosine to three places, as compared with the errors generated by perturbing an angle measurement by a 1 or 0.1% error in the distance measurements used to estimate it.

In practice, both distance measurements used for an angle estimate would be made with the same clocks, and little error would be seen except for that caused by a simple chip counting error.

INDEX

WIDENER UNIVERSITY WOLFGRAM LIBRARY CHESTER, PA.